Introduction to Quantum Mechanics in Chemistry

Mark A. Ratner
Northwestern University

George C. Schatz
Northwestern University

PRENTICE HALL
Upper Saddle River, New Jersey 07458

Library of Congress Cataloging-in-Publication Data

Schatz, George C.,
    An introduction to quantum mechanics in chemistry / George C. Schatz, Mark A Ratner,
      p.   cm.
    Includes bibliographical references and index.
    ISBN 0-13-895491-7
    1. Quantum chemistry. I. Ratner, Mark A, 1942. II. Title.
    QD462.S328   2000
    541.2′8—dc21   00-029130
                   CIP

Executive Editor: **John Challice**
Cover Designer: **Bruce Kenselaar**
Manufacturing Manager: **Trudy Pisciotti**
Manufacturing Buyer: **Beth Sturla**
Production supervision/composition: **Prepare, Inc.**
Cover Caption: Ground $^1A_1$ and lowest excited $^1B_1$ state potential energy curves of biphenyl H_5C_6-C_6H_5 as a function of the inter-ring torsion angle. The ground state minimum occurs for a twisted geometry, and the highest occupied molecular orbital (HOMO) is depicted on the right. The excited state minimum is planar, and the corresponding HOMO is on the left. Because of this change in torsion angle, absorption from the ground state (long arrow on right) creates an excited state that twists before fluorescence (long arrow on left) occurs. See problem 14.13 for further details.

© 2001 by Prentice-Hall, Inc.
Upper Saddle River, New Jersey 07458

Printed in the United States of America

ISBN 0-13-895491-7

Prentice-Hall International (UK) Limited, *London*
Prentice-Hall of Australia Pty. Limited, *Sydney*
Prentice-Hall Canada Inc., *Toronto*
Prentice-Hall Hispanoamericana, S.A., *Mexico*
Prentice-Hall of India Private Limited, *New Delhi*
Prentice-Hall of Japan, Inc., *Tokyo*
Pearson Education Asia Pte. Ltd., *Singapore*
Editora Prentice-Hall do Brasil, Ltda., *Rio de Janeiro*

4 2019

To our children: Stacy, Daniel, Paul, Albert, and Jonathan

Contents

Preface

Quantum mechanics, particularly quantum chemistry, is a crucial part of the language of modern chemical science. Terms such as *pπ–dπ interaction, symmetry–forbidden reactions, bond-order/bond-energy relationships, hypervalency,* and *exchange repulsion* are typical of those that arise in contemporary discussions of chemical structure and properties. This language is a powerful one with which the contemporary chemical scientist needs to be familiar.

The language starts with quantum mechanics, and that brings with it some formal and mathematical impediments. Most chemists do not intend to be theorists or even traditional physical chemists. For them, many of the niceties and formal elegances of quantum mechanics are really irrelevant, since their aim is a more utilitarian one: to use quantum chemistry to understand the molecular problems or the materials problems that arise in their own research and understanding. Within the past 15 years, the broad applicability and availability of appropriate modeling tools has made quantum chemical techniques part of the arsenal of most chemists. Indeed, more electronic structure calculations are published by people who do not call themselves physical chemists than by people who do!

Given this situation, we felt that it was appropriate to write a text focused on the language of quantum chemistry and the tools that it makes available. The approach is straightforward: It attempts to avoid all unnecessary complexity, detail, and formalism. The book is not written for theorists; rather, it is intended to allow all chemical scientists to become familiar with the language of quantum chemistry and with the use of many of its most important tools.

The book is designed to provide an integrated approach to the conceptual development of quantum chemistry and to its application in current research questions. It is intended to be modest and straightforward, easily completed in its entirety either in a one-semester formal course (at the advanced undergraduate or beginning graduate level) or as a self-study document that can be completed in about 50 hours of reading, problem solving, and computer exercises.

The in-chapter exercises are intended to demonstrate problem-solving methods. We recommend trying to solve them before looking at the solution that is given. (Answers, to odd-numbered exercises are provided in the appendix to this text. Full solutions are available to instructors by contacting Prentice Hall and requesting ISBN 0-13-015487-3.) The problems at the end of each chapter are of differing levels of difficulty; some (for example, Problems 2.12, 6.5, 8.8, 8.11, 13.4, 14.11, and 15.6) contain important thematic material.

The computer exercises (especially those in Chapter 14) are an important part of the book, just as, to most chemists, computational applications are arguably the most important contribution of quantum chemistry. The methods discussed in Chapters 11, 12, and 14 are available through the use of a large number of commercial and freeware codes. Such software permits the chemist to answer, more accurately and efficiently, many of the questions involving molecular structure and response that arise in understanding the behavior of molecules.

The book is intended for chemists, materials scientists, and chemical engineers who wish to learn the language of quantum chemistry and the computational methods that it provides. The volume can also serve as a bare-bones introduction for those who intend to pursue quantum chemistry more deeply, perhaps supplemented by some of the texts that are discussed in the bibliographies at the end of each chapter. Our more advanced book, *Quantum Mechanics in Chemistry* (Englewood Cliffs, NJ-Prentice Hall, 1993), is organized so that one can jump directly from Chapter 14 of the present book to Chapter 3 of that one, so as to provide enough material for an entire year's course on quantum chemistry.

It is our aim in this text to provide an introduction to quantum chemistry that can be used with ease (and, we hope, with some pleasure) by most chemists.

We are grateful to Margaret and to Nancy for allowing us to spend even more time than usual in the completion of this book, and to John Challice and his colleagues at Prentice Hall for inspiration in writing and help in assembling the book. We also thank Fred Northrup for the spectra in Chapter 15; Janet Goranson, who suffered through several nearly fatal word-processor upgrades, for her expert typing; our students, who suffered through several early versions of the book, for their useful and challenging suggestions. We thank Brian Hoffman, Northwestern Unversity; John Head, University of Hawaii; W. Vern Hicks, Jr., Northern Kentucky University; and Duane Swank, Pacific Lutheran University, for their careful review of the manuscript for this book.

Mark A. Ratner
ratner@mercury.chem.nwu.edu

George C. Schatz
schatz@chem.nwu.edu

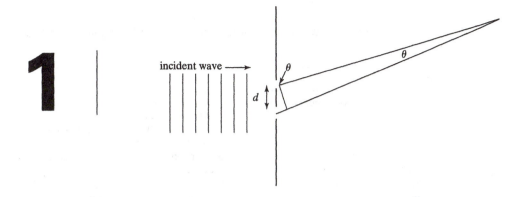

Introduction and Background to Quantum Mechanics

Understanding molecular behavior through the use of quantum mechanical ideas requires a fundamental understanding of quantum mechanics itself. In this chapter, we discuss some of the aims of theoretical chemistry, mention key concepts from classical physics, including both particle and wave behavior, and briefly summarize some of the early history of quantum mechanics. Ideas such as wave–particle duality, the uncertainty principle, and wave mechanics underlie the work in the remainder of the book.

1.1 AIM OF THEORETICAL CHEMISTRY

Prior to the development of quantum mechanics, molecular chemistry was essentially an empirical science. Elements and compounds were categorized into groups (e.g., the periodic table, alkanes, alkenes, alkynes, etc.), and there were empirical rules concerning the interrelationships of different chemical properties (structural, thermodynamic, reactivity, etc.). The underlying physical basis for these rules was largely unknown, and this ignorance hindered attempts to generalize the rules to different kinds of molecules.

Quantum mechanics changed molecular chemistry to a more exact science in which most molecular behavior can be understood on the basis of one unified concept: the Schrödinger equation. In this picture, the empirical rules can be derived by studying the behavior of solutions of the Schrödinger equation, and this, in principle, enables the rules to be generalized. The following table

Empirical rule	Quantum property
Pauling's rule (relates bond energy and length)	Occupation numbers of bonding and antibonding molecular orbitals approximately determine both bond energy and bond length
Arrhenius formula (chemical reactions are activated)	Barriers on potential-energy surfaces determine activation energy
Linear free-energy relations (Hammett rule)	Inductive electron density changes in aromatic systems determine substituent effects

gives some empirical rules and their underlying rationalization, based on quantum mechanics.

It has been suggested (by Dirac) that quantum mechanics changes the practice of chemistry to an exercise in applied mathematics, wherein it is only solutions to the Schrödinger equation that need be determined to understand any chemical problem. This statement is true more in principle than in practice for most problems, however, as the computational complexity associated with solving the Schrödinger equation is often quite formidable. But it also means that the development of accurate and efficient methods for solving the Schrödinger equation for molecular systems is an important research topic in chemistry and that understanding quantum mechanics is an essential part of one's chemical education. One significant consequence of this research and education in the chemical applications of quantum mechanics has been the emergence of several software companies whose survival depends on the ability of quantum methods to solve chemical problems that are important to industry.

The detailed history of the discovery of quantum mechanics is now largely decoupled from its modern practice, but there are key concepts that are best understood in a historical context. This chapter describes these concepts, as well as background material from classical physics that is relevant to understanding molecular quantum mechanics. In addition, the chapter serves as an introduction to the basic concepts of quantum mechanics that are defined more rigorously in Chapter 2. Our intention in this development is to keep the level of mathematics as low as possible—ideally requiring nothing beyond what is covered in undergraduate calculus courses. However, a few topics go beyond this level, so in Appendix A we give a short primer on those subjects: complex variables; differential equations; and matrices, determinants, and eigenvalues.

1.2 KEY CONCEPTS FROM CLASSICAL PHYSICS

To understand quantum mechanics, it is necessary first to understand classical mechanics and classical wave theory. Here we give just the essentials of each, plus some simple examples that also show up in quantum mechanics.

1.3 CLASSICAL MECHANICS

1.3.1 Potential-Energy Functions

The motions we are interested in involve particles moving under the influence of forces that may be derived from a potential-energy function that we denote by V. This situation could involve two particles interacting with each other subject to a *coulomb* potential

$$V(r) = \frac{q_1 q_2}{4\pi\varepsilon_0 r},\tag{1.1}$$

where r is the distance between the particles, q_1 and q_2 are the charges on each particle, and ε_0 is the permittivity of free space. [*Note*: Throughout this text, we use SI units. The corresponding CGS formula would omit the factor $1/(4\pi\varepsilon_0)$.] A *gravitational* potential would have the same form, except that it is always attractive and the charges are replaced by the masses of the particles. For either coulomb or gravitational potentials, the potential appropriate for many particles is just the sum of all the two-particle potentials.

Another example would be a *harmonic oscillator*, which can be thought of as a single particle moving in a parabolic potential. For motion in one dimension, the potential is

$$V(x) = \frac{1}{2} kx^2,\tag{1.2}$$

where x is the displacement of the particle from equilibrium and k is a parameter known as the force constant. This potential is appropriate for describing the motion of a mass oscillating at the end of a spring or for describing the motion of atoms vibrating in a molecule or crystal.

EXERCISE 1.1 For the CO molecule, the force constant is 0.0185 N/m. If the molecule is displaced by 1.0 pm (0.01 Å) from its equilibrium position, how much vibrational potential energy is present in the molecule?

Answer $V = \frac{1}{2}(0.0185 \text{ N/m})(1.0 \times 10^{-12} \text{ m})^2 = 9.3 \times 10^{-27} \text{ J} = 0.00557 \text{ J/mol.}$

1.3.2 Newton's Second Law

Newton's second law consists of an equation that determines the evolution of the coordinate and velocity of a particle (or particles) with time, given initial values for the coordinate and velocity. The usual expression for Newton's law (or Newton's equation) for a single particle moving in one dimension is

$$ma = F,\tag{1.3}$$

where m is the mass of the particle, a is its acceleration, and F is the instantaneous force on the particle. The acceleration is related to the velocity v and the coordinate x of the particle by

$$a = \frac{dv}{dt} = \frac{d^2x}{dt^2},$$ (1.4)

and the force is related to the potential energy V by

$$F = -\frac{dV}{dx}.$$ (1.5)

Sometimes we will use the momentum p of the particle rather than its velocity. In Cartesian coordinates, the momentum is related to the velocity by $p = mv$, so that Newton's equation may be written $dp/dt = F$.

By substituting the formulas that relate a and F to x into Newton's equation, we obtain a second-order ordinary differential equation for the evolution of the coordinate x with time:

$$\frac{d^2x}{dt^2} = -\frac{1}{m}\frac{dV(x)}{dx}.$$ (1.6)

This equation is not fully defined until the functional form of V is specified, but once it is, the evolution of x is completely determined by solving the equation.

EXERCISE 1.2   Solve Newton's equation for a harmonic oscillator.

Answer   The force in this case [from Eqs. (1.2) and (1.5)] is $F = -kx$, so the equation to be solved is

$$\frac{d^2x}{dt^2} + \omega^2 x = 0,$$ (1.7)

where

$$\omega = (k/m)^{1/2}$$ (1.8)

is a constant. Equation (1.7) is a differential equation that we shall encounter over and over again in this book. It is not difficult to show (by explicit substitution) that its general solution is of the form (see Appendix A.2 for further discussion)

$$x(t) = A \cos \omega t + B \sin \omega t,$$ (1.9)

where A and B are constants that are determined by the initial conditions in the problem. Since there are two undetermined parameters, there must be two initial conditions, which means that both the initial coordinate x and initial velocity v must be specified to determine the subsequent evolution of the motion with time.

1.3.3 Generalization to Three Dimensions and Many Particles

It is easy to generalize the development just given to three dimensions, provided that we use Cartesian coordinates. If we let $\mathbf{r}$ stand for the vector $x\hat{\mathbf{i}} + y\hat{\mathbf{j}} + z\hat{\mathbf{k}}$ that locates the particle relative to some arbitrary origin, then the acceleration is given by

$$\mathbf{a} = \frac{d\mathbf{v}}{dt} = \frac{d^2\mathbf{r}}{dt^2}, \qquad (1.10)$$

and the force is

$$\mathbf{F} = -\nabla V, \qquad (1.11)$$

where

$$\nabla = \hat{\mathbf{i}}\,\frac{d}{dx} + \hat{\mathbf{j}}\,\frac{d}{dy} + \hat{\mathbf{k}}\,\frac{d}{dz}$$

is the gradient operator. Newton's equation is still $\mathbf{F} = m\mathbf{a}$, but now this is a set of three ordinary differential equations, one for each Cartesian coordinate. It is possible to convert these differential equations to other coordinate systems, such as spherical polar coordinates, if desired, but it is not possible to write down Newton's equations for arbitrary coordinates in the same simple form that one has in Cartesian coordinates. The generalization of classical mechanics to arbitrary coordinate systems requires additional development that is outside the scope of the book.

The generalization of classical mechanics to treat many particles is also simple, provided that we stick to Cartesian coordinates. In this case the vector pointing to the ith particle is denoted $\mathbf{r}_i$, and the force on the ith particle (including that due to the presence of all other particles in the system) is $\mathbf{F}_i = -\nabla_i V$. Newton's equation thus turns into coupled ordinary differential equations of the form

$$\mathbf{F}_i = m_i\mathbf{a}_i, \qquad (1.12)$$

where m_i is the mass of the ith particle.

1.3.4 Energy Conservation and the Classical Hamiltonian

For a single particle moving in one dimension, let us define the classical Hamiltonian function as

$$H = \frac{p^2}{2m} + V(x). \qquad (1.13)$$

The first term on the right side of this expression is the *kinetic energy T* (i.e., $T = p^2/2m$), so clearly, H is the sum of the kinetic energy and the potential energy. This, of course, is just the total energy of the particle.

If one takes the time derivative of H and converts everything to coordinate time derivatives using the chain rule [i.e., $dV/dt = (dV/dx)(dx/dt)$], then, as long as V depends on x and not explicitly on time, we get

$$\frac{dH}{dt} = \frac{dx}{dt}\left[m\frac{d^2x}{dt^2} + \frac{dV}{dx} \right]. \tag{1.14}$$

The term in the square brackets on the right-hand side vanishes, provided that x satisfies Newton's equation [from Eq. (1.6)], which means that H is a constant of the motion (i.e., H doesn't change with time). But to say that H is constant is just to say, classically, that energy is conserved:

$$H = E = \text{constant for systems with } V \text{ independent of time.} \tag{1.15}$$

Systems for which Eq. (1.15) holds are called *conservative* systems.

To illustrate energy conservation for the harmonic oscillator, we simply substitute the expression for $x(t)$ from Eq. (1.9) into Eq. (1.13), then calculate $p = m\,dx(t)/dt$, and substitute into H to get

$$H = E = \frac{1}{2}m\omega^2(A^2 + B^2). \tag{1.16}$$

This expression depends only on constants and thus is itself a constant. Often, it is convenient to express the time dependence of the coordinates in a form in which the conserved energy appears explicitly in the solution. Thus, for the harmonic oscillator, we can write, as an alternative to Eq. (1.9), the formula

$$x(t) = \sqrt{\frac{2E}{m\omega^2}}\cos(\omega t + \delta), \tag{1.17}$$

where δ is a phase factor that is determined from the initial conditions.

EXERCISE 1.3   Suppose that the harmonic oscillator is damped (for example, by frictional forces due to the environment). Then the total force will be $F = -\nabla V(x) - \xi\,dx/dt$, where ξ is a friction coefficient.

1. Write the Newton equation of motion.
2. Show that $x(t) = x_0 e^{\alpha t}$ is a solution of the equation.
3. Solve for α as a function of ξ, m, and k.
4. Describe what the motion is like for $\xi = 0$, for $\xi^2 = 4mk$, and for $\xi^2 > 4mk$.
5. Is energy conserved? Note that the friction force is dependent on time.

Answers

1. $m\,d^2x/dt^2 = -kx - \xi\,dx/dt$.
2. Trying $x(t) = x_0 e^{\alpha t}$, we have $mx_0\alpha^2 e^{\alpha t} = -kx_0 e^{\alpha t} - \xi\alpha x_0 e^{\alpha t}$. This is a solution if $m\alpha^2 = -k - \xi\alpha$.
3. From the quadratic formula, we obtain

$$\alpha = \frac{-\xi \pm \sqrt{\xi^2 - 4mk}}{2m}.$$

4. If $\xi = 0, \alpha = i\omega$, where ω is given by Eq. (1.8). This is the usual harmonic motion (i.e., $x = x_0 e^{i\omega t}$). If $\xi^2 = 4mk$, $\alpha = -\xi/2m$ and $x = x_0 e^{-\xi t/2m}$. This is called critical damping: The oscillator never oscillates, but just decays smoothly to rest ($x = 0$ at $t \rightarrow \infty$). If $\xi^2 > 4mk$, the lower sign is the only appropriate root. (Otherwise the displacement of the system grows without limit.) This is the overdamped limit and again decays smoothly (with no oscillation) to rest.

5. No, the energy of the system is not conserved. This is what usually happens for time-dependent forces.

For one-dimensional problems, we can use the expression for H to generate an exact solution of Newton's equation. To do this, we substitute $p = m\, dx/dt$ into Eq. (1.13), invoke Eq. (1.15), and rearrange terms to get

$$m\frac{dx}{\sqrt{2m(E - V(x))}} = dt. \tag{1.18}$$

If this equation is now integrated over the coordinate x and time t, with the initial condition that the initial coordinate is x_0 and the initial time is t_0, then we obtain

$$m\int_{x_0}^{x} \frac{dx}{\sqrt{2m(E - V(x))}} = t - t_0. \tag{1.19}$$

This reduces the process of solving Newton's equation to integrating the expression under the integral sign on the left-hand side of Eq. (1.19). Sometimes the resulting equation is called an "implicit solution," as the dependence of x on t is only contained implicitly in Eq. (1.19).

EXERCISE 1.4   Find the integral for a harmonic oscillator, and show that it reduces to the same form as Eq. (1.17). The following integral will be useful:

$$\int \frac{dx}{\sqrt{1 - x^2}} = \sin^{-1} x.$$

Answer   For a harmonic oscillator, the potential $V(x) = \frac{1}{2}m\omega^2 x^2$. So Eq. (1.19) becomes

$$t - t_0 = m\int_{x_0}^{x} \frac{dx}{\sqrt{2m\left[E - \frac{1}{2}m\omega^2 x^2\right]}}$$

$$= m\int_{x_0}^{x} \frac{dx}{(2mE)^{\frac{1}{2}}\left[1 - \frac{m\omega^2}{2E}x^2\right]^{\frac{1}{2}}}.$$

Let $A = (m\omega^2/2E)^{1/2}$, $y = Ax$, and $dy = A\, dx$. Then we have

$$t - t_0 = \left(\frac{m}{2E}\right)^{\frac{1}{2}}\int_{Ax_0}^{Ax} \frac{dy}{A\sqrt{1 - y^2}} = \left(\frac{m}{2E}\right)^{\frac{1}{2}}\frac{1}{A}\sin^{-1} y\Big|_{Ax_0}^{Ax},$$

or

$$t - t_0 = \left(\frac{m}{2E}\frac{2E}{m\omega^2}\right)^{\frac{1}{2}}[\sin^{-1} Ax - \sin^{-1} Ax_0] = \frac{1}{\omega}\left[\sin^{-1} Ax - \sin^{-1} Ax_0\right].$$

Now define $\delta = -\pi/2 - \omega t_0 + \sin^{-1} Ax_0$ (a constant). Then $\omega t + \delta + \pi/2 = \sin^{-1} Ax$, and we get

$$x = \frac{1}{A}\sin\left(\frac{\pi}{2} + \omega t + \delta\right),$$

which reduces to Eq. (1.17).

It is possible to generalize the treatment just given to three dimensions and to many particles, provided that the potential function allows the problem to be separated into many one-dimensional ones (i.e., the function is a so-called *separable* potential). In three dimensions, this can sometimes be accomplished by using coordinate systems other than Cartesian. For nonseparable potentials, it is generally not possible to obtain analytical solutions to the equations of motion, but numerical solutions are usually feasible. In fact, the process of integrating the classical equations for the motions of atoms in molecular systems such as proteins (generally known as *molecular dynamics calculations*) is a multimillion-dollar industry.

1.4   CLASSICAL WAVE THEORY

1.4.1 Wave Amplitude

Waves are part of our common experience. If you splash the water in a pool, a traveling wave will be set up that will propagate away from you. One way to describe this motion would be to specify the perpendicular displacement of the surface of the water from its equilibrium position as a function of time and as a function of the distance parallel to the surface. We call this displacement the *wave amplitude*, and if we use the symbol $A(x, t)$ for this amplitude at position x and time t, then a *traveling wave* might be described using the formula

$$A(x, t) = A_0 \cos(kx - \omega t). \tag{1.20}$$

Here, A_0 is a constant that determines the maximum value of the wave amplitude, k is a constant related to the wavelength λ via the equation

$$k = \frac{2\pi}{\lambda}, \tag{1.21}$$

and ω is another constant that is related to the period of oscillation, τ, via the formula

$$\omega = \frac{2\pi}{\tau}. \tag{1.22}$$

We also use the circular frequency $\nu = \omega/2\pi = 1/\tau$.

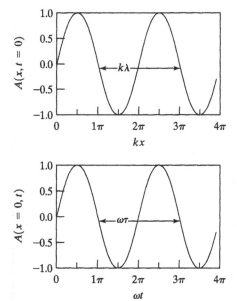

Figure 1.1 Wave amplitude $A(x, t)$ as a function of x for fixed t and as a function of t for fixed x.

Figure 1.1 illustrates the appearance of the wave amplitude in Eq. (1.20). One way to think about this formula is to consider the motion of the wave for a position of constant *phase*. A surfer experiences what this means by positioning the surfboard at a position where the wave amplitude is fixed. In this case, the argument of the cosine function in Eq. (1.20) is set equal to a constant; that is,

$$kx - \omega t = \text{constant,} \tag{1.23}$$

or

$$x = \left(\frac{\omega}{k}\right) t + \text{constant.} \tag{1.24}$$

Equation (1.24) indicates that the wave moves to increasing x with a constant velocity known as the phase velocity $v_p = \omega/k = \nu\lambda$. For electromagnetic waves (e.g., light), the phase velocity is just the speed of light, c.

1.4.2 Superposition and Diffraction

Although the wave amplitude $A(x, t)$ is often associated with the motions of particles, the amplitude has certain properties that are quite different from the properties of particle motion. In particular, if two wave trains collide, the resulting amplitude will be the coherent superposition of the individual amplitudes. In other words, if we denote the individual amplitudes by A_1 and A_2, then the resulting amplitude will be $A = A_1 + A_2$. Obvious examples occur with water waves.

One consequence of this principle of superposition is the diffraction of light in slit experiments. Figure 1.2 shows a beam of light incident on two slits that are

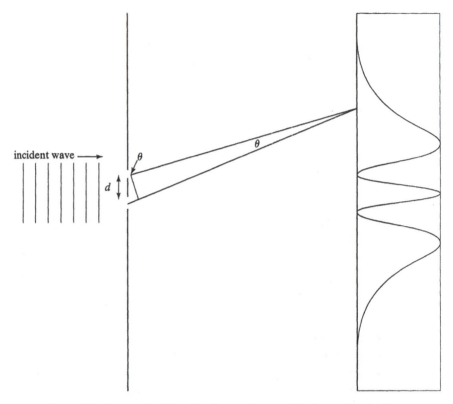

Figure 1.2 Schematic of double-slit experiment, with plane waves incident on slit at left and producing diffraction pattern on screen at right.

separated by a distance d. On the other side of the slits is a screen, on which a diffraction pattern is observed. A formula for the positions of the maxima may be derived by calculating the path lengths between the slits and the screen. Whenever the difference between the lengths equals a multiple of the wavelength, constructive interference occurs. Simple geometry shows that the difference in path lengths is $d \sin \theta$, so the positions of the maxima are given by

$$n\lambda = d \sin \theta. \tag{1.25}$$

1.4.3 Standing Waves

The principle of superposition can be used to define waves in problems where boundary conditions must be imposed. For a violin string, for example, the wave amplitude is forced to vanish at the fixed ends of the string. To accomplish this, we sum up traveling waves [Eq. (1.20)] moving in opposite directions, as in

$$A(x, t) = a \cos(kx - \omega t) + b \cos(kx + \omega t). \tag{1.26}$$

Now we require that $A(x = 0, t) = A(x = L, t) = 0$, where L is the distance between the ends of the string. This leads to the equations

$$(a + b) \cos(\omega t) = 0 \tag{1.27}$$

and

$$a \cos(kL - \omega t) + b \cos(kL + \omega t) = 0. \tag{1.28}$$

The first equation is satisfied if $a = -b$. The second can then be reduced to $2a \sin kL \sin \omega t = 0$ by using the trigonometric identity $\cos(A - B) = \cos A \cos B + \sin A \sin B$. The equation can be satisfied by requiring that $\sin kL = 0$, which in turn requires that $kL = n\pi$, where $n = 1, 2, 3$, etc. As a result, $L = (n/2)\lambda$, which means that only wave amplitudes in which a half-integral number of wavelengths fits between the ends of the string are allowed. (The $n = 0$ solution corresponds to zero displacement everywhere and is possible, but not interesting.) The resulting expression for the wave amplitude is

$$A(x, t) = 2a \sin(\omega t) \sin\left(\frac{n\pi x}{L}\right), \tag{1.29}$$

where the constant a is still arbitrary, as it determines the overall amplitude of oscillation.

1.4.4 Energy Associated with Wave Amplitude

For an oscillating string, each small segment of the string can be considered to undergo periodic oscillation about an equilibrium point, just like a harmonic oscillator. This means that the energy in the string is given by an expression that is essentially the same as a harmonic oscillator—namely, there is a kinetic energy that is proportional to $(dA/dt)^2$ and a potential energy proportional to A^2—a result that holds also for other kinds of waves (water waves, electromagnetic waves, etc.).

1.4.5 Wave Equation

For the traveling wave defined by Eq. (1.20), it is easy to show that the wave amplitude satisfies the partial differential equation

$$\frac{\partial^2 A}{\partial x^2} = \frac{1}{v_p^2} \frac{\partial^2 A}{\partial t^2}, \tag{1.30}$$

provided that the constant v_p equals ω/k. This constant is just the phase velocity, and the equation is called the wave equation. It is the appropriate differential equation that must be satisfied by any wave amplitude (whether it be of traveling waves or not) for many kinds of wave phenomena.

1.5   EARLY HISTORY OF QUANTUM MECHANICS

Quantum mechanics was developed in response to several key experiments that were in conflict with the predictions of classical physics. There were two distinct classes of these experiments, some in which it was found that light could not be described exclusively using wave theory and others in which it was found that particles possessed wavelike properties. We briefly describe a few of these experiments.

1.6   PARTICLE NATURE OF LIGHT

1.6.1 Blackbody Radiation
(M. Planck, *Verh. Deut. Phys. Ges. 2*, 237–45, 1900)

Blackbody radiation is concerned with the distribution of frequencies of light emitted by a heated solid. This distribution can depend on the materials that make up the solid and on the properties of the surface, but becomes independent of materials and surface properties in the ideal limit of a *blackbody* (a cavity that emits radiation). Measurement of the emitted intensity as a function of frequency produces a result shown schematically in Figure 1.3, in which there is a peak at intermediate frequencies whose precise location depends only on the temperature of the blackbody. Classical electromagnetic theory was able to explain the increase in intensity with frequency only at low frequencies. This relationship arises from the increasing number of wavelengths that can satisfy the boundary conditions inside the blackbody (see Section 1.4.3) as the frequency is increased. However, at high frequencies, the observed distribution falls to zero, but classical theory continues to predict a rise in intensity.

In 1900, Planck developed a theory that correctly explained the observed distribution of frequencies. In his theory, he postulated that the energy seen in the blackbody spectrum comes in discrete quanta of magnitude $h\nu$, where h is now called *Planck's constant*. From classical statistical mechanics, the probability of

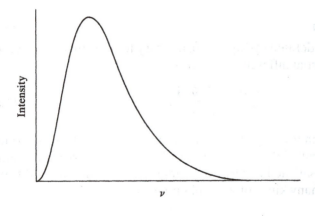

Figure 1.3 Typical blackbody emission spectrum.

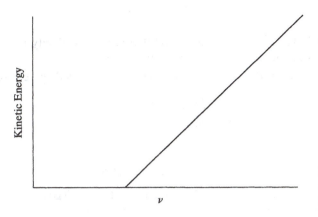

Figure 1.4 Kinetic energy of photoelectrons vs. frequency of light.

each quantum is determined by the Boltzmann distribution, which varies as $\exp(-h\nu/k_BT)$, where k_B is Boltzmann's constant. Planck's theory predicts exponential decay of the frequency distribution at high frequencies, in agreement with observation.

1.6.2 Photoelectric Effect
(A. Einstein, *Ann. Phys. 17*, 132–48, 1905)

The photoelectric effect (discovered by Heinrich Hertz in 1887) refers to the emission of electrons from the surface of a metal irradiated with light. Figure 1.4 shows the observed electron kinetic energy as a function of the irradiating light frequency ν. The figure indicates that there is a certain threshold frequency ν_0 below which there is no emission and that the kinetic energy increases linearly with frequency above this threshold. The quantity $h\nu_0$ is called the *work function*. The kinetic energy is not sensitive to the intensity of the light, a result that violates classical theory.

Einstein realized that photoemission could be explained if one assumes that each electron absorbs one quantum of energy from the radiation field, with the quantum of energy equal to $h\nu$, where h is Planck's constant. The basic formula is

$$\text{K.E.} = h\nu - h\nu_0, \tag{1.31}$$

where $h\nu_0$, the work function, represents the minimum energy needed for electrons to escape from the surface.

EXERCISE 1.5   In a photoelectric effect experiment, if the photocurrent under a certain set of conditions is measured as J amps, and the electrons have velocity v, how will J and v change if

1. The intensity of the incident light is doubled, with no change in frequency?

2. The frequency of the incident light is doubled, but the intensity remains constant? (Assume monochromatic radiation.)

Answers

1. According to Eq. (1.31), the kinetic energy, and therefore the velocity, is unchanged if the intensity doubles. The current J will double.

2. If we label the original and the doubled condition 1 and 2, respectively, then we have

$$\text{K.E.}(1) = h\nu_1 - h\nu_0$$

$$\text{K.E.}(2) = h\nu_2 - h\nu_0.$$

These equations imply that

$$\text{K.E.}(2) - \text{K.E.}(1) = h(\nu_2 - \nu_1) = h(2\nu_1 - \nu_1) = h\nu_1,$$

so the number of electrons will not change, but the kinetic energy of each will go up by $h\nu_1$.

1.7 WAVE NATURE OF PARTICLES

1.7.1 Atomic Spectra and the Bohr Model

In the late 1800s, spectroscopists began to measure the emission spectra of atoms. They noticed that the spectra consisted of discrete lines, often in regular patterns. The spectrum of the hydrogen atom was especially simple, and in particular, Balmer noted in 1885 that the emission frequency of one of the groups of lines could be represented using the simple formula

$$\nu = \text{const}\left(\frac{1}{2^2} - \frac{1}{n^2}\right). \tag{1.32}$$

Balmer's spectrum was successfully explained by Bohr in 1911, using a simple, essentially classical model for the hydrogen atom in which it is imagined that the electron orbits around the proton in circular orbits, with the constraint that the only possible values of the orbital angular momentum are

$$\ell = n\hbar, \tag{1.33}$$

where $\hbar = h/2\pi$ and $n = 1, 2, 3, \ldots$. Spectral transitions are associated with discontinuous changes in the value of n, with the energy difference ΔE between the initial and final states related to the frequency ν of the light by $\Delta E = h\nu$. Balmer's formula, Eq. (1.32), is thus to be interpreted as representing the difference in energy between two possible energy levels in which the final value of n is 2 and the initial value is $n = 3, 4, 5, \ldots$.

Although Bohr's theory (or better yet, Sommerfeld's improved version of Bohr's theory) works spectacularly for the hydrogen atom and other one-electron atoms like He^+, and works well for pseudo one-electron atoms like Li and Na, it fails seriously for atoms such as He, in which the two electrons interact strongly. This shortcoming shows that the Bohr theory is incomplete, and we now know that Bohr theory is an approximation to the correct formulation of quantum mechanics.

1.7.2 De Broglie Waves and Electron Diffraction

In 1923, de Broglie, in his Ph.D. thesis, postulated that electrons and other particles have waves associated with them and that the wavelength of these waves is given by $\lambda = h/p$, where p is the momentum of the particle. By assuming that an integral number of wavelengths must fit into Bohr's circular orbits, one can re-derive his formulas for the energy levels of the hydrogen atom. Thus, de Broglie's theory is compatible with Bohr's, but actually goes much further. If the particles have waves associated with them, then it must be possible to observe particle diffraction similar to the diffraction of light described in Section 1.4.2. Such particle diffraction, however, had not been observed at the time that de Broglie wrote his thesis, but shortly thereafter (1927) Davisson and Germer observed the diffraction of electrons, and in 1932 Stern observed He and H_2 diffraction, proving that particles showed the wavelike property of diffraction.

1.8   UNCERTAINTY PRINCIPLE

In 1925, Heisenberg realized that a consequence of the de Broglie waves is the uncertainty principle, which states that it is impossible to measure the momentum and position of a particle simultaneously to arbitrary precision. To show how the principle works, we consider again the double-slit experiment of Section 1.4.2 (see Figure 1.2), this time imagining that particles rather than light are undergoing diffraction. If, originally, the particle is moving along the z-axis, the diffraction introduces a nonzero momentum in the x direction (i.e., perpendicular to the direction of the incident beam). Since diffraction changes the *direction*, but not the magnitude, of the momentum, the range of p_x values, Δp_x, is found from the equation

$$p_x = p_0 \sin \theta = \frac{\Delta p_x}{2}, \tag{1.34}$$

where the last equality arises because it is just as likely for the particle to be scattered up as down. It is reasonable to choose the angle θ so that it refers to the position of the first diffraction minimum outside the central beam. From the figure, this is half the angle associated with the $n = 1$ maximum in Eq. (1.25); that is,

$$\sin \theta \cong \frac{\lambda/2}{d}, \tag{1.35}$$

where d is the size of the slit. If this is now substituted into the previous equation, we find that

$$\Delta p_x \cong \frac{p_0 \lambda}{d}. \tag{1.36}$$

Defining Δx as the uncertainty in the x-position of the particle (which is just d), we find that the uncertainty product is

$$(\Delta x)(\Delta p_x) \cong p_0 \lambda = h, \tag{1.37}$$

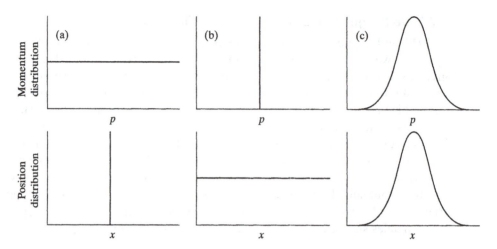

Figure 1.5 Position (x) and momentum (p) distributions, including the limiting cases in which (a) x is specified exactly, (b) p is specified exactly, and (c) both x and p are uncertain.

where we have substituted the de Broglie formula in the second equality. A more rigorous evaluation of the uncertainty product leads to the inequality (the Heisenberg uncertainty principle)

$$\Delta x \Delta p_x \geq \frac{\hbar}{2}. \tag{1.38}$$

Figure 1.5 illustrates the complementary roles of the uncertainties in the momentum and position implied by the uncertainty relationship. In Part (a), we consider the case in which the position is specified exactly, which means that the momentum is not determined. Part (b) shows the reverse situation, and Part (c) shows the in-between case in which there is uncertainty in both x and p.

1.9 DISCOVERY OF QUANTUM MECHANICS

1.9.1 Schrödinger Wave Mechanics and Heisenberg Matrix Mechanics

By 1925, the idea that particles have wavelike properties had been established, but the meaning and form of the wave amplitude, or what we shall now call the wavefunction ψ, was unclear. While on a Christmas holiday in 1925–26, Schrödinger discovered the wave equation that bears his name [E. Schrödinger, *Ann. Phys.* 79, 361–76 (1926)] and solved it for the hydrogen atom, showing that it reproduced the result of Bohr theory, but without making the *ad hoc* assumptions of the Bohr model. In Section 1.10 we present a qualitative motivation for the functional form of the Schrödinger equation, and then in Chapter 2 we present the detailed theory, starting with the postulates of quantum mechanics.

In spite of Schrödinger's success, the physical meaning of ψ was still unclear at the time of his first work, and it remained for Born and the Copenhagen Interpretation to postulate that ψ is a probability amplitude, which means that $|\psi|^2$ is the probability density for finding the particle at a particular location. About the same time that Schrödinger was developing "wave mechanics," Heisenberg developed a matrix approach to the evaluation of measureable quantities (so-called matrix mechanics) that was quickly recognized to be equivalent to wave mechanics. Since the Heisenberg formalism is not needed in this book, we will not present it.

1.9.2 Relativistic Quantum Mechanics

Despite the success of wave mechanics and matrix mechanics, it was realized from the beginning that these theories were incomplete in the sense that they are based on the nonrelativistic formula for the relation between the kinetic energy and the momentum (i.e., $E = p^2/2m$). Several researchers, including Schrödinger, attempted to develop a relativistic generalization of wave mechanics that would be based on the relativistic formula $E = \left(p^2c^2 + m^2c^4\right)^{1/2}$, where m is the particle's rest mass, but the square root in this formula causes serious difficulties in developing a wave equation that is similar in structure to the Schrödinger equation. Eventually, Dirac (1929) realized that the way around this problem is to "take the square root" by introducing an additional degree of freedom into the description of the wavefunction. This degree of freedom is what we now call *spin*, and the resulting relativistic wave equation is called the *Dirac equation*. It is interesting to note that the application of the Dirac equation to the motions of electrons, subject only to coulomb interactions between them, correctly describes all the properties of electrons that we are aware of. The corresponding description of protons, neutrons, and other heavy particles requires the introduction of nuclear forces.

We will not need to develop the Dirac equation in this book, but certain results from it, such as the properties of electron spin and the spin–orbit interaction (the magnetic coupling between electron spin and orbital angular momentum), will be used where needed.

1.10   CONCEPTS IN QUANTUM MECHANICS

The history that we have just presented demonstrates that particles must be described as waves, with an associated wavefunction ψ that behaves in many respects like the amplitude of a classical wave, exhibiting diffraction phenomena and satisfying boundary conditions that lead to discrete energy levels that show up in spectroscopy. There are, however, some important distinctions between quantum waves and classical waves, and these are responsible both for the difference between the Schrödinger equation and the classical wave equation and for the relationship between the energy in the wave and the amplitude of the wave.

Earlier, we noted that in classical wave theory, solutions of the wave equation for a traveling wave were of the form $\cos(kx - \omega t)$. In quantum mechanics, if we want $|\psi|^2$ to stand for the probability density, then the spatial oscillations associated with $\cos(kx - \omega t)$ would not be appropriate for a uniform beam of free particles moving in a vacuum. Instead, the functional form of ψ that works for free particles is

$$\psi \approx \exp[i(kx - \omega t)] \quad (i^2 = -1), \tag{1.39}$$

for in this case $|\psi|^2$ is simply a constant, which is physically what one would expect. (See Appendix A.1 for a discussion of the meaning of the complex exponential and other issues related to the use of complex numbers.)

Assuming that Eq. (1.39) is correct (which it is), the question then arises as to how to interpret the constants k and ω. For that, we use two formulas we developed in the previous section, namely,
(1) de Broglie's formula $\lambda = h/p$, which relates wavelength to momentum and may here be rewritten

$$p = \frac{h}{\lambda} = \hbar k \quad (\hbar = h/2\pi) \tag{1.40}$$

and (2) the Planck–Einstein relationship $E = h\nu$, which, in the present application, becomes

$$E = \hbar\omega. \tag{1.41}$$

If these two formulas are now substituted into Eq. (1.39), we obtain

$$\psi \approx \exp\left[\frac{i}{\hbar}(px - Et)\right], \tag{1.42}$$

in which only the physical properties E and p of the particle appear. This expression for ψ may now be used to develop a wave equation. Since, for a free particle, E and p are related by $E = p^2/2m$, by simple substitution we can show that the first time derivative of Eq. (1.42) is proportional to the second derivative of that equation with respect to the coordinate x; specifically,

$$i\hbar \frac{\partial \psi}{\partial t} = \frac{-\hbar^2}{2m} \frac{\partial^2 \psi}{\partial x^2}. \tag{1.43}$$

This is the Schrödinger equation for a free particle. Note that the formula involves a first derivative with respect to time and a second derivative with respect to the coordinate x. This differs from the classical wave equation, (Eq. 1.30), which involves second derivatives with respect to both t and x, but it is a natural consequence of the relationship $E = p^2/2m$ that is appropriate for the motion of a free particle. In fact, the classical wave amplitude $\cos(kx - \omega t)$ is physically inappropriate for a free particle, as it is not compatible with the interrelationship between E and p. This means, then, that for a free particle, we are forced to use complex wavefunctions for ψ, and there are a number of consequences of this constraint that pertain to the nature of the functions that represent acceptable solutions of the Schrödinger equation (in so-called *Hilbert* space). In addition, we see that there is a natural association in quantum mechanics between certain differential expressions and certain dynamical variables, notably, (1) between the kinetic energy $p^2/2m$ and the expression $(-\hbar^2/2m)(\partial^2/\partial x^2)$ and also (2) between the energy E and the expression $i\hbar(\partial/\partial t)$. These differential expressions are examples of quantum mechanical *operators*. What we mean by that will be defined in the next chapter, but we note that the association demonstrated here between operators and dynamical variables will play an important role throughout quantum mechanics.

EXERCISE 1.6   If we formally replace t by it in the free-particle Schrödinger equation and define $D = \hbar/2m$, we obtain a different result, one that is familiar from classical physical chemistry.

1. What dimensions (units) does D have?
2. A solution of the modified Schrödinger equation is

$$\psi(x) = \frac{ae^{-x^2/4Dt}}{\sqrt{Dt}}.$$

   Plot $\psi(x)$ for $t = 0.05, 0.1$, and 1.0, taking $D = 1$ and $a = 1$.
3. Given the units of D and the behavior of ψ, what is the new equation called? (*Hint*: The letter is D, not a or b.)

Answer

1. $D = $ (energy)(time)/mass $= $ (length)^2/time
2. See Figure 1.6.

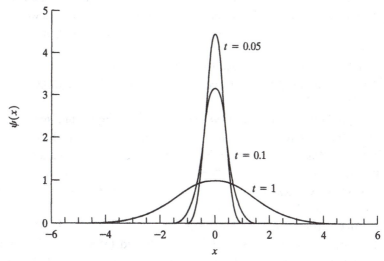

Figure 1.6   Plot of $\psi(x)$ versus x for the parameters in Exercise 1.6.

> **3.** It is the diffusion equation, sometimes called Fick's law. The replacement $t \to it$ in this fashion actually leads to a useful way to find solutions of the Schrödinger equation that are of very high accuracy.

The "derivation" of quantum mechanics given here is really just a plausibility argument. In fact, there is no derivation of quantum mechanics from some other theory. Instead, quantum mechanics is based on several assumptions, which we will call postulates. Chapter 2 presents these postulates, thus giving us the formal theory that underlies quantum mechanics. But it should be noted that the only justification for the postulates is that they lead to predictions about the properties of atoms and molecules and about spectroscopy, kinetics, and other physical phenomena that are in accord with experiment. There have been many tests of quantum mechanics over the years—many with extraordinary precision—and at this point there is no indication of error in any predictions. Thus, we can truly say that quantum mechanics works!

SUGGESTED READINGS

The following are standard quantum chemistry textbooks that give additional introductory material:

P. W. Atkins and R. S. Friedman, *Molecular Quantum Mechanics*, 3rd ed. (Oxford, New York, 1997).

C. E. Dykstra, *Quantum Chemistry and Molecular Spectroscopy* (Prentice Hall, Englewood Cliffs, NJ, 1992).

C. E. Dykstra, *Introduction to Quantum Chemistry* (Prentice Hall, Englewood Cliffs, NJ, 1994).

R. L. Flurry, Jr., *Quantum Chemistry* (Prentice-Hall, Englewood Cliffs, NJ, 1983).

W. H. Flygare, *Molecular Structure and Dynamics* (Prentice-Hall, Englewood Cliffs, NJ, 1978).

M. W. Hanna, *Quantum Mechanics in Chemistry* (Benjamin/Cummings, Menlo Park, CA, 1981).

H. F. Hameka, *Quantum Mechanics* (Wiley, New York, 1981).

W. Hehre, L. Radom, P. vR. Schleyer and J. A. Pople, *Ab Initio Molecular Orbital Theory* (Wiley, New York, 1986).

M. Karplus and R. N. Porter, *Atoms and Molecules* (Benjamin, New York, 1970).

W. Kauzmann, *Quantum Chemistry* (Academic Press, New York, 1957).

I. N. Levine, *Quantum Chemistry*, 5th ed. (Prentice Hall, Upper Saddle River, NJ, 2000).

J. P. Lowe, *Quantum Chemistry*, 2nd ed. (Academic Press, New York, 1993).

P. S. C. Matthews, *Quantum Chemistry of Atoms and Molecules* (Cambridge University Press, Cambridge, U.K., 1986).

D. A. McQuarrie, *Quantum Chemistry* (University Science Books, Mill Valley, CA, 1983).

F. L. Pilar, *Elementary Quantum Chemistry*, 2nd ed. (McGraw-Hill, New York, 1990).

H. L. Strauss, *Quantum Mechanics, an Introduction* (Prentice-Hall, Englewood Cliffs, NJ, 1968).

PROBLEMS

1.1 The vibrational motion of the H_2 molecule can be modeled roughly as a harmonic oscillator. If the frequency $\bar{\nu}(\equiv \nu/c)$ of H_2 is 4,395.2 cm^{-1}, evaluate the force constant k. (The appropriate mass is the reduced mass (see problem 3) $\mu = m_a m_b/(m_a + m_b) = m/2$ where $m_a = m_b = m$ is the hydrogen atom mass.) Also, compute the force acting at the minimum of the potential-energy curve and at the turning points (where the kinetic energy vanishes), assuming that the oscillator energy is 2,197.6 cm^{-1} (the lowest energy that the quantum oscillator may have). Finally, calculate the largest value of the vibrational velocity for the energy just given, and compare it with the average thermal velocity of H_2, (defined by $\left(m v_{th}^2/2\right) = \frac{3}{2}k_B T$, where k_B is Boltzmann's constant and m is the mass of H_2).

1.2 (a) Consider the classical motion of a single particle in three dimensions. The kinetic energy of the particle is

$$T = \frac{1}{2}m(\dot{x}^2 + \dot{y}^2 + \dot{z}^2) = \frac{\mathbf{p}^2}{2m},$$

where $\dot{x} = dx/dt$, etc. Now introduce spherical polar coordinates

$$x = r \sin\theta \cos\phi,$$
$$y = r \sin\theta \sin\phi,$$

and

$$z = r \cos\theta.$$

Show that

$$T = \frac{1}{2}m(\dot{r}^2 + r^2(\dot{\theta}^2 + \sin^2\theta\dot{\phi}^2)).$$

(b) Now introduce the angular-momentum vector $\boldsymbol{\ell}$ of the particle; that is,

$$\boldsymbol{\ell} = \mathbf{r} \times \mathbf{p}.$$

Show that

$$\ell^2 = \ell_x^2 + \ell_y^2 + \ell_z^2 = m^2 r^4(\dot{\theta}^2 + \sin^2\theta\dot{\phi}^2).$$

(c) Combine the results of (a) and (b) to show that

$$\mathbf{p}^2 = p_r^2 + \frac{\ell^2}{r^2},$$

where $p_r = m\dot{r}$ is called the *radial momentum* of the particle.

1.3 Consider a diatomic molecule with atoms labeled a and b and with a classical Hamiltonian given by

$$H = \frac{1}{2}m_a(\dot{x}_a^2 + \dot{y}_a^2 + \dot{z}_a^2) + \frac{1}{2}m_b(\dot{x}_b^2 + \dot{y}_b^2 + \dot{z}_b^2) + V(r),$$

where $r = \left[(x_a - x_b)^2 + (y_a - y_b)^2 + (z_a - z_b)^2\right]^{1/2}$ is the distance between the atoms and $\mathbf{r}_a = (x_a, y_a, z_a)$ and $\mathbf{r}_b = (x_b, y_b, z_b)$ are vectors that locate each atom.

(a) Show, by using the variables $\mathbf{R} = (X, Y, Z)$ and $\mathbf{r} = (x, y, z)$ defined by

$$\mathbf{R} = \frac{m_a \mathbf{r}_a + m_b \mathbf{r}_b}{m_a + m_b}$$

and

$$\mathbf{r} = \mathbf{r}_a - \mathbf{r}_b,$$

that H becomes

$$H = \frac{1}{2} M(\dot{X}^2 + \dot{Y}^2 + \dot{Z}^2) + \frac{1}{2} \mu(\dot{x}^2 + \dot{y}^2 + \dot{z}^2) + V(r),$$

where

$$M = m_a + m_b$$

and

$$\mu = \frac{m_a m_b}{(m_a + m_b)}.$$

(b) What are Newton's equations in terms of X, Y, Z, x, y and z? How has the transformation to center-of-mass coordinates uncoupled these equations of motion? This transformation also works in the quantum description of two or more particles.

1.4 Using the de Broglie and Planck–Einstein formulas, and considering (a) a helium atom with a kinetic energy of 0.025 eV (i.e., thermal energy at 300 K) and (b) a baseball (mass = 500 g) moving at 49 m/s, evaluate and interpret the wavelength and frequency associated with the wave motion of each. How do these results compare with the size of an atom $(1\ \text{Å})$ and the diameter of a nucleus $(1 \times 10^{-5}\ \text{Å})$?

1.5 If a given crystal with fixed spacing d between the atomic planes is studied both by electron diffraction and by neutron diffraction, and if the electrons and neutrons have the same velocity, how will the scattering angles θ differ in the two experiments?

1.6 The X rays used for both scientific and medical applications are usually produced by making high-velocity electrons collide with metallic targets. For electrons accelerated in a field of 75,000 volts,

(a) compute the wavelength of the X rays produced.

(b) if this same kinetic energy were to be transferred to a proton, how fast would that proton be traveling?

2

$$i\hbar \frac{\partial}{\partial t} \Psi = \hat{H}\Psi$$

Quantum Theory

Since our overall aim is to understand the use of quantum mechanics in chemistry, formal quantum mechanics will be dealt with in the most straightforward way: by postulate. All the postulates except one are given in this chapter, and such crucial concepts as the Schrödinger equation, eigenvalues, expectation values, operator properties, and measurement theory are introduced.

2.1 POSTULATES OF QUANTUM MECHANICS

The postulates of quantum mechanics are simply the basic assumptions that define the theory. There is no unique list, as quantum mechanics may be developed in many different ways. However, the treatment we set forth next is common, and it follows logically from the historical development of the theory described in Chapter 1.

2.2 DEFINITION OF Ψ AND $|\Psi|^2$

The first postulate defines what we mean by the wavefunction. We begin by considering a single particle. Associated with this particle is a wavefunction $\Psi(x, y, z, t)$ that depends on the position x, y, z of the particle and on the time t. We then have the first postulate.

Postulate 1: $\Psi(x, y, z, t)$ is a well-behaved, square integrable function, and $|\Psi(x, y, z, t)|^2 \, dx \, dy \, dz$ is the probability of finding the particle between the points (x, y, z) and $(x + dx, y + dy, z + dz)$ at time t.

By square integrable, we mean that $\int |\Psi(x, t)|^2\, dx\, dy\, dz$ is finite, where the limits of integration are over all physically meaningful values of x, y, z, which typically means that $-\infty < x < \infty$, $-\infty < y < \infty$, and $-\infty < z < \infty$. "Well behaved" means that Ψ is a single-valued function; generally, Ψ is also a smooth and continuous function. We will get more specific about what this means later, but for now we note that wavefunctions with discontinuities are not permitted, nor can we have solutions that diverge in such a way that $|\Psi(x, y, z, t)|^2\, dx\, dy\, dz$ cannot be integrated.

Let us generalize the preceding statements to many particles. In that case, Ψ becomes a function of the coordinates of all the particles, which we take to be $x_1, y_1, z_1, x_2, y_2, z_2, \ldots$. The integration volume is $dx_1\, dy_1\, dz_1\, dx_2\, dy_2\, dz_2 \ldots$, which we abbreviate by $d\tau$. Later on, when we introduce spin coordinates, we will add them to the list of coordinates. The second part of Postulate 1 now states that $|\Psi(x_1, y_1, z_1, x_2, y_2, z_2, \ldots, t)|^2\, d\tau$ is the probability of finding particle 1 between the points (x_1, y_1, z_1) and $(x_1 + dx_1, y_1 + dy_1, z_1 + dz_1)$, particle 2 between the points (x_2, y_2, z_2) and $(x_2 + dx_2, y_2 + dy_2, z_2 + dz_2)$, etc., at time t. If the particles are indistinguishable, then the probability density refers to any particle rather than a specific particle.

It is conventional to define the integral over all space of $|\Psi|^2$ to be unity. Wavefunctions that satisfy this formula are said to be *normalized*. The general normalization condition is

$$\int |\Psi^2|\, d\tau = 1. \tag{2.1}$$

EXERCISE 2.1   We saw in the last chapter that the wavefunction $\Psi(x, t) = e^{i(\omega t - kx)}$ is appropriate for a free particle. Show that this wavefunction cannot be normalized.

Answer   For a function $\Psi(x, t)$ to be normalizable, we require $\int |\Psi(x, t)|^2\, dx$ to be finite. In the case of the plane-wave solution, we have

$$\Psi(x, t) = e^{i(\omega t - kx)},$$

$$\Psi^*(x, t) = e^{-i(\omega t - kx)},$$

and

$$\Psi^*\Psi = 1,$$

so that

$$\int_{-\infty}^{\infty} |\Psi(x, t)|^2\, dx = \int_{-\infty}^{\infty} dx.$$

This integral is not finite, so Ψ cannot be normalized. In spite of this, Ψ is still a physically meaningful wavefunction.

2.3   OPERATORS

2.3.1 Definition of Operators Corresponding to Observables

In Chapter 1, we mentioned that dynamical variables are associated with quantum mechanical operators. In this and the next three sections, we define this relationship in more detail. We begin by defining the connection between observables and operators.

> **Postulate 2:** For every observable property, there is a linear Hermitian operator that is obtained by expressing the classical form in Cartesian coordinates x and momenta p_x and making the replacements

$$x \rightarrow \hat{x} \tag{2.2a}$$

and

$$p_x \rightarrow \hat{p}_x = \frac{\hbar}{i} \frac{\partial}{\partial x}. \tag{2.2b}$$

Here we have put a caret ($^\wedge$) on x to emphasize that it is an operator. (We will omit the caret if the context makes its presence obvious.) In some cases it is necessary to write the classical form in a symmetrical way in order to generate an operator that is Hermitian; we shall illustrate this with examples.

Let us define the terms in Postulate 2. First, an *operator* is a symbol that describes what to do to what follows it. The differential d/dx is an example of an operator. When it operates on a function $f(x)$, it tells us to take the derivative of $f(x)$. The coordinate $\hat{x}$ can be thought of as a multiplicative operator when it multiplies the function $f(x)$ in the expression $\hat{x}f(x)$. Table 2.1 gives some additional examples.

TABLE 2.1   EXAMPLES OF OPERATIONS

Operator	Quantity	Result
$\partial/\partial x$	$f(x)$	$\partial f/\partial x$
$\hat{x}$	$f(x)$	$xf(x)$
c (a constant)	$f(x)$	$cf(x)$
$\partial/\partial x$	$x^2 + a$	$2x$
$\hat{x}$	$x^2 + a$	$x^3 + ax$
$\int dx$	$x^2 + a$	$x^3/3 + ax + c$
$x+$	$x^2 + a$	$x^2 + x + a$
$\sqrt{}$	x	$x^{1/2}$

A *linear* operator $\hat{B}$ is an operator that satisfies the (linearity) property that

$$\hat{B}(\phi + c\psi) = \hat{B}\phi + c\hat{B}\psi, \tag{2.3}$$

where ϕ and ψ are arbitrary functions and c is a constant.

A *Hermitian* operator B satisfies the equation

$$\int_{\text{all space}} \psi(B\phi)^* \, d\tau = \int_{\text{all space}} \phi^* B\psi \, d\tau, \tag{2.4}$$

where ϕ and ψ are well behaved, but otherwise arbitrary, wavefunctions that satisfy boundary conditions which are appropriate for the problem being considered. An important property of Hermitian operators that will be used later is that

$$\int \psi^* B\psi \, d\tau = \int \psi(B\psi)^* \, d\tau = \text{a real number.} \tag{2.5}$$

EXERCISE 2.2   **a.** Is x Hermitian?

Answer   This result can easily be proved by direct substitution.

   b. Is p_x Hermitian?

Answer   The proof is complicated, but it works like this (here we omit integrations over the coordinates y and z, as they carry through trivially): Integrating by parts, we obtain

$$\int_{-\infty}^{\infty} \phi^*(x)(-i\hbar)\frac{\partial}{\partial x}\psi(x)\,dx = (-i\hbar)\phi^*(x)\psi(x)\Big|_{-\infty}^{\infty} - \int_{\infty}^{\infty}(-i\hbar)\frac{\partial\phi^*}{\partial x}\cdot\psi\,dx. \tag{2.6a}$$

If ϕ and ψ are wavefunctions associated with a particle that is bound, then these wavefunctions must vanish at $x \to \infty$ to be square integrable. Thus, we must have

$$\int_{-\infty}^{\infty} \phi^*(x)(-i\hbar)\frac{\partial}{\partial x}\psi(x)\,dx = -\int_{-\infty}^{\infty}\psi(-i\hbar)\frac{\partial}{\partial x}\phi^*\,dx$$

$$= \int_{-\infty}^{\infty}\psi\left((-i\hbar)\frac{\partial}{\partial x}\right)^*\phi^*\,dx \tag{2.6b}$$

$$= \int\psi\left[(-i\hbar)\frac{\partial}{\partial x}\phi\right]^*\,dx.$$

This means that $-i\hbar(\partial/\partial x)$ is Hermitian, but $(\partial/\partial x)$ is not.

   c. Is $xp_x x$ Hermitian?

Answer   This proof is more complicated, but it works through in much the same way as for p_x. Note that the nonsymmetrical form of the same operator, xxp_x, is not Hermitian.

2.3.2 Other Properties of Operators Corresponding to Observables

   1. *Operator multiplication is associative and distributive.* This means that when a product of three operators, say, $\hat{A}\hat{B}\hat{C}$, operates on a function $f(x)$, we can write $\hat{A}\hat{B}\hat{C}(x) = \hat{A}(\hat{B}[\hat{C}f(x)])$. As an example, suppose that

$$\hat{A} = x, \hat{B} = 2, \hat{C} = \frac{\partial}{\partial x},$$

and

$$f = x^2.$$

Then $\hat{A}\hat{B}\hat{C}f = x \cdot 2 \cdot \partial/\partial x \cdot x^2 = x \cdot 2 \cdot 2x = 4x^2.$

2. *Operator multiplication is not necessarily commutative.* In other words, the operators $\hat{A}\hat{B}$ and $\hat{B}\hat{A}$ may give different results when applied to a function $f(x)$.

Here it is convenient to define the *commutator* of two operators as

$$[\hat{A}, \hat{B}] \equiv \hat{A}\hat{B} - \hat{B}\hat{A}. \tag{2.7}$$

Then if $\hat{A}\hat{B}f = \hat{B}\hat{A}f$ for all f, we say that $\hat{A}$ and $\hat{B}$ commute. The following are examples of commutative and noncommutative operators.

$[x, x] = 0$; (commutative)

$[x, \partial/\partial x] \neq 0$; (noncommutative)

If $\hat{A} = \sqrt{\ }, \hat{B} = 1-$;

$\hat{A}\hat{B}x = \sqrt{1 - x}$ and $\hat{B}\hat{A}x = 1 - \sqrt{x}$; so $[\hat{A}, \hat{B}] \neq 0$. (noncommutative)

2.3.3 Examples of Important Operators in Quantum Mechanics

Among the most important operators in quantum mechanics are those that refer to the energy of individual particles. Using the prescription of Postulate 2, we find that the kinetic-energy operator in one dimension is given by

$$\frac{\hat{p}_x^2}{2m} \rightarrow \frac{1}{2m}\left(\frac{\hbar}{i}\frac{\partial}{\partial x}\right)^2 = -\frac{\hbar^2}{2m}\frac{\partial^2}{\partial x^2}. \tag{2.8a}$$

In three dimensions, we have

$$\frac{\hat{p}_x^2 + \hat{p}_y^2 + \hat{p}_z^2}{2m} \rightarrow \frac{-\hbar^2}{2m}\left(\frac{\partial^2}{\partial x^2} + \frac{\partial^2}{\partial y^2} + \frac{\partial^2}{\partial z^2}\right) = \frac{-\hbar^2}{2m}\nabla^2, \tag{2.8b}$$

where ∇^2 is the Laplacian operator defined by the term in parentheses on the left. If this expression for the kinetic-energy operator is added to that for the potential-energy operator (a coordinate-dependent operator), we obtain the following expression for the Hamiltonian operator [analogous to the classical Hamiltonian of Eq. (1.13)]:

$$\hat{H} = \frac{-\hbar^2}{2m}\nabla^2 + V(x, y, z). \tag{2.9a}$$

For a system that consists of n particles, each of mass m_i, interacting via a potential V, we have

$$H = \frac{-\hbar^2}{2}\sum_{j=1}^{n}\frac{\nabla_j^2}{m_j} + V. \tag{2.9b}$$

Another set of important operators in quantum mechanics are those corresponding to the angular momentum ℓ. For a single particle, the classical angular momentum is $\boldsymbol{\ell} = \mathbf{r} \times \mathbf{p}$. The quantum analogue is easily constructed by replacing $\mathbf{r}$ and $\mathbf{p}$ by their nonclassical counterparts. Table 2.2 gives some specific examples.

TABLE 2.2 CLASSICAL AND QUANTUM
MECHANICAL OPERATORS

Classical	Quantum mechanical
$\ell_x = yp_z - zp_y$	$\hat{\ell}_x = -i\hbar\left(y\dfrac{\partial}{\partial z} - z\dfrac{\partial}{\partial y}\right)$
$\ell_y = zp_x - xp_z$	$\hat{\ell}_y = -i\hbar\left(z\dfrac{\partial}{\partial x} - x\dfrac{\partial}{\partial z}\right)$
$\ell_z = xp_y - yp_x$	$\hat{\ell}_z = -i\hbar\left(x\dfrac{\partial}{\partial y} - y\dfrac{\partial}{\partial x}\right)$

EXERCISE 2.3 Prove directly that x commutes with ℓ_x and that ℓ_x commutes with p_x, but that p_x does not commute with x.

Answer We have $[x, \ell_x] = [x, yp_z - zp_y]$. Then, using $[A, BC] = [A, B]C + B[A, C]$, we obtain $[x, \ell_x] = [x, y]p_z + y[x, p_z] - [x, z]p_y - z[x, p_y] = 0$. Similarly, one can show that $[p_x, \ell_x] = [p_x, yp_z - zp_y] = [p_x, y]p_z + y[p_x, p_z] - [p_x, z]p_y - z[p_x, p_y] = 0$. By contrast (when acting on a function ψ) the evaluation of $[p_x, x]$ yields $[p_x, x]\psi = \hbar/i[d/dx(x\psi) - x\, d/dx\, \psi] = \hbar/i\psi$, which implies that $[p_x, x] = \hbar/i \neq 0$.

2.4   TIME-DEPENDENT AND TIME-INDEPENDENT SCHRÖDINGER EQUATIONS

Postulate 3: The wavefunction $\Psi(x, t)$ is obtained by solving the equation

$$i\hbar\,\frac{\partial}{\partial t}\,\Psi = \hat{H}\Psi, \tag{2.10}$$

which is known as the *time-dependent Schrödinger equation.*

This is a partial differential equation in spatial coordinates and time that determines the evolution of Ψ, and it says that the action of the Hamiltonian operator on the wavefunction gives the same result as the operator $i\hbar\partial/\partial t$. We derived the same equation heuristically in Section 1.10.

In the special circumstance where the Hamiltonian is independent of time, the operator on the right-hand side of Eq. (2.10) depends only on the spatial coordinates, while the operator on the left-hand side depends only on time. In that case, it is possible to solve the time-dependent Schrödinger equation by a method known as *separation of variables*. In this method, the wavefunction is written as a product of a temporal and a spatial part:

$$\Psi(x, t) = \psi(x)f(t). \tag{2.11a}$$

Upon substituting this into the Schrödinger equation, dividing both sides by $\psi(x)f(t)$, and canceling where possible, one obtains

$$i\hbar \frac{1}{f(t)} \frac{df(t)}{dt} = \frac{H\psi(x)}{\psi(x)}. \tag{2.11b}$$

In this equation, the left-hand side depends on time only, while the right-hand side depends on the coordinate x only. Since these are independent variables in quantum mechanics, the left- and right-hand sides can be equal only if they are both equal to a constant. If we call the constant E, then Eq. (2.11b) reduces to the two equations

$$i\hbar \frac{df}{dt} = Ef \tag{2.12}$$

and

$$H\psi = E\psi. \tag{2.13}$$

Rearranging Eq. (2.12) into the form

$$\frac{df}{f} = \frac{-i}{\hbar} E\, dt,$$

followed by quadrature and then exponentiation (and ignoring the integration constant) yields

$$f(t) = e^{-iEt/\hbar}.$$

Overall, then, we have

$$\Psi(x,t) = e^{-iEt/\hbar} \psi(x).$$

Equation (2.13) is called the *time-independent Schrödinger equation.* It is an example of an *eigenvalue–eigenfunction* equation, whereby applying the Hamiltonian operator to the wavefunction gives the parameter E (the eigenvalue) times the wavefunction (the eigenfunction).

Since the Hamiltonian operator corresponds to the energy of the system, the constant E can be thought of as the total energy. In addition, note that the form of $f(t)$ yields

$$\Psi^*\Psi = \psi^*\psi. \tag{2.14}$$

This means that the probability density is independent of time for this solution, so the wavefunction $\psi(x)$ can be thought of as the *stationary state* associated with the Hamiltonian.

EXERCISE 2.4   Suppose that, for some particular problem, the time-dependent wavefunction is

$$\Psi(x, t) = \psi_1 e^{-iE_1 t/\hbar} + \psi_2 e^{-iE_2 t/\hbar},$$

with ψ_1 and ψ_2 real. Is the probability $\Psi^*\Psi = P(x)$ time independent? If $E_1 = E_2$, is $P(x)$ time independent?

Answer   We have

$$\Psi^*\Psi = (\psi_1 e^{iE_1 t/\hbar} + \psi_2 e^{iE_2 t/\hbar})(\psi_1 e^{-iE_1 t/\hbar} + \psi_2 e^{-iE_2 t/\hbar})$$

$$= \psi_1^2 + \psi_2^2 + 2\psi_1\psi_2 \cos[(E_1 - E_2)t/\hbar].$$

The system described by this equation oscillates in time. But if $E_1 = E_2$, then $\Psi^*\Psi = \psi_1^2 + \psi_2^2 + 2\psi_1\psi_2 = $ constant.

2.5 EIGENVALUES

Many kinds of eigenvalues and eigenfunctions play a significant role in quantum mechanics, and it is important to understand what their properties are. Postulate 4 tells us about the connection of eigenvalues to measurements.

>   **Postulate 4:** If ψ_a is an eigenfunction of the operator $\hat{A}$ with eigenvalue a, then if we measure the property A for a system whose wavefunction is ψ_a, we always get a as the result.

Postulate 4 implies that if a molecule is in an eigenstate of the Hamiltonian operator $\hat{H}$, then the only energy that can be measured for the system is the eigenvalue E associated with that eigenstate.

The eigenfunctions of the momentum operator are another example of an eigenvalue–eigenvector system. The eigenvalue equation in this case is

$$\hat{p}_x \psi_\ell = p_\ell \psi_\ell, \tag{2.15}$$

where $\hat{p}_x$ is taken to be the x component of the momentum operator [from Eq. (2.2b)] and p_ℓ is its corresponding eigenvalue. The equation to be solved is

$$-i\hbar \frac{\partial}{\partial x} \psi_\ell = p_\ell \psi_\ell, \tag{2.16}$$

and by substitution, one can verify that the eigenfunctions are

$$\psi_\ell = A e^{ip_\ell x/\hbar}, \tag{2.17}$$

where A is a constant. Note that $\psi^*\psi = A^*A$. These momentum eigenfunctions are called *plane waves*, and they have the property that $|\psi_\ell|^2$ is a constant, independent of x. This means that the coordinate x is completely undetermined, which is what the uncertainty principle requires. There is an analogous set of eigenfunctions of the coordinate operator that are completely localized in space, but delocalized in momentum.

EXERCISE 2.5 One of the simplest quantum systems is the harmonic oscillator, whose ground-state wavefunction is $\psi = Ae^{-\lambda x^2}$. Is this wavefunction an eigenfunction of the momentum? Is it an eigenfunction of the kinetic energy?

Answer Use explicit expressions for the operators to obtain

$$\hat{p}\psi = -i\hbar\frac{d\psi}{dx} = -i\hbar A(-2\lambda x)e^{-\lambda x^2}$$

and

$$\hat{p}^2\psi = (-i\hbar)^2 A(-2\lambda)\frac{d}{dx}\left(xe^{-\lambda x^2}\right)$$

$$= 2\lambda A\hbar^2\left[-2\lambda x^2 e^{-\lambda x^2} + e^{-\lambda x^2}\right],$$

so ψ is not an eigenfunction of either $\hat{p}_x$ or $\hat{p}_x^2$. Since the kinetic energy operator is $\hat{p}_x^2/2m$, neither is ψ an eigenfunction of the kinetic energy.

2.6 EXPECTATION VALUES

Postulate 4 tells us what we will measure when we consider eigenfunctions associated with a dynamical variable of interest. In the more general case where the wavefunctions are not eigenfunctions of the desired operator, we must calculate the average (or expectation) value. If the observable depends on coordinates, the expression for the average value is the same as in classical theory, namely, the integral of the probability density times the quantity to be averaged. The formula is

$$\langle x\rangle = \frac{\displaystyle\int x|\psi(x)|^2\,dx}{\displaystyle\int |\psi(x)|^2\,dx}. \tag{2.18}$$

But what about other expectation values? This is where we turn to Postulate 5:

Postulate 5: The average (or expectation) value of an observable A is given by

$$\langle A\rangle = \frac{\displaystyle\int \psi^*\hat{A}\psi\,d\tau}{\displaystyle\int \psi^*\psi\,d\tau}, \tag{2.19}$$

where $\hat{A}$ is the operator associated with the observable. Evidently, if ψ is an eigenfunction of the operator $\hat{A}$, then the expectation value is just the eigenvalue.

The sixth and last postulate arises only for systems with several identical particles and is given in Chapter 8.

EXERCISE 2.6 For the harmonic oscillator of Exercise 2.5, evaluate $\langle x \rangle$ and $\langle p \rangle$. Interpret your results.

Answer First,

$$\langle x \rangle = \int_{-\infty}^{\infty} A^2 e^{-2\lambda x^2} x \, dx = 0.$$

This integral vanishes because the integrand is an odd function.

Next,

$$\langle p_x \rangle = -i\hbar A^2 \int_{-\infty}^{\infty} e^{-\lambda x^2} \frac{d}{dx} e^{-\lambda x^2} \, dx$$

$$= -i\hbar A^2 \int_{-\infty}^{\infty} e^{-2\lambda x^2} (-2\lambda x) \, dx = 0.$$

This integral also vanishes because the integrand is an odd function.

Thus, for a harmonic oscillator, the averaged position and the averaged momentum both vanish. This makes sense, as the velocity and position of a classical spring would both *average* to zero.

2.7 PROPERTIES OF THE TIME-INDEPENDENT SCHRÖDINGER EIGENFUNCTIONS

2.7.1 Hermitian Operators

Let the operator $\hat{A}$ satisfy

$$\hat{A}\psi = a\psi, \tag{2.20}$$

where a is an eigenvalue. From the definition of Hermitian operators,

$$\int \psi(\hat{A}\phi)^* \, d\tau = \int \phi^* \hat{A}\psi \, d\tau. \tag{2.21}$$

Now choose $\phi = \psi$. Then

$$\int \psi(\hat{A}\psi)^* \, d\tau = \int \psi^* \hat{A}\psi \, d\tau. \tag{2.22}$$

If we substitute from the eigenvalue equation, we find that

$$\int \psi a^* \psi^* \, d\tau = \int \psi^* a\psi \, d\tau. \tag{2.23}$$

Or, taking the constant out of the integral, we obtain

$$a^* \int \psi\psi^* \, d\tau = a \int \psi^* \psi \, d\tau, \tag{2.24}$$

which requires that $a^* = a$, so a is real. This proves that all Hermitian operators (including the Hamiltonian) have real eigenvalues. A similar proof leads to the conclusion that *only* Hermitian operators have real expectation values.

2.7.2 Orthogonality of Eigenfunctions

Suppose that we have two different eigenfunctions of the same Hamiltonian. That is,

$$\hat{H}\psi_1 = E_1\psi_1 \tag{2.25a}$$

and

$$\hat{H}\psi_2 = E_2\psi_2, \tag{2.25b}$$

where E_1 and E_2 might be the same or different. Now consider the integrals

$$\int \psi_2^* \hat{H}\psi_1 \, d\tau = E_1 \int \psi_2^* \psi_1 \, d\tau \tag{2.26a}$$

and

$$\int \psi_1 (H\psi_2)^* \, d\tau = E_2^* \int \psi_1 \psi_2^* = E_2 \int \psi_2^* \psi_1 \, d\tau, \tag{2.26b}$$

where we have used the fact that E_2 is real in Eq. (2.26b). If, now, Eq. (2.26b) is subtracted from Eq. (2.26a), we find that

$$(E_1 - E_2) \int \psi_2^* \psi_1 \, d\tau = \int \psi_2^* H\psi_1 \, d\tau - \int \psi_1(H\psi_2)^* \, d\tau = 0, \tag{2.27}$$

since H is Hermitian. This implies that

$$0 = (E_1 - E_2) \int \psi_1 \psi_2^* \, d\tau. \tag{2.28}$$

Two functions ψ_1 and ψ_2 for which $\int \psi_1 \psi_2^* \, d\tau = 0$ are said to be *orthogonal*. Equation (2.28) says that eigenfunctions of the same operator with different eigenvalues must be orthogonal. It also states that eigenfunctions with *degenerate* (equal) eigenvalues need not be orthogonal, although we are free to make them so if we like. Wavefunctions that are both normalized and orthogonal are called *orthonormal*.

2.7.3 Simultaneous Eigenfunctions

Suppose we have an operator $\hat{G}$ and a wavefunction ψ that is an eigenfunction of $\hat{H}$. Is ψ an eigenfunction of both $\hat{G}$ and $\hat{H}$? To answer this question, we first assume that ψ is an eigenfunction of both $\hat{G}$ and $\hat{H}$, so that $\hat{G}\psi = g\psi$ and $\hat{H}\psi = E\psi$. Now let us examine the application of $\hat{G}\hat{H}$ and $\hat{H}\hat{G}$ to ψ. This gives us

$$\hat{G}\hat{H}\psi = \hat{G}[\hat{H}\psi] \tag{2.29}$$

$$= \hat{G}E\psi = E\hat{G}\psi = Eg\psi \tag{2.30}$$

and

$$\hat{H}\hat{G}\psi = \hat{H}g\psi = g\hat{H}\psi = Eg\psi,$$

so that

$$[\hat{G}\hat{H} - \hat{H}\hat{G}]\psi = [\hat{G}, \hat{H}]\psi = 0. \tag{2.31}$$

This tells us that *simultaneous eigenfunctions are associated with commuting operators*, which is a fundamental theorem of quantum mechanics.

EXERCISE 2.7

  a. Apply the preceding theorem to a free particle, using

$$\hat{H} = -\frac{\hbar^2}{2m}\frac{\partial^2}{\partial x^2} \tag{2.32}$$

  and

$$\hat{G} \equiv \hat{p} = -i\hbar\frac{\partial}{\partial x}. \tag{2.33}$$

Answer By inspection, we have $[\hat{H}, \hat{G}] = 0$, so simultaneous eigenfunctions are possible. Indeed, e^{ikx} is a simultaneous eigenfunction, with $p = \hbar k$ and $E = p^2/(2m)$.

  b. Is $\cos(kx)$ a simultaneous eigenfunction of the operators $\hat{H}$ and $\hat{p}$?

Answer Clearly, $\cos(kx)$ is an eigenfunction of $\hat{H}$, but not of $\hat{p}$.

The previous example illustrates that even when operators commute, we can (sometimes) find non-simultaneous eigenfunctions. However, when this happens, the function can always be decomposed into linear combinations of simultaneous eigenfunctions. Thus, $\cos(kx)$ is a linear combination of the two momentum eigenfunctions:

$$\cos(kx) = \frac{1}{2}\left(e^{ikx} + e^{-ikx}\right). \tag{2.35}$$

Note that one can measure the eigenvalues g and g' to infinite precision simultaneously if $[\hat{G}, \hat{G}'] = 0$. Thus, the uncertainty principle holds only for non-commuting operators. For the operators $\hat{p}_x$ and $\hat{x}$, we have

$$[\hat{p}_x, \hat{x}] = -i\hbar \neq 0, \tag{2.36}$$

which means that x and p may *not* have simultaneous eigenfunctions. Of course, this is consistent with the uncertainty relationship $\Delta x\,\Delta p \geq \hbar/2$.

2.7.4 Differentiability of the Wavefunction

For a particle moving in one dimension, the Schrödinger equation (2.13) is

$$\frac{-\hbar^2}{2m}\frac{d^2\psi}{dx^2} + V(x)\psi = E\psi, \tag{2.37}$$

which can be rearranged to give

$$\frac{d^2\psi}{dx^2} = \frac{2m}{\hbar^2}\left(V(x) - E\right)\psi. \tag{2.38}$$

If this equation is integrated over x between $x = a$ and $x = b$, we obtain

$$\frac{d\psi}{dx}\bigg|_{x=b} - \frac{d\psi}{dx}\bigg|_{x=a} = \int_a^b \frac{2m}{\hbar^2}(V(x) - E)\psi \, dx. \tag{2.39}$$

In the limit such that b approaches a, the integral on the right-hand side goes to zero (provided that $V(x)$ is finite). This means that

$$\lim_{b \to a} \frac{d\psi}{dx}\bigg|_{x=b} = \frac{d\psi}{dx}\bigg|_{x=a}. \tag{2.40}$$

In other words $d\psi/dx$ is *continuous*. Similarly, one can show that $\psi(x)$ is both *continuous* and *smooth*. Exceptions to these statements arise only for geometries wherein $V(x)$ can become infinite, as in an infinite hard wall potential.

2.7.5 Completeness

One final property of the Schrödinger equation is that it is possible to expand any well-behaved function (that is defined over the same range of coordinates and that satisfies the same boundary conditions) exactly in terms of the complete set of eigenfunctions of $\hat{H}\psi = E\psi$. This property is called *completeness*, and it actually applies to the eigenfunctions of any Hermitian operator. It means that for an arbitrary, but well-behaved, function $f(x)$, the expansion

$$f(x) = \sum_n c_n \psi_n(x) \tag{2.41}$$

is exact, provided that the ψ_n's are eigenfunctions of the Schrödinger equation. If the ψ_n's are orthonormal, then the coefficient c_n in Eq. (2.41) may be obtained by multiplying that equation by $\psi_m^*(x)$ and integrating, yielding

$$c_n = \int \psi_n^*(x) f(x) \, dx. \tag{2.42}$$

Note that if $f(x)$ is normalized, then $\langle f | f \rangle = \sum_n |c_n|^2 = 1$. If we consider Eq. (2.41) as decomposing the function $f(x)$ into its component parts, then $|c_n|^2$ can be thought of as the probability of finding state n in the function f, and the sum over all states of this probability is unity.

SUGGESTED READINGS

The postulates of quantum mechanics are discussed in all the standard quantum chemistry textbooks (cited at the end of Chapter 1). In addition, a more rigorous treatment is given in a number of physics-oriented quantum mechanics books, including the following:

G. Baym, *Lectures on Quantum Mechanics* (Benjamin/Cummings, London, 1981).

C. Cohen-Tannoudji, B. Diu, and F. Laloe, *Quantum Mechanics* (Wiley, New York, 1976).

A. S. Davydov, *Quantum Mechanics* (Pergamon, Oxford, 1976).

P. A. M. Dirac, *The Principles of Quantum Mechanics* (Oxford University Press, Oxford, 1947).

E. Merzbacher, *Quantum Mechanics* (3rd ed.) (Wiley, New York, 1998).

D. Rapp, *Quantum Mechanics* (Holt, Rinehart and Winston, New York, 1971).

L. I. Schiff, *Quantum Mechanics* (McGraw-Hill, New York, 1968).

PROBLEMS

2.1 Which of the following operators is Hermitian?

   (a) i

   (b) $*$ (complex conjugate)

   (c) e^{ix}

   (d) $x(d/dx)x$

   (e) $x\dfrac{d}{dx} + \dfrac{d}{dx}x$

   (f) $-ix\dfrac{\partial}{\partial y}$ (in two dimensions)

   (g) $\dfrac{d^2}{dx^2}$ (in one dimension)

   (h) $-i\hbar\dfrac{\partial}{\partial r}$ (in spherical polar coordinates)

2.2 (a) Given the three operators $A, B,$ and $C,$ show that $[A, BC] = B[A, C] + [A, B]C.$

   (b) Given the three operators $A, B,$ and $C,$ show that $[A,[B, C]] = [B,[A, C]]$ if A and B commute.

2.3 What is $[p_x, x^2]$? The final expression should contain a single operator.

2.4 Prove that any real wavefunction has a vanishing average momentum.

2.5 True or False

   (a) Nondegenerate eigenfunctions of the same operator are orthogonal.

   (b) All Hermitian operators are real.

   (c If two operators commute with a third, they will commute with each other.

   (d) If two operators commute, they must have the same eigenfunctions.

   (e) $\dfrac{d\psi}{dx}$ must be continuous as long as the potential $V(x)$ is finite.

   (f) If a wavefunction is simultaneously the eigenfunction of two operators, then that wavefunction will also be an eigenfunction of the product of those operators.

2.6 Show that if one requires that the wavefunction

$$\psi = Ae^{ikx} + Be^{-ikx}$$

and its derivative be real at some value of x, then $A = B^*$ and $B = A^*$. Show that this result implies that ψ is real for all x.

2.7 Consider the following functions:

$$\psi_1 = 1;$$
$$\psi_2 = x;$$
$$\psi_3 = x^2 + bx + c;$$
$$\psi_4 = x^3 + dx^2 + ex + f.$$

(a) Show that ψ_1 and ψ_2 are orthogonal over the interval $[-1, 1]$.

(b) Find b, c, d, e, and f such that ψ_1, ψ_2, ψ_3, and ψ_4 are orthogonal on the interval $[-1, 1]$.

(c) Normalize the functions in (b) on the interval $[-1, 1]$.

2.8 The ground vibrational state of a symmetric diatomic molecule A_2 can be approximated as

$$\psi_0 = A_0 \exp\left\{-\frac{1}{2\hbar}\mu\omega x^2\right\},$$

where μ, the reduced mass, is $\frac{1}{2}M_A$, A_0 is an arbitrary normalization constant, and x is the stretching distance ($x = 0$ at equilibrium). The ground state energy is $E = \frac{1}{2}\hbar\omega$.

(a) What value of x is the most probable? Interpret this classically.

(b) At the classical turning point, the momentum vanishes. Find the value of x_c, the classical turning point. *Hint:* See Eq. 1.17.

(c) Compute the relative probabilities of finding the diatomic molecule at $x = 0$ and at $x = x_c$.

2.9 A plane wave traveling along the x-axis in the positive direction has the (non-normalized) wavefunction

$$\psi(x) = e^{ipx/\hbar}.$$

Prove directly that this wavefunction has an expectation value p for the operator $\hat{p}_x$ and that the probability distribution of the function is constant all along the x-axis. Finally, compute the uncertainty $\Delta p = [\langle p_x^2\rangle - \langle p_x\rangle^2]^{1/2}$ in the momentum of the wave. (*Hint:* Don't try to evaluate the normalization integral, which is infinite, but carry it along in the calculation, and eventually it should cancel.)

2.10 For operators A and B that commute, prove that $\Delta A\Delta B = 0$. (The Δ notation is defined in Problem 2.9.) Prove that if A and B do not commute, then $\Delta A\Delta B \neq 0$. The interpretation of these relationships is that two commuting observables can each be measured exactly and simultaneously, but if such observables do not commute, there will be an uncertainty.

2.11 The time-dependent Schrödinger equation suggests that the operator H is equivalent to the operator $i\hbar\,\partial/\partial t$. Use this equivalence to demonstrate that (at least heuristically)

$$[H, t] = i\hbar.$$

The interpretation of this equation is that there is a heuristic time–energy uncertainty relation $\Delta E\Delta t \sim \hbar$, and this relation can be used to interpret the time associated with typical quantum processes. If a spectral line has a width ΔE of 0.2 eV, estimate Δt, the lifetime of the excited state involved in the transition.

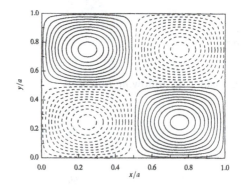

3

Particle-in-Box Models

The particle-in-box models are among the simplest examples of solutions to the time-independent Schrödinger equation. The properties of these models are useful in understanding systems ranging from conduction in metals to optical properties of polymers. We will consider the one-dimensional (1D) and two-dimensional (2D) box models in detail and the three-dimensional (3D) box briefly; then we illustrate the application of these models to a chemical problem via the free-electron molecular orbital method.

3.1 PARTICLE IN A ONE-DIMENSIONAL BOX

3.1.1 Eigenvalues and Eigenfunctions

Figure 3.1 shows the potential-energy function $V(x)$ (or, simply, the potential) for the one-dimensional box. A mathematical expression for the potential is

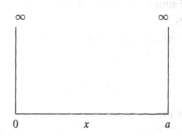

Figure 3.1 Potential for one-dimensional box.

$$V(x) = \begin{cases} 0, & 0 \le x \le a \\ \infty, & \text{otherwise} \end{cases}, \tag{3.1}$$

and the Hamiltonian is

$$H = \frac{-\hbar^2}{2m} \frac{d^2}{dx^2} + V. \tag{3.2}$$

Outside the box, the Schrödinger equation is

$$\frac{-\hbar^2}{2m} \frac{d^2}{dx^2} \psi = (E - \infty)\psi. \tag{3.3}$$

A finite second derivative requires that $\psi(x)$ be zero outside the box. Thus, we have

$$\psi = 0 \quad \text{outside.} \tag{3.4}$$

Inside the box, the Schrödinger equation is

$$-\frac{\hbar^2}{2m} \frac{d^2}{dx^2} \psi = E\psi. \tag{3.5}$$

From Chapter 1 (Eq. 1.9) and Appendix A (Section A.2), we know that the solution of this equation is

$$\psi = A \sin \alpha x + B \cos \alpha x, \tag{3.6}$$

where A and B are constants and

$$\alpha = \left(\frac{2mE}{\hbar^2} \right)^{1/2}. \tag{3.7}$$

By continuity of the wavefunction, $\psi(x = 0)$ must vanish, as must $\psi(x = a)$. This condition requires that

(1) $B = 0$, to satisfy the boundary condition at $x = 0$ (3.8)

and

(2) $\alpha = \dfrac{n\pi}{a}$, to satisfy the boundary condition at $x = a$. (3.9)

Here n can in principle be any integer ($n = 0, \pm 1, \pm 2, \ldots$). However, note that the solution which satisfies the boundary conditions is

$$\psi = A \sin \frac{n\pi x}{a}, \tag{3.10}$$

so $n = 0$ implies that $\psi = 0$, which is not a physically useful solution. Also, negative values of n lead to solutions that are identical to those for positive values except for a difference in sign. Such solutions are therefore not independent of the positive n solutions, which means that only $n = 1, 2, 3, \ldots$ gives unique solutions.

Normalization of the wavefunction implies that

$$1 = \int_0^a \psi^*\psi \, dx = A^2 \int_0^a \sin^2 \frac{n\pi x}{a} \, dx. \tag{3.11}$$

To evaluate this integral, we use

$$\sin^2 y = \frac{1}{2}(1 - \cos 2y), \tag{3.12}$$

which leads to

$$1 = A^2 \left\{ \frac{x}{2} \Big|_0^a - \frac{\sin(2n\pi x/a)}{2n\pi/a} \Big|_0^a \right\} \tag{3.13}$$

$$= A^2(a/2), \tag{3.14}$$

so that

$$A = \sqrt{\frac{2}{a}}. \tag{3.15}$$

Thus, the complete solution is

$$\psi = \begin{cases} 0, & \text{outside box} \\ \sqrt{\frac{2}{a}} \sin \frac{n\pi x}{a}, & \text{inside box} \end{cases}, \tag{3.16}$$

where $n = 1, 2, 3, \ldots$

To find the energy, we substitute back into the Schrödinger equation to get

$$\frac{-\hbar^2}{2m} \frac{d^2}{dx^2} A \sin \frac{n\pi x}{a} = E A \sin \frac{n\pi x}{a}, \tag{3.17}$$

or in other words,

$$E = \frac{(n\pi\hbar)^2}{2ma^2} = \frac{n^2 h^2}{8ma^2}. \tag{3.18}$$

The resulting energy levels are depicted in Figure 3.2.

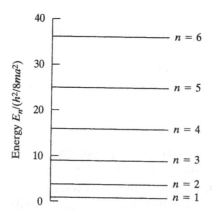

Figure 3.2   Energy levels of particle in box.

Note that the spacing of the energy levels gets larger as the energy increases. Note also that the ground-state energy is nonzero. This so-called *zero-point energy* is an important feature of quantum mechanics that is a natural consequence of the uncertainty principle, as we show later.

EXERCISE 3.1 Show that, although the $n = 0$ function $\psi_0 = 0$ satisfies the Schrödinger equation, it is not normalizable, so it is not an acceptable wavefunction.

Answer If $\psi_0 = 0$, then

$$\int_{-\infty}^{\infty} |\psi_0|^2 \, dx = \int 0 \, dx = 0.$$

For the function to be normalizable, this integral must be finite. Since the integral vanishes, the wavefunction $\psi_0 = 0$ is not normalizable, so it is not allowed. (Physically, $|\psi_0|^2 = 0$ means that the probability that the particle is *anywhere* is zero.)

3.1.2 Expectation Values

To find the average value of x, we simply substitute the wavefunctions just obtained into Eq. (2.19). This gives

$$\langle x \rangle = \frac{\int \psi^* x \psi \, dx}{\int \psi^* \psi \, dx} = A^2 \int_0^a \sin^2 \frac{n\pi x}{a} x \, dx, \tag{3.19}$$

or

$$\langle x \rangle = \frac{a}{2}. \tag{3.20}$$

Similarly, one can show that

$$\langle x^2 \rangle = \frac{a^2}{3}, \tag{3.21}$$

which means that the uncertainty in the position x of the particle (here taken to be the standard deviation) is given by

$$\Delta x = \sqrt{\langle x^2 \rangle - \langle x \rangle^2} = \sqrt{\langle (x - \langle x \rangle)^2 \rangle} = \sqrt{\frac{a^2}{12}}. \tag{3.22}$$

Now let us calculate the expectation value of the momentum. If Eq. (2.2b) is inserted into Eq. (2.19), the result is

$$\langle p_x \rangle = \frac{\hbar}{i} A^2 \int_0^a \sin \frac{n\pi x}{a} \frac{d}{dx} \sin \frac{n\pi x}{a} \propto \int_0^a \sin \frac{n\pi x}{a} \cos \frac{n\pi x}{a} \, dx$$

$$\propto \int_0^a \sin \frac{n\pi x}{a} \, d \sin \frac{n\pi x}{a} \propto \sin^2 \frac{n\pi x}{a} \Big|_0^a \tag{3.23}$$

$$= 0,$$

which means that the particle is moving with equal probability from right to left and from left to right. The corresponding p^2 expectation value gives

$$\langle p_x^2 \rangle = 2m\langle H \rangle = 2m \frac{n^2 h^2}{8ma^2} = \frac{n^2 h^2}{4a^2}, \tag{3.24}$$

and the uncertainty in the momentum is

$$\Delta p_x = \sqrt{\langle p_x^2 \rangle - \langle p_x \rangle^2} = \sqrt{(p_x - \langle p_x \rangle)^2} = \frac{hn}{2a}. \tag{3.25}$$

If Eqs. (3.22) and (3.25) are multiplied together, we obtain the following expression for the uncertainty product:

$$(\Delta x)(\Delta p) \cong \frac{hn}{2a} \cdot \frac{a}{\sqrt{12}} = \frac{nh}{4\sqrt{3}} \geq \frac{\hbar}{2}. \tag{3.26}$$

This is another verification of the uncertainty principle, and it shows why even the ground state must have nonzero energy (i.e., zero-point energy).

3.1.3 Orthogonality

The overlap between the two lowest eigenfunctions is given by

$$\begin{aligned}
\int_0^a \psi_1^* \psi_2 \, dx &= \int_0^a A^2 \sin \frac{\pi x}{a} \sin \frac{2\pi x}{a} \, dx \\
&= \int_0^a A^2 2 \sin \frac{\pi x}{a} \sin \frac{\pi x}{a} \cos \frac{\pi x}{a} \, dx \\
&= \int_0^a 2A^2 \sin^2 \frac{\pi x}{a} \, d\left(\sin \frac{\pi x}{a} \right) \frac{a}{\pi} \\
&\propto \sin^3 \frac{\pi x}{a} \Big|_0^a = 0.
\end{aligned} \tag{3.27}$$

So these eigenfunctions are **orthogonal**. In general, one can show that

$$\int_0^a \psi_i^* \psi_j \, dx = \delta_{ij},$$

which is in accord with a theorem proved in Chapter 2 (Section 2.7.2).

3.1.4 Wavefunctions and Nodes

Plots of the first four eigenfunctions are given in Figure 3.3. Note that the nth eigenfunction $\psi_n(x)$ has a wavelength $\lambda = 2a/n$, so that there are $n/2$ periods within the box. A *node* of the wavefunction occurs whenever $\psi_n(x) = 0$, so that there are $n - 1$ nodes associated with quantum state n (not counting the nodes at $x = 0$ and $x = a$, which are associated with the boundary conditions).

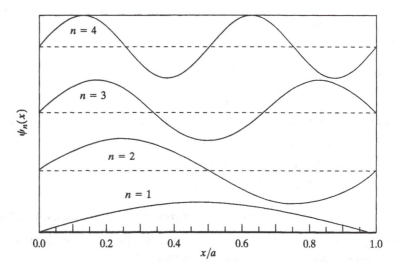

Figure 3.3 Particle-in-box eigenfunctions for $n = 1, \ldots, 4$. (For clarity, successive eigenfunctions have been shifted up.)

Note that the number of nodes increases by one for each successive quantum state. In addition, each node of the nth state is located between nodes of the $(n + 1)$st state. These results hold for any solution to the one-dimensional Schrödinger equation.

The nodal structure of eigenfunctions of the Schrödinger equation is crucial to making those functions orthogonal. In addition, more nodes lead to shorter de Broglie wavelengths and thus to higher kinetic energies, so the growth in nodes with increasing energy is a natural consequence of how kinetic energy is defined in quantum mechanics.

Plots of the square of the wavefunction $|\psi_n(x)|^2$, are presented in Figure 3.4 for $n = 1, \ldots, 4$. The plots show that the lowest energy state is localized mostly

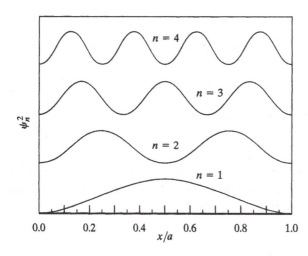

Figure 3.4 Probability densities associated with the $n = 1, \ldots, 4$ particle-in-box eigenfunctions. Successive states are shifted up for clarity.

near the center of the well, while for $n = 2$, the probability that the particle is in the center is zero.

3.1.5 Correspondence Principle

It is not difficult to show that

(1) As $n \to \infty$, the locally averaged probability density $|\psi|^2$ is the same for all x.
(2) As $m \to \infty$, the energy spacing $\to 0$.
(3) As $h \to 0$, the energy spacing $\to 0$.
(4) As $a \to \infty$, the energy spacing $\to 0$.

These are all examples of the **Bohr correspondence principle**, which states that

As $\hbar \to 0$, or as sizes become macroscopic, the predictions of quantum mechanics become those of classical mechanics.

Of course, $\hbar$ is not a variable in the real world, so quantum and classical results are different. However, the Bohr correspondence principle serves as a useful guide in the development of approximate quantum theories, such as semiclassical theories that lie in-between quantum and classical mechanics.

3.2   PARTICLE IN A TWO-DIMENSIONAL BOX

In two dimensions, the Hamiltonian is an obvious generalization of the 1D Hamiltonian, namely,

$$H_{\text{inside}} = \frac{-\hbar^2}{2m} \left(\frac{\partial^2}{\partial x^2} + \frac{\partial^2}{\partial y^2} \right) \tag{3.28}$$

and

$$H_{\text{out}} = H_{\text{in}} + \infty, \tag{3.29}$$

We take the width of the box to be a in the x-direction and b in the y-direction. The solution outside is easily shown to be

$$\psi_{\text{out}} = 0. \tag{3.30}$$

To determine the solution inside, we take advantage of the fact that

$$H(x, y) = h_x(x) + h_y(y), \tag{3.31}$$

where h_x does not depend on y and h_y does not depend on x. A Hamiltonian that satisfies this property is called a **separable** Hamiltonian, and its solutions may be written (as in Section 2.4) in the product form

$$\psi = X(x)Y(y). \tag{3.32}$$

To prove this, we simply substitute the assumed solution into the Schrödinger equation to obtain

$$\frac{-\hbar^2}{2m}\left(\frac{\partial^2}{\partial x^2} + \frac{\partial^2}{\partial y^2}\right)XY = EXY, \tag{3.33}$$

which can be rearranged to give

$$\frac{-\hbar^2}{2m}\left\{Y\frac{\partial^2 X}{\partial x^2} + X\frac{\partial^2 Y}{\partial y^2}\right\} = XYE \tag{3.34}$$

and then

$$\frac{-\hbar^2}{2m}\left\{\frac{1}{X}\frac{\partial^2 X}{\partial x^2} + \frac{1}{Y}\frac{\partial^2 Y}{\partial y^2}\right\} = E. \tag{3.35}$$

Now, as x is varied, only the first term in brackets will vary. But to satisfy the equation, that term must be constant. Also, as y is varied, the same is true of the second term. These two conditions imply that

$$\begin{cases} \dfrac{-\hbar^2}{2m}\dfrac{1}{X}\dfrac{\partial^2 X}{\partial x^2} = \text{const} \equiv \varepsilon_x & (3.36a) \\[4mm] \dfrac{-\hbar^2}{2m}\dfrac{1}{Y}\dfrac{\partial^2 Y}{\partial y^2} = \varepsilon_y & (3.36b) \end{cases}$$

where

$$\varepsilon_x + \varepsilon_y = E. \tag{3.37}$$

Equation (3.36a) may be rearranged to read

$$\varepsilon_x X = \frac{-\hbar^2}{2m}\frac{\partial^2 X}{\partial x^2}, \tag{3.38}$$

which is just the Schrödinger equation for the one-dimensional box. The solution $\psi(x, y)$ may thus be written using the 1D results already given, namely,

$$\begin{cases} \psi(x, y) = XY = \sqrt{\dfrac{4}{ab}}\sin\dfrac{n_x \pi x}{a}\sin\dfrac{n_y \pi y}{b} & (3.39) \\[4mm] E = \dfrac{h^2}{8m}\left(\dfrac{n_x^2}{a^2} + \dfrac{n_y^2}{b^2}\right) & (3.40) \end{cases}$$

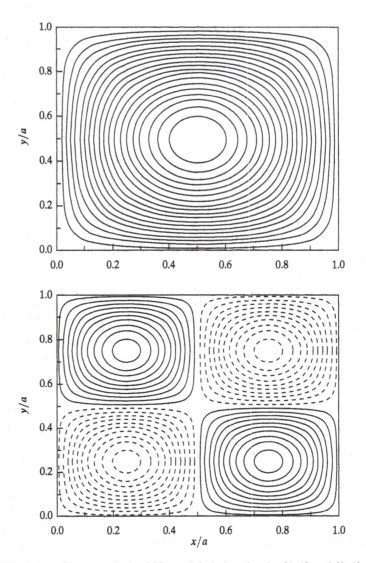

Figure 3.5 Eigenfunctions of 2D particle in box for the $(1, 1)$ and $(2, 2)$ states.

If the box is **square**, then $a = b$, and the expression for the energy is

$$E = \frac{h^2}{8ma^2} \left(n_x^2 + n_y^2 \right). \tag{3.41}$$

Table 3.1 gives energies for the lowest few states of the square box. Figure 3.5 shows contours of the $(n_x, n_y) = (1, 1)$ and $(2, 2)$ eigenfunctions. Note that many of the states are *degenerate*, which means that they have the same energy.

TABLE 3.1 LOWEST ENERGY LEVELS OF SQUARE BOX

n_x	n_y	$E(h^2/8ma^2)$
1	1	2
1	2	5
2	1	5
2	2	8
1	3	10
3	1	10
3	2	13
2	3	13
1	4	17
4	1	17
3	3	18
2	4	20
4	2	20
1	5	26
5	1	26

3.3 PARTICLE IN A THREE-DIMENSIONAL BOX

The solution to the particle in a three-dimensional box is completely analogous to what we have just done for the 2-D box. Again the Hamiltonian is separable, so the wavefunction can be written as a product of one-dimensional solutions and the energy as a sum of one-dimensional energies. We leave as an exercise to the reader to show that the analogues of Eqs. (3.39) and (3.40) are, respectively,

$$\psi(x, y, z) = \sqrt{\frac{8}{abc}} \sin \frac{n_x \pi x}{a} \sin \frac{n_y \pi y}{b} \sin \frac{n_z \pi z}{c} \qquad (3.42)$$

and

$$E_{n_x n_y n_z} = \frac{h^2}{8m} \left(\frac{n_x^2}{a^2} + \frac{n_y^2}{b^2} + \frac{n_z^2}{c^2} \right). \qquad (3.43)$$

EXERCISE 3.2 What are the energy and the degeneracy (the number of states) of the lowest-lying degenerate state for a particle in a cubic box of volume a^3?

Answer

$$E_{n_x, n_y, n_z} = \frac{h^2}{8ma^2} \left(n_x^2 + n_y^2 + n_z^2 \right),$$

so $E_{111} = (h^2/8ma^2) \cdot 3$ and $E_{112} = E_{121} = E_{211} = (h^2/8ma^2) \cdot (4 + 1 + 1) = (h^2/8ma^2) \cdot 6$. This says that the first excited state $(n_x, n_y, n_z) = (1, 1, 2), (1, 2, 1), (2, 1, 1)$ is triply degenerate.

3.4   FREE-ELECTRON MOLECULAR ORBITAL MODEL

The free-electron molecular orbital (FEMO) imodel s a simple model of the π electronic states in conjugated organic molecules that is based on particle-in-box eigenfunctions. It was originally developed in the 1950s by Kuhn, Bayliss, Platt, and Simpson. The model assumes that the π electrons in a conjugated molecule can be separated from the σ electrons and that the σ framework is frozen. In addition, all interelectronic interactions are neglected, and the effective potential acting on each π electron is assumed to be given by a particle-in-a-box potential. For linear or nearly linear molecules, the box can be taken to be one dimensional. This immediately leads to an expression for the energy levels of the π electrons given by Eq. (3.18). These energy levels will be occupied in accordance with the Pauli principle (see Chapter 8), namely, that two electrons (one with up spin and one with down spin) will go into each level. The absolute energies of these electrons are not by themselves of interest, but the *differences* in energy between adjacent levels can be compared by means of optical absorption measurements. Thus, if n is the quantum number of the highest occupied energy level, then the energy difference would be

$$\Delta E = E_{n+1} - E_n. \tag{3.44}$$

For a **linear** box (representing butadiene, hexatriene, etc.), the energy difference is thus given by

$$\Delta E = \frac{h^2}{8ma^2} \left[(n+1)^2 - n^2 \right] = (2n+1) \frac{h^2}{8ma^2}. \tag{3.45}$$

Application to Butadiene   Butadiene is a four-carbon molecule with the formula $CH_2{=}CH{-}CH{=}CH_2$. The carbon–carbon bond lengths are 1.34 Å for $C{=}C$ and 1.46 Å for $C{-}C$, so the overall length of the carbon backbone is $a = 1.46 + 1.34 + 1.34 = 4.14$ Å. Since there are four π electrons, the $n = 1$ and $n = 2$ levels are doubly occupied. This means that the lowest possible excitation is from $n = 2$ to $n = 3$. The energy associated with this transition is

$$\Delta E = E_3 - E_2 = \frac{5h^2}{8ma^2} = h\nu = h\frac{c}{\lambda}, \tag{3.46}$$

which can be rearranged to

$$\frac{5h^2}{8ma^2} = \frac{hc}{\lambda}, \tag{3.47}$$

or

$$\lambda = \frac{(hc)(8ma^2)}{5h^2} \qquad \lambda = 113.0 \text{ nm}. \tag{3.48}$$

A more reasonable box length allows the wavefunction to be extended a bit beyond the outer C atoms, say, by half a bond on each end. This leads to

$$a \cong 2 \cdot \tfrac{1}{2} \cdot 1.34 + 4.14 = 5.48 \text{ Å} \tag{3.49}$$

and to

$$\frac{1}{\lambda_{max}} = 50482 \text{ cm}^{-1}, \tag{3.50}$$

or

$$\lambda_{max} = 198.1 \text{ nm}. \tag{3.51}$$

Table 3.2 summarizes these results and compares them with experiment. It turns out that the estimate which accounts for the end correction is almost perfect.

TABLE 3.2   FEMO ESTIMATES OF ABSORPTION
WAVELENGTH IN BUTADIENE

Wavelength	Small box	Big box	Observed
λ_{max}	113 nm	198.1 nm	210.0 nm

An even better result is obtained if we choose $a = 1.4(k + 2)$ Å where k is the number of double bonds. For $k = 2$,

$$a = 5.6 \text{ Å} \tag{3.52}$$

and

$$\lambda_{max} = 207.0 \text{ nm}. \tag{3.53}$$

When the same formula is applied to octatetraene (C_8H_{10}), we find that

$$\lambda_{max} = \begin{cases} 271 \text{ nm (calculated)} \\ 290 \text{ nm (observed)} \end{cases}. \tag{3.54}$$

The high accuracy of these comparisons is somewhat accidental, but the model does demonstrate an important qualitative result, namely, that for linear polyenes, the absorption wavelengths get longer as the length of the chain is increased.

EXERCISE 3.3   Optical transitions from the highest occupied level to the lowest empty one are usually strong, but other transitions can also be. Using $a = 1.40(k + 2)$ Å, compute λ_{max} for transitions from the second-highest occupied orbital to the lowest empty one, and from the highest occupied to the second-lowest empty one, for butadiene. In which region of the electromagnetic spectrum do these transitions occur?

Answer   For butadiene, $a = 1.40(2 + 2) = 5.60$ Å and

$$E_n = \frac{n^2h^2}{8ma^2} = \frac{\pi^2n^2\hbar^2}{2ma^2}.$$

By direct substitution, we find that $E_n = 115.6n^2$ kJ/mol. The $n = 1$ and $n = 2$ levels are full, as there are 4π electrons. So $E_3 - E_2 = E_{\text{lowest empty}} - E_{\text{highest full}} = 5 \cdot 115.6 = 578$ kJ/mol. Similarly, $E_3 - E_1 = 924$ kJ/mol and $E_4 - E_2 = 1,386$ kJ/mol. Using $\Delta E = hc/\lambda$ to

TABLE 3.3   FEMO PREDICTIONS OF WAVELENGTHS ASSOCIATED
WITH TRANSITIONS IN BUTADIENE.

Upper quantum number	Lower quantum number	λ (nm)
3	2	207
3	1	129
4	2	86.3

convert these energy differences to λ_{max} yields the results shown in Table 3.3. All of these
transitions occur in the ultraviolet. (As a point of reference for Chapter 15, we note that
the 3–1 and 4–2 transitions are examples of forbidden transitions.)

SUGGESTED READINGS

Box problems are described in most quantum chemistry and quantum mechanics books
(cited at the end of Chapters 1 and 2, respectively). The FEMO model is discussed in
I. N. Levine, *Quantum Chemistry*, 5th ed. (Prentice Hall, Upper Saddle River, NJ, 2000).

PROBLEMS

3.1 Draw
   (a) the $n = 10$ state of the 1D particle in box.
   (b) the corresponding probability density.
   (c) the dependence of the particle-in-box eigenvalue on the mass m.

3.2 Write down an expression for the average value of the square of the displacement
   of a particle in a one-dimensional box from the middle of the box. No evaluation of
   integrals is necessary, but indicate what limits are used on any integrals.

3.3 Consider a one-dimensional-box potential such that the bottom of the box has a
   step at the midpoint, as shown in the following diagram, where the step height is V_s:

   (a) What boundary conditions must be applied to the solutions of the Schrödinger
   equation at $x = 0$, $x = a$, and $x = a/2$?
   (b) What is the solution of the Schrödinger equation for $x < a/2$ that satisfies the
   $x = 0$ boundary condition, but not that at $x = a/2$?
   (c) What is the solution of the Schrödinger equation for $x > a/2$ that satisfies the
   $x = a$ boundary condition, but not that at $x = a/2$? Assume that $E > V_s$.
   (d) Take the solutions from (b) and (c) and apply the boundary conditions appropriate for $x = a/2$. It should be possible to manipulate the resulting equations
   into a single equation that determines the eigenvalues. What is this equation?
   It is not necessary to solve it.

3.4 For the particle-in-box problem, why does the energy eigenvalue increase as the number of nodes in the corresponding eigenfunction increases? (*Hint:* Think about de Broglie's formula.) Also, why must each eigenfunction have a different number of nodes?

3.5 Solve the Schrödinger equation for a particle in a half-infinite box, shown in the following diagram:
   (a) Write down solutions for $x < 0, 0 \leq x \leq L$, and $x > L$. Consider $E < V_0$ only.
   (b) Apply boundary conditions at $x = 0$, $x = L$, and $x = \infty$, generating a formula for the energy. You do not need to solve this formula explicitly to determine the energy as a function of the quantum number.
   (c) Use a calculator or computer to solve numerically for the four lowest energy levels (in eV) for the case $L = 1$ Å, $m = 1$ g/mol, and $V_0 = 4$ eV. Compare your result with the corresponding infinite-box energies (i.e., same parameters, except that $V_0 = \infty$).

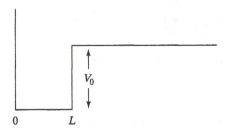

3.6 Consider a particle in a two-dimensional box having sides L_x and L_y with $L_x = 3L_y$. Take the potential to be infinite outside the box and zero inside.
   (a) What are the energy eigenvalues?
   (b) Calculate the energies and degeneracies of the first 10 states.
   (c) Show contours for the first four eigenfunctions.

3.7 Consider a particle in a three-dimensional cubic box. Which of the following wavefunctions are eigenfunctions of the Hamiltonian?

$$\psi = \frac{1}{\sqrt{2}} (\psi_{1,2,3} - \psi_{3,1,2});$$

$$\psi = \psi_{1,1,1} - \psi_{2,2,2}.$$

3.8 Impurities in solids can sometimes be described by a particle-in-a-box model. Suppose He is substituted for Xe, and assume a particle-in-a-cubic-box model, the length of whose sides is equal to the atomic diameter of Xe (≈ 2.62 Å). Compute the lowest excitation energy for the He atom's motion.

3.9 Suppose that you wanted to synthesize a polyene that has its lowest excitation energy at a wavelength of 600 nm. What would the structure of this molecule be, according to FEMO?

3.10 An alternative FEMO model that can be applied to molecules like benzene involves using a two-dimensional box and assuming that the area of the box is the same as the cross-sectional area of the molecule. Apply this model to determining the lowest excitation energy of benzene. Assume that the C–C bond distance is 1.4 Å. How does your result compare with the measured wavelength of 200 nm?

4

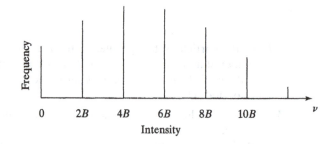

Rigid-Rotor Models and Angular-Momentum Eigenstates

Rotational energy is one of the important forms of energy that molecules possess. Rotations also characterize the symmetry of spherical species, such as atoms. The understanding of rigid-rotor models and angular-momentum eigenstates will be useful both in rotational problems themselves and in understanding atomic orbitals.

4.1 MOTIONS OF A DIATOMIC MOLECULE: SEPARATION OF THE CENTER OF MASS

In Chapter 3, we discussed some applications of the Schrödinger equation to electrons. Of course, that equation also holds for atomic nuclei, and this is what we now study. Consider the motions of a diatomic molecule AB, where the two atoms have masses m_a and m_b, respectively. The quantum mechanical Hamiltonian governing the motion of the two atoms is

$$H = -\frac{\hbar^2}{2m_a}\nabla_a^2 - \frac{\hbar^2}{2m_b}\nabla_b^2 + V(r), \qquad (4.1)$$

where $V(r)$ depends only on the distance r between the atoms (in the absence of external fields). Because V depends only on r, it is possible to simplify the problem greatly by transforming to center-of-mass coordinates. These are defined (compare Problem 1.3) as follows:

$$\mathbf{R} = \frac{m_a \mathbf{r}_a + m_b \mathbf{r}_b}{m_a + m_b}; \qquad (4.2a)$$

$$\mathbf{r} = \mathbf{r}_a - \mathbf{r}_b. \qquad (4.2b)$$

Here, $\mathbf{R}$ locates the center of mass of the AB system, while $\mathbf{r}$ is the vector between A and B whose magnitude is r. It is not hard to solve these equations for $\mathbf{r}_a$ and $\mathbf{r}_b$; we have

$$\mathbf{r}_a = \mathbf{R} + \frac{m_b}{m_a + m_b} \mathbf{r}; \qquad (4.3a)$$

$$\mathbf{r}_b = \mathbf{R} - \frac{m_a}{m_a + m_b} \mathbf{r}. \qquad (4.3b)$$

If the chain rule is now applied to convert the derivatives with respect to $\mathbf{r}_a$ and $\mathbf{r}_b$ that appear in the Laplacian operators in Eq. (4.1) to derivatives with respect to $\mathbf{R}$ and $\mathbf{r}$, the Hamiltonian becomes

$$H = \frac{-\hbar^2}{2M} \nabla_{\mathbf{R}}^2 - \frac{\hbar^2}{2\mu} \nabla_{\mathbf{r}}^2 + V(r), \qquad (4.4)$$

where $M = m_a + m_b$ is the total mass of the diatomic molecule and μ, defined by

$$\frac{1}{\mu} = \frac{1}{m_a} + \frac{1}{m_b}, \qquad (4.5)$$

is called the *reduced mass* of the molecule. The first term on the right-hand side of Eq. (4.4) represents the kinetic energy associated with motions of the center of mass of the molecule. Since the potential V does not depend on $\mathbf{R}$, the Hamiltonian is separable into a part that depends only on R and a part that depends only on r. The R-dependent part (i.e., that governing motion of the center of mass) is identical to the Hamiltonian for a free particle. We will ignore this term in what follows, as we are interested here just in the internal states (vibration and rotation) of the diatomic molecule. Thus, we need only consider the Hamiltonian governing *relative* motions, namely,

$$H = \frac{-\hbar^2}{2\mu} \nabla_{\mathbf{r}}^2 + V(r). \qquad (4.6)$$

4.2 RIGID-ROTOR MODEL IN TWO DIMENSIONS

For motions in two dimensions according to Eq. (4.6), the Laplacian is

$$\nabla^2 = \frac{\partial^2}{\partial x^2} + \frac{\partial^2}{\partial y^2}. \qquad (4.7)$$

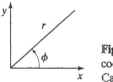

Figure 4.1 Polar coordinates related to Cartesian coordinates.

Because V depends only on the scalar quantity r, it is convenient to convert Eq. (4.7) to polar coordinates. For planar motions, these are the coordinates r and ϕ (Figure 4.1), and the conversion equations are

$$x = r \cos \phi \tag{4.8a}$$

and

$$y = r \sin \phi. \tag{4.8b}$$

By repeated application of the chain rule, the Laplacian operator can be converted to

$$\nabla^2 = \frac{1}{r} \frac{\partial}{\partial r} r \frac{\partial}{\partial r} + \frac{1}{r^2} \frac{\partial^2}{\partial \phi^2}, \tag{4.9}$$

where the radial term can also be expressed as $(\partial^2/\partial r^2) + (1/r)(\partial/\partial r)$. The Schrödinger equation for the diatomic internal wavefunction, denoted $\chi(r, \phi)$, is

$$\left\{ \frac{-\hbar^2}{2\mu} \left[\frac{1}{r^2} \frac{\partial^2}{\partial \phi^2} + \frac{1}{r} \frac{\partial}{\partial r} r \frac{\partial}{\partial r} \right] + V(r) \right\} \chi(r, \phi) = E \chi(r, \phi). \tag{4.10}$$

The first term in the braces in this equation refers to angular motion (i.e., rotation) of the molecule, while the second and third terms govern the molecule's radial motion (vibration).

We now assume that the molecule is a *rigid rotor*, in which the distance r is fixed. This means that we can omit the radial kinetic-energy term, and since the potential energy depends only on r, and that is a constant, we can ignore it. As a result, the only contribution to the energy comes from rotational (ϕ) motion, and the Hamiltonian is

$$H_{\text{rotor}} = \frac{-\hbar^2}{2\mu r^2} \frac{\partial^2}{\partial \phi^2}. \tag{4.11}$$

If we now use the symbol ψ for the rotational wavefunction (the relationship of ψ to the wavefunction χ will be described in Chapter 5), the Schrödinger equation is

$$\frac{-\hbar^2}{2\mu r^2} \frac{d^2 \psi}{d\phi^2} = E \psi. \tag{4.12}$$

It is convenient to write this as

$$\frac{d^2 \psi}{d\phi^2} = -m^2 \psi, \tag{4.13}$$

where the dimensionless constant m is defined by

$$m^2 = \frac{2\mu E r^2}{\hbar^2}, \text{ or } E = \frac{\hbar^2 m^2}{2\mu r^2}, \tag{4.14}$$

which means that

$$E = \frac{\hbar^2 m^2}{2I}, \tag{4.15}$$

where $I \equiv \mu r^2 = \sum_{i=a,b} m_i r_i^2$ is the moment of inertia of the molecule. In the summation, r_i is the distance of atom i from the center of mass of the molecule.

The differential equation (4.13) is the same as was discussed in Chapter 1, and its solutions can be expressed in terms of $\sin(m\phi)$ and $\cos(m\phi)$. In the present case, it is convenient to rewrite these solutions in terms of complex exponentials instead as

$$\psi = A e^{im\phi} + B e^{-im\phi}, \tag{4.16}$$

where A and B are constants. The boundary condition on the wavefunction is that it must be periodic in ϕ; that is,

$$\psi(\phi) = \psi(\phi + 2\pi), \tag{4.17}$$

which means that

$$A e^{im(\phi + 2\pi)} + B e^{-im(\phi + 2\pi)} = A e^{im\phi} + B e^{-im\phi}. \tag{4.18}$$

This equation can be satisfied for any A and B if

$$e^{im2\pi} = e^{-im2\pi} = 1, \tag{4.19}$$

which requires that m be an integer:

$$m = 0, \pm 1, \pm 2, \pm 3, \ldots$$

There is redundancy in allowing m to have both positive and negative values and in including both terms in Eq. (4.16), so the convention is to use the solution

$$\psi = A e^{im\phi}, \tag{4.20}$$

with m allowed to have positive, negative, and zero integer values; that is,

$$m = 0, \pm 1, \pm 2, \ldots$$

Since m is constrained to have integer values, the energy given by Eq. (4.15) is quantized, and in fact, the form of this expression is familiar from the particle-in-box model, where (from Eq. 3.18) the energy is also proportional to the square of a quantum number. Note, however, that with the rigid rotor, the quantum number m can have positive, negative, and zero values. The case of $m = 0$ gives $E = 0$, which means that there is no zero-point energy. Why? One way to rationalize this is to assert that the uncertainty in the angular coordinate $\Delta\phi$ is infinite, so the uncertainty in its conjugate momentum can be zero (leading to zero energy) and still satisfy the condition that $\Delta\phi\Delta p_\phi \neq 0$. Note also that the states with $m \neq 0$ are all doubly degenerate. Physically, this corresponds to the two possible directions of rotor motion (clockwise and counterclockwise). Another property of the solutions is that

$$\psi\psi^* = A^2 = \text{constant, for all } m, \tag{4.21}$$

which means that there is equal probability of finding the rotor with any value of ϕ.

To normalize the wavefunction, we have

$$\int d\tau \psi^*\psi = 1 = \int_0^{2\pi} d\phi \, e^{im\phi}e^{-im\phi}A^2 = 2\pi A^2, \tag{4.22}$$

so

$$A = \frac{1}{\sqrt{2\pi}}, \tag{4.23}$$

which means that

$$\psi = \frac{1}{\sqrt{2\pi}} e^{im\phi}. \tag{4.24}$$

EXERCISE 4.1 If the H_2 molecule rotates in the plane of a crystalline surface (in a chemisorption situation), it can be approximated as a two-dimensional rigid rotor. Calculate, in cm^{-1}, the lowest energy transition for such a system.

Answer From Eq. (4.15), the energy levels of the 2-D rotor are

$$E_n = \frac{m^2\hbar^2}{2I} = \frac{m^2\hbar^2}{2\mu r^2}.$$

For H_2, $r = r_0 = 0.7416$ Å, μ is one-half the mass of the hydrogen atom, and the lowest energy transition is between $m = 0$ and $m = 1$. Plugging in the constants from Appendix D, we obtain

$$\Delta E = 0.727 \text{ kJ/mol} = 60.8 \text{ cm}^{-1}.$$

4.3 THREE-DIMENSIONAL RIGID ROTOR

The three-dimensional case is more or less the same as the two-dimensional one, but more difficult. Here we introduce spherical polar coordinates (see Figure 4.2), whereby

$$x = r\sin\theta\cos\phi, \tag{4.25a}$$

$$y = r\sin\theta\sin\phi, \tag{4.25b}$$

and

$$z = r\cos\theta. \tag{4.25c}$$

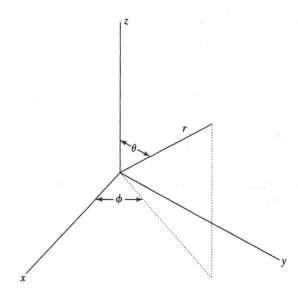

Figure 4.2 Definition of spherical polar coordinates.

In terms of these coordinates, the integration element is $d\tau = r^2 dr \sin\theta d\theta d\phi$, and the Laplacian is

$$\nabla^2 = \frac{1}{r^2}\frac{\partial}{\partial r} r^2 \frac{\partial}{\partial r} + \frac{1}{r^2 \sin\theta}\frac{\partial}{\partial \theta} \sin\theta \frac{\partial}{\partial \theta} + \frac{1}{r^2 \sin^2\theta}\frac{\partial^2}{\partial \phi^2}, \qquad (4.26)$$

where the radial term can also be written as $(\partial^2/\partial r^2) + [(2/r)(\partial/\partial r)]$ or as $(1/r)(\partial^2/\partial r^2)r$. For a **rigid** rotor, r is fixed, so that

$$\nabla^2 = \frac{1}{r^2}\left\{\frac{1}{\sin\theta}\frac{\partial}{\partial\theta}\sin\theta\frac{\partial}{\partial\theta} + \frac{1}{\sin^2\theta}\frac{\partial^2}{\partial\phi^2}\right\}. \qquad (4.27)$$

This gives us the Schrödinger equation

$$\frac{-\hbar^2}{2I}\left\{\frac{1}{\sin\theta}\frac{\partial}{\partial\theta}\sin\theta\frac{\partial}{\partial\theta} + \frac{1}{\sin^2\theta}\frac{\partial^2}{\partial\phi^2}\right\}\psi(\phi,\theta) = E\psi(\phi,\theta). \qquad (4.28)$$

To solve this equation, we will use a variant of the method of separation of variables. Note that an alternative approach based on raising and lowering operators is also possible, as outlined at the end of this chapter.

To solve Eq. (4.28), we note that if the Hamiltonian operator is multiplied by $\sin^2\theta$, it is separable. This suggests that we try a solution of the form

$$\psi = \Theta(\theta)\Phi(\phi). \qquad (4.29)$$

Substituting into the Schrödinger equation and multiplying by $\sin^2\theta$, we get

$$\frac{-\hbar^2}{2I}\left\{\sin\theta\frac{\partial}{\partial\theta}\sin\theta\frac{\partial\Theta}{\partial\theta}\Phi + \Theta\frac{\partial^2\Phi}{\partial\phi^2}\right\} = E\sin^2\theta\Phi\Theta, \qquad (4.30)$$

and if this is now divided by ψ, we obtain

$$\frac{-\hbar^2}{2I}\left\{\frac{1}{\Theta}\sin\theta\frac{\partial}{\partial\theta}\sin\theta\frac{\partial\Theta}{\partial\theta} + \frac{1}{\Phi}\frac{\partial^2\Phi}{\partial\phi^2}\right\} = E\sin^2\theta. \tag{4.31}$$

Invoking the separation-of-variables procedure (see Chapter 2, Section 2.4), we may write the ϕ part of this equation as

$$\frac{-\hbar^2}{2I}\frac{1}{\Phi}\frac{\partial^2\Phi}{\partial\phi^2} = \text{constant} = +\lambda, \tag{4.32}$$

where λ is a constant. This equation is the same as the 2D rigid-rotor Schrödinger equation, so the solutions are

$$\Phi = \frac{e^{im\phi}}{\sqrt{2\pi}}, \tag{4.33}$$

where

$$m^2 = \frac{\lambda 2I}{\hbar^2} \quad \text{and} \quad \lambda = \frac{\hbar^2 m^2}{2I}. \tag{4.34}$$

The equation for Θ, when combined with Eq. (4.34), gives

$$\frac{-\hbar^2}{2I}\left\{\frac{1}{\sin\theta}\frac{\partial}{\partial\theta}\sin\theta\frac{\partial}{\partial\theta} - \frac{m^2}{\sin^2\theta}\right\}\Theta = E\Theta. \tag{4.35}$$

This is a much more challenging differential equation to solve than that for the 2D rotor. Rather than solve this equation explicitly, we simply give some of the solutions and verify that they work. By substituting into the Schrödinger equation, one can verify that

a. $\Theta = 1$ is a solution if $m = 0$ and $E = 0$.

b. $\Theta = \cos\theta$ is also a solution. When we substitute into the Schrödinger equation, we have

$$\left(\frac{-1}{\sin\theta}\frac{\partial}{\partial\theta}\sin^2\theta - \frac{m^2}{\sin^2\theta}\cos\theta\right) \stackrel{?}{=} \frac{-2IE}{\hbar^2}\cos\theta, \tag{4.36}$$

which reduces to

$$\left(-2\cos\theta - \frac{m^2\cos\theta}{\sin^2\theta}\right) \stackrel{?}{=} \frac{-2IE}{\hbar^2}\cos\theta. \tag{4.37}$$

This is clearly an equality if $m = 0$ and

$$E = \frac{\hbar^2}{I}. \tag{4.38}$$

c. $\Theta = \sin\theta$ is also a solution. In this case, substitution into the Schrödinger equation gives

$$\left(\frac{1}{\sin\theta}\frac{\partial}{\partial\theta}\sin\theta\cos\theta - \frac{m^2}{\sin^2\theta}\sin\theta\right) \stackrel{?}{=} \frac{-2IE\sin\theta}{\hbar^2}, \tag{4.39}$$

which can be satisfied if

$$m = \pm 1$$

and

$$E = \frac{\hbar^2}{I}. \tag{4.40}$$

The general solution of Eq. (4.35) is

$$\Theta = S_{\ell m} = (-1)^m\left[\frac{2\ell + 1}{2}\frac{(\ell - |m|)!}{(\ell + |m|)!}\right]^{1/2} P_\ell^{|m|}(\cos\theta), \qquad m \geq 0, \tag{4.41}$$

and

$$\Theta = \left[\frac{2\ell + 1}{2}\frac{(\ell - |m|)!}{(\ell + |m|)!}\right]^{1/2} P_\ell^{|m|}(\cos\theta), \qquad m < 0,$$

where the quantum numbers ℓ and m are given by

$$\ell = 0, 1, 2, \ldots, \qquad m = 0, \pm 1, \ldots, \pm\ell.$$

In Eq. (4.41), the term in brackets is a normalization factor, and the P_ℓ^m are called *associated Legendre polynomials*. The following are some specific examples of these polynomials:

$$P_0^0 = 1, \qquad\qquad P_2^0 = \tfrac{1}{2}(3\cos^2\theta - 1),$$
$$P_1^0 = \cos\theta, \qquad\qquad P_2^1 = \tfrac{1}{2}\sin\theta\cos\theta,$$
$$P_1^1 = \sin\theta, \qquad\qquad P_2^2 = \sin^2\theta.$$

The first three on this list are the solutions (a)–(c) just discussed.

The energy eigenvalues in this case are independent of m and are given by

$$E_\ell = \frac{\hbar^2}{2I}\ell(\ell + 1). \tag{4.42}$$

Note that, just as with the 2D rotor, there is no zero-point energy associated with rotational motion.

The normalization condition associated with the solutions $S_{\ell m}(\theta)$ is

$$\int_0^\pi \sin\theta \, d\theta \, S_{\ell m}^* S_{\ell m} = 1. \tag{4.43}$$

4.4 SPHERICAL HARMONICS

The combined eigenstates (the product of the θ and ϕ parts) for the 3-D rigid rotor are called the spherical harmonics and are given by

$$Y_{\ell m} = \frac{e^{im\phi}}{\sqrt{2\pi}} S_{\ell m}(\theta). \tag{4.44a}$$

The following are some specific examples of spherical harmonics:

$$Y_{00} = \frac{1}{\sqrt{4\pi}};$$

$$Y_{1-1} = \sqrt{\frac{3}{8\pi}}\, e^{-i\phi} \sin\theta; \quad Y_{10} = \sqrt{\frac{3}{4\pi}} \cos\theta; \quad Y_{11} = -\sqrt{\frac{3}{8\pi}}\, e^{i\phi} \sin\theta;$$

$$Y_{2-1} = \sqrt{\frac{13}{8\pi}} \sin\theta \cos\theta e^{-i\phi}; \quad Y_{20} = \sqrt{\frac{5}{16\pi}} (3\cos^2\theta - 1); \tag{4.44b}$$

$$Y_{21} = -\sqrt{\frac{13}{8\pi}} \sin\theta \cos\theta e^{i\phi}.$$

The $Y_{\ell m}$ have some interesting properties:

1. They are orthonormal; that is,

$$\int_0^{2\pi} \int_0^{\pi} Y_{\ell m}^* Y_{\ell' m'} \sin\theta \, d\theta \, d\phi = \delta_{\ell\ell'}\delta_{mm'}, \tag{4.45}$$

   where δ_{ij} is called the *Kronecker* δ and has the values 1 if $i = j$ and 0 if $i \neq j$.

2. They satisfy the Schrödinger equation

$$\frac{-\hbar^2}{2I} \left\{ \frac{1}{\sin\theta} \frac{\partial}{\partial\theta} \sin\theta \frac{\partial}{\partial\theta} + \frac{1}{\sin^2\theta} \frac{\partial^2}{\partial\phi^2} \right\} Y_{\ell m} = E Y_{\ell m}, \tag{4.46}$$

   with energy given by Eq. (4.42).

3. They also satisfy the eigenvalue equation

$$\frac{\partial^2 Y_{\ell m}}{\partial\phi^2} = -m^2 Y_{\ell m}, \tag{4.47}$$

   which will prove important in our later discussion of angular-momentum properties.

EXERCISE 4.2 Are $Y_{11} \pm Y_{1-1}$ eigenfunctions of the 2-D rigid-rotor Hamiltonian? Are they eigenfunctions of the 3-D rigid-rotor Hamiltonian? What is the energy of each of these states?

Answer

$$H_{2D} = -\frac{\hbar^2}{2\mu r^2} \frac{\partial^2}{\partial\phi^2} \quad [\text{Eq. (4.11)}];$$

$$Y_{\ell m} = e^{im\phi} \frac{S_{\ell m}(\theta)}{\sqrt{2\pi}} \quad [\text{Eq. (4.44a)}].$$

Direct substitution gives

$$-\frac{\hbar^2}{2\mu r^2}\frac{\partial^2}{\partial\phi^2}\left[Y_{11} \pm Y_{1-1}\right] = -\frac{\hbar^2}{2\mu r^2}\frac{\partial^2}{\partial\phi^2}\left(e^{i\phi}S_{11} \pm e^{-i\phi}S_{1-1}\right)/\sqrt{2\pi}$$

$$= -\frac{\hbar^2}{2\mu r^2}\left(-e^{i\phi}S_{11} \mp e^{-i\phi}S_{1-1}\right)/\sqrt{2\pi}$$

$$= \frac{\hbar^2}{2\mu r^2}\left(Y_{11} \pm Y_{1-1}\right),$$

so $Y_{11} \pm Y_{1-1}$ is an eigenfunction of H_{2D}. For H_{3D}, we use the result in Eq. (4.46) to get

$$H_{3D}\left(Y_{11} \pm Y_{1-1}\right) = \frac{\hbar^2}{2\mu r^2}(1)(2)\left(Y_{11} \pm Y_{1-1}\right),$$

where the factor $(1)(2)$ comes from $\ell(\ell + 1)$ for the case $\ell = 1$. This shows that the combinations $Y_{11} \pm Y_{1-1}$ are eigenfunctions of H for both 2D and 3D rigid rotors. The energies are $\hbar^2/2I$ and $2\hbar^2/2I$, respectively. The 3D rotor has more energy because it can move in more directions.

4.5 ROTATIONAL SPECTRA

The quantum numbers normally used in describing rotational spectra of diatomic molecules are called JM (rather than ℓ, m), so for the purposes of this discussion, we switch notation. Later on, when we consider electronic wavefunctions, ℓ, m are the symbols most commonly used, so we will switch back. In terms of J, the energy levels are given by

$$E_J = \frac{J(J + 1)\hbar^2}{2I}, \tag{4.48}$$

and the energy change in making the transition from state J'' to J' is

$$\frac{E' - E''}{hc} = B\left[(J' + 1)(J') - (J'' + 1)J''\right], \tag{4.49}$$

where

$$B = \frac{h}{8\pi^2 I c} \tag{4.50}$$

is called the *rotational constant*.

Selection Rules   Only *some* transitions are permitted. We will see in Chapter 15 that (1) only molecules with a nonzero dipole moment ($\mu \neq 0$) have a rotational spectrum and (2) the intensity is proportional to the square of the matrix elements $x_{J'M',J''M''}$, $y_{J'M',J''M''}$, $z_{J'M',J''M''}$. By "matrix elements," we mean integrals involving a product of two Y_{JM}'s multiplied by x, y, or z. The general form of these integrals is (note that M' and M'' are quantum numbers while μ is a dipole matrix element, $\mu = er$)

$$
\mu_{J'M', J''M''} = e \int_0^{2\pi} d\phi \int_0^\pi \sin\theta \, d\theta \, Y_{J'M'}^* \begin{Bmatrix} x \\ y \\ z \end{Bmatrix} Y_{J''M''}
$$

$$
= e \int_0^{2\pi} d\phi \int_0^\pi d\theta \sin\theta \begin{Bmatrix} r\cos\phi\sin\theta \\ r\sin\phi\sin\theta \\ r\cos\theta \end{Bmatrix} S_{J'M'}^* S_{J''M''} e^{-iM'\phi} e^{iM''\phi},
$$

(4.51)

where we have used Eqs. (4.25). Upon further substitution, the ϕ part of this integral becomes

$$
\mu_{J'M', J''M''} \propto \int_0^{2\pi} d\phi \begin{Bmatrix} \cos\phi \\ \sin\phi \\ 1 \end{Bmatrix} e^{i(M''-M')\phi}
$$

(4.52)

$$
= 0,
$$

unless $M'' - M' = \pm 1, 0$.

From Eq. (4.52), we see that $\mu_{J'M', J''M''} = 0$, unless

$$
M'' = M' \text{ or } M'' = M' \pm 1,
$$

(4.53)

which defines a *selection rule* for rotational transitions. The analogous θ integration needed to complete the evaluation of Eq. (4.51) can be done, but is much harder. Here we just give the resulting selection rule, which is

$$
\mu_{J'M'J''M''} = 0, \text{ unless } J' - J'' = \pm 1.
$$

(4.54)

If we are interested in light *absorption*, then the only possible final state J' starting from J'' is $J' = J'' + 1$. The frequency associated with this transition is

$$
\nu_{J''+1, J''} = B(J'' + 1)(J'' + 2) - B(J'' + 1)(J''),
$$

(4.55a)

or

$$
\nu_{J''+1, J''} = 2B(J'' + 1).
$$

(4.55b)

The spectrum thus should consist of a series of equally spaced lines in which the spacing is 2B. This is pictured schematically in Figure 4.3. The relative intensities

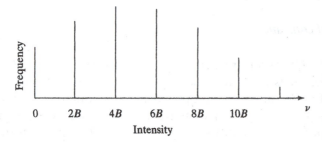

Figure 4.3 Schematic of a rotational spectrum for a molecule with a thermal distribution of rotational states.

of the lines are determined by the magnitude of $\mu_{J'M'J''M''}$, along with the initial-state population (which is usually determined by the Boltzmann distribution). The measurement of rotational spectra provides a powerful method for determining bond distances in diatomics (as derived from the value of B using Eq. (4.50) and the relation $I = \mu r^2$).

EXERCISE 4.3   Several unstable species can be recognized spectroscopically by their pure rotational spectrum. Compute the rotational spacing $2B$ for the OH radical $(r_e = r = 0.97\text{Å})$ and for the OD radical. Do you think that these can be differentiated experimentally?

Answer   Substitute into $B = h/(8\pi^2 Ic)$ and $I = \mu r_e^2$ to get $2B(\text{OH}) = 37.5 \text{ cm}^{-1}$ and $2B(\text{OD}) = 20.0 \text{ cm}^{-1}$.

4.6   ANGULAR MOMENTUM

One way to write the kinetic energy for a diatomic molecule is as

$$\frac{-\hbar^2 \nabla^2}{2\mu} = \frac{-\hbar^2}{2\mu} \left\{ \frac{1}{r} \frac{\partial^2}{\partial r^2} r + \frac{1}{r^2 \sin\theta} \frac{\partial}{\partial\theta} \sin\theta \frac{\partial}{\partial\theta} + \frac{1}{r^2 \sin^2\theta} \frac{\partial^2}{\partial\phi^2} \right\} \quad (4.56a)$$

$$= \frac{\left(P_r^2 + \dfrac{\hat{\ell}^2}{r^2} \right)}{2\mu}, \quad (4.56b)$$

where P_r is the radial momentum operator and $\hat{\ell}^2$ is the operator associated with the square of the angular momentum of the rotor. Equation (4.56b) comes from classical mechanics (Chapter 1, Problem 2), but it applies equally well to quantum mechanics. $\hat{\ell}^2$ is is defined by

$$\hat{\ell}^2 = \hat{\ell}_x^2 + \hat{\ell}_y^2 + \hat{\ell}_z^2. \quad (4.57)$$

Comparing Eqs. (4.56a) and (4.56b), we see that

$$\hat{\ell}^2 = -\hbar^2 \left\{ \frac{1}{\sin\theta} \frac{\partial}{\partial u} \sin\theta \frac{\partial}{\partial\theta} + \frac{1}{\sin^2\theta} \frac{\partial^2}{\partial\phi^2} \right\}, \quad (4.58)$$

which means that the rigid-rotor Hamiltonian is

$$H_{\text{rotor}} = \frac{\hat{\ell}^2}{2I}, \quad (4.59)$$

and from Eq. 4.46 the spherical harmonics satisfy the relation

$$\hat{\ell}^2 Y_{\ell m} = \ell(\ell + 1)\hbar^2 Y_{\ell m}. \quad (4.60)$$

The spherical harmonics are also eigenfunctions of other angular-momentum operators. These operators were introduced in Section 2.3.3, and the z component is given by

$$\hat{\ell}_z = \frac{\hbar}{i}\left(x\frac{\partial}{\partial y} - y\frac{\partial}{\partial x}\right)$$

$$= \frac{\hbar}{i}\frac{\partial}{\partial \phi}. \tag{4.61}$$

If this operator is now applied to the spherical harmonics, we obtain

$$\hat{\ell}_z Y_{\ell m} = \hat{\ell}_z\{A_{\ell m}P_\ell^m(\theta)e^{im\phi}\}$$

$$= \hbar m Y_{\ell m}, \tag{4.62}$$

so the spherical harmonics are eigenfunctions of both $\hat{\ell}^2$ and $\hat{\ell}_z$. They are also eigenfunctions of $\hat{\ell}_z^2$ (by property 3 noted in Section 4.4).

The x and y components of ℓ are given by

$$\hat{\ell}_x = \frac{\hbar}{i}\left(y\frac{\partial}{\partial z} - z\frac{\partial}{\partial y}\right) = \frac{-\hbar}{i}\left(\sin\phi\frac{\partial}{\partial\theta} + \cot\theta\cos\phi\frac{\partial}{\partial\phi}\right) \tag{4.63}$$

and

$$\hat{\ell}_y = \frac{\hbar}{i}\left(z\frac{\partial}{\partial x} - x\frac{\partial}{\partial z}\right) = \frac{-\hbar}{i}\left(\cos\phi\frac{\partial}{\partial\theta} - \cot\theta\sin\phi\frac{\partial}{\partial\phi}\right). \tag{4.64}$$

Now, consider the commutator of these two operators. From Eqs. (4.63) and (4.64), we find that

$$[\hat{\ell}_x, \hat{\ell}_y] = \frac{\hbar^2}{i^2}\left[y\frac{\partial}{\partial z} - z\frac{\partial}{\partial y}, z\frac{\partial}{\partial x} - x\frac{\partial}{\partial z}\right] \tag{4.65}$$

$$= \frac{\hbar}{i}[yp_z - zp_y, zp_x - xp_z].$$

To proceed from here, we invoke the two operator identities

$$[A, BC] = ABC - BCA = ABC + BAC - BAC - BCA$$

$$= [A, B]C + B[A, C] \tag{4.66}$$

and

$$[A + B, C] = [A, C] + [B, C], \tag{4.67}$$

and we also invoke the commutator

$$[x, p_x] = i\hbar. \tag{4.68}$$

(See Section 2.3.2.) This leads to

$$[yp_z, xp_z] = 0 \quad \text{and} \quad [zp_y, zp_x] = 0, \tag{4.69}$$

so that

$$[\hat{\ell}_x, \hat{\ell}_y] = [yp_z, zp_x] + [zp_y, xp_z]$$
$$= yp_x[p_z, z] + xp_y[z, p_z] = i\hbar[xp_y - yp_x] \qquad (4.70)$$
$$= i\hbar\hat{\ell}_z.$$

Similar commutation formulas can be developed for any cyclic permutation of the indices x, y, and z, so that $[\hat{\ell}_z, \hat{\ell}_x] = i\hbar\hat{\ell}_y$. One can also show that

$$[\hat{\ell}_x, \hat{\ell}^2] = [\hat{\ell}_y, \hat{\ell}^2] = [\hat{\ell}_z, \hat{\ell}^2] = 0. \qquad (4.71)$$

We conclude that $\hat{\ell}^2$ and $\hat{\ell}_z$ commute. From Chapter 2 (Section 2.7.3), we know that such commuting operators can have simultaneous eigenfunctions, and indeed, the spherical harmonics are these eigenfunctions.

EXERCISE 4.4   Although $\hat{\ell}_z$ and $\hat{\ell}_x$ both commute with $\hat{\ell}^2$, they do not commute with one another and therefore are not expected to have simultaneous eigenfunctions. Show directly that Y_{11} is not an eigenfunction of $\hat{\ell}_x$.

Answer

$$\begin{array}{lll} Y_{11} = Ae^{i\phi}\sin\theta & & \text{[Eq. (4.46)]};\\[4pt] \hat{\ell}_x = yp_z - zp_y & & \text{(Table 2.2)};\\[4pt] Y_{11} = A(\cos\phi + i\sin\phi)\sin\theta & & \text{[Eq. (A.4)]};\\[4pt] x = r\sin\theta\cos\phi; & & \\[4pt] y = r\sin\theta\sin\phi & & \text{(Problem 1.2)}; \end{array}$$

So

$$Y_{11} = \frac{A}{r}(x + iy);$$

$$\hat{\ell}_x Y_{11} = \frac{A}{r}\frac{\hbar}{i}\left(y\frac{\partial}{\partial z} - z\frac{\partial}{\partial y}\right)(x + iy)$$

$$= \frac{A\hbar}{ir}(0 - zi) = -\frac{A\hbar}{r}z.$$

So $\hat{\ell}_x Y_{11}$ is not a constant times Y_{11}.

4.7 DIRAC NOTATION

It is convenient to write the integrals that appear in quantum mechanics using the following *Dirac notation*:

$$\int \phi^* \hat{A}\psi \, d\tau \equiv \langle\phi|\hat{A}|\psi\rangle. \qquad (4.72)$$

The $\langle\phi|$ that appears in this formula is termed a "bra," and the $|\psi\rangle$ is called a "ket." Note that the complex conjugation operation is implicitly assumed in writing the bra. For Hermitian operators, $\langle\phi|\hat{A}|\psi\rangle = \langle\hat{A}\phi|\psi\rangle$.

In Dirac notation, the following are some commonly encountered integrals:

$$\text{Scalar product:} \quad \int \phi_1^* \phi_2 \, dx = \langle\phi_1|\phi_2\rangle.$$

$$\text{Orthogonality:} \quad \langle\phi_1|\phi_2\rangle = 0.$$

$$\text{Normalization:} \quad \langle\phi_1|\phi_1\rangle = 1.$$

EXERCISE 4.5   Evaluate

$$\langle Y_{1,0}|Y_{2,0}\rangle, \langle Y_{1,0} + \sqrt{2}Y_{2,0}|Y_{1,0} - \sqrt{2}Y_{2,0}\rangle, \langle Y_{4,3}|Y_{4,4} + \sqrt{13}Y_{4,3}\rangle$$

explicitly.

Answer

$$\langle Y_{1,0}|Y_{2,0}\rangle = 0 \quad \text{(eigenfunctions with different eigenvalues)};$$
$$\langle Y_{1,0} + \sqrt{2}Y_{2,0}|Y_{1,0} - \sqrt{2}Y_{2,0}\rangle = \langle Y_{1,0}|Y_{1,0}\rangle - 2\langle Y_{2,0}|Y_{2,0}\rangle = -1;$$
$$\langle Y_{4,3}|Y_{4,4} + \sqrt{13}Y_{4,3}\rangle = \sqrt{13}.$$

4.8 RAISING AND LOWERING ANGULAR-MOMENTUM OPERATORS

Consider the operators

$$\hat{\ell}_+ = \hat{\ell}_x + i\hat{\ell}_y \tag{4.73a}$$

and

$$\hat{\ell}_- = \hat{\ell}_x - i\hat{\ell}_y. \tag{4.73b}$$

It is not difficult to show from Eq. (4.70) that

$$[\hat{\ell}_+, \hat{\ell}_z] = -\hbar\hat{\ell}_+. \tag{4.74}$$

If the operator $\hat{\ell}_z$ is applied to $\hat{\ell}_+|\ell m\rangle$ (here we switch to Dirac notation, wherein $|\ell m\rangle = Y_{\ell m}$), we get

$$\begin{aligned}
\hat{\ell}_z\hat{\ell}_+|\ell m\rangle &= \hat{\ell}_+\hat{\ell}_z|\ell m\rangle + [\hat{\ell}_z, \hat{\ell}_+]|\ell m\rangle \\
&= \hbar m\hat{\ell}_+|\ell m\rangle + \hbar\hat{\ell}_+|\ell m\rangle \\
&= \hbar(m+1)\hat{\ell}_+|\ell m\rangle.
\end{aligned} \tag{4.75}$$

This implies that $\hat{\ell}_+|\ell m\rangle$ is an eigenfunction of $\hat{\ell}_z$ with eigenvalue $m + 1$. It is also an eigenfunction of $\hat{\ell}^2$ (this is easily shown), so it must be that

$$\hat{\ell}_+|\ell m\rangle \propto |\ell m + 1\rangle. \tag{4.76}$$

The proportionality is used to allow for the possibility that $\hat{\ell}_+|\ell m\rangle$ is not normalized. Let us introduce a proportionality constant $c_{\ell m}$ into Eq. (4.76) so that it may be written

$$\hat{\ell}_+|\ell m\rangle = c_{\ell m}|\ell m + 1\rangle. \tag{4.77}$$

A similar formula can be developed for the quantity $\hat{\ell}_-|\ell m\rangle$, except that here the resulting eigenfunction is $|\ell m - 1\rangle$. Thus, we can write

$$\hat{\ell}_-|\ell m\rangle = d_{\ell m}|\ell m - 1\rangle, \tag{4.78}$$

where $d_{\ell m}$ is a constant. We see that the operator $\hat{\ell}_+$ serves to *raise* the quantum number m, while $\hat{\ell}_-$ lowers m. For this reason, we call them *raising* and *lowering* operators, respectively.

To determine the proportionality constants, we consider the integral

$$\langle \ell m|\hat{\ell}_-\hat{\ell}_+|\ell m\rangle = \langle \ell m|(\hat{\ell}_x - i\hat{\ell}_y)(\hat{\ell}_x + i\hat{\ell}_y)|\ell m\rangle \tag{4.79a}$$

$$= \langle \ell m|\hat{\ell}^2 - \hat{\ell}_z^2 - \hbar\hat{\ell}_z|\ell m\rangle, \tag{4.79b}$$

where we invoked Eqs. (4.57) and (4.70) in deriving Eq. (4.79b). The evaluation of this last equation is straightforward using Eqs. (4.60) and (4.62) and gives

$$\langle \ell m|\hat{\ell}^2 - \hat{\ell}_z^2 - \hbar\hat{\ell}_z|\ell m\rangle = \hbar^2[\ell(\ell + 1) - m(m + 1)]. \tag{4.80}$$

Now let us use the Hermitian nature of $\hat{\ell}_x$ and $\hat{\ell}_y$ to rewrite our original expression in Eqs. (4.79a). The complete integral may be written

$$\int Y_{\ell m}^*(\hat{\ell}_x - i\hat{\ell}_y)\hat{\ell}_+ Y_{\ell m}\, d\tau = \int [(\hat{\ell}_x + i\hat{\ell}_y)Y_{\ell m}]^*\hat{\ell}_+ Y_{\ell m}\, d\tau$$

$$= \int (\hat{\ell}_+ Y_{\ell m})^*\hat{\ell}_+ Y_{\ell m}\, d\tau$$

$$= |c_{\ell m}|^2 \int Y_{\ell m+1}^* Y_{\ell m+1}\, d\tau \tag{4.81}$$

$$= |c_{\ell m}|^2.$$

In the second line of Eq. (4.81), we have switched the ℓ_- operator so that it operates on $Y_{\ell m}^*$. In so doing, the factor of i inside the square brackets changes sign (which means that the Hermitian conjugate of ℓ_- is ℓ_+). The final form of Eq. (4.81) follows from Eq. (4.77) and its complex conjugate.

If we assume that the $c_{\ell m}$'s are all real and positive, then we can combine Eq. (4.81) with Eq. (4.80) to give

$$c_{\ell m} = \hbar\sqrt{\ell(\ell + 1) - m(m + 1)}. \tag{4.82}$$

A similar derivation gives

$$d_{\ell m} = \hbar\sqrt{\ell(\ell + 1) - m(m - 1)}. \tag{4.83}$$

Combining these with Eqs. (4.77) and (4.78) gives

$$\hat{\ell}_{\pm}|\ell m\rangle = \hbar\sqrt{\ell(\ell+1) - m(m \pm 1)}\,|\ell m \pm 1\rangle, \qquad (4.84)$$

which is a very useful formula for evaluating integrals involving the operators $\hat{\ell}_x$ and $\hat{\ell}_y$, and any multiples or products thereof. Equation (4.84) also shows that $|m| \le \ell$: Trying to raise the system from the state with $m = \ell$ gives zero, as does trying to lower it from $m = -\ell$.

These angular-momentum results are very useful whenever problems have angular asymmetry that leads to coupling between rotational states with different values of m. Two examples in which these results are used are the rotation of polyatomic molecules and selection rules for magnetic resonance spectroscopy.

EXERCISE 4.6 For some molecules with heavy atoms, a relativistic correction enters the Hamiltonian that can be written $H' = K\mathbf{j}_1 \cdot \mathbf{j}_2$ with K a constant and $\mathbf{j}_1$, $\mathbf{j}_2$ the total angular momenta for two electrons. Show that

$$H' = K\left(j_{z1}j_{z2} + \tfrac{1}{2}[j_{+1}j_{-2} + j_{-1}j_{+2}]\right).$$

Answer The $\mathbf{j}$'s are vector operators, so $\mathbf{j}_1 \cdot \mathbf{j}_2 = j_{1x}j_{2x} + j_{1y}j_{2y} + j_{1z}j_{2z}$. Defining $j_{1\pm} = j_{1x} \pm ij_{1y}$, we have

$$j_{+1}j_{-2} + j_{-1}j_{+2} = \left(j_{1x} + ij_{1y}\right)\left(j_{2x} - ij_{2y}\right) + \left(j_{1x} - ij_{1y}\right)\left(j_{2x} + ij_{2y}\right)$$
$$= 2\left(j_{1x}j_{2x} + j_{1y}j_{2y}\right).$$

Substituting back into the expression for $\mathbf{j}_1 \cdot \mathbf{j}_2$ yields the desired expression for H'. This "trick" works with all angular momenta.

SUGGESTED READINGS

The description of angular momentum presented here may be found in all of the books cited at the end of Chapter 2. A very thorough treatment of angular momentum is in R. N. Zare, *Angular Momentum* (Wiley-Interscience, New York, 1988). Another specialized book is M. E. Rose, *Angular Momentum* (John Wiley and Sons, New York, 1988).

PROBLEMS

4.1 Draw each of the following curves using the format in the graph that follows:
   (a) the $\ell = 1$, $m = 0$ wavefunction of the 3D rigid rotor.
   (b) the $\ell = 0$, $m = 0$ probability density of the 3D rigid rotor.
   (c) the $\ell = 1$, $m = 0$ probability density of the 3D rigid rotor.

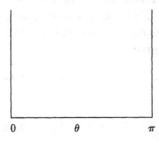

4.2 Suppose that we use a FEMO-type model to describe π electron motion in benzene. Here we assume that each electron (mass $= m$) moves as a particle in a ring of radius a. Let the polar angle be ϕ. The Schrödinger equation is

$$-\frac{\hbar^2}{2ma^2}\frac{d^2}{d\phi^2}\psi = E\psi.$$

(a) What are the eigenvalues and eigenfunctions? (*Hint:* Think about the ϕ-dependent part of the solution to the rigid-rotor problem.)

(b) Show which energy levels are occupied for the ground configuration of the benzene cation $C_6H_6^+$. What is the spin multiplicity? (The multiplicity is $2n_u + 1$, where n_u is the number of unpaired spins.)

(c) What is the projection of the electronic orbital angular momentum along the z-axis (i.e., perpendicular to the molecular plane) in the highest occupied orbital?

(d) If $C_6H_6^+$ is placed in a magnetic field B in the z-direction, a perturbation

$$V = \gamma B\ell_z$$

is added to the Hamiltonian of each electron, where ℓ_z is the orbital angular-momentum operator and γ is a constant. In first-order perturbation theory (see Chapter 5, Section 5.4), the energy level changes as a result of this perturbation are just $\langle\psi|V|\psi\rangle$. Evaluate this expression.

4.3 Single-walled carbon nanotubes can be approximated by a particle-on-a-cylindrical-surface model. Suppose the cylinder has length ℓ and radius a, with the z-axis along the cylinder.

(a) Write the kinetic energy of the electron in terms of the length z, the radius a, and the radial angle ϕ of the cylinder.

(b) Combining ideas from the particle-in-a-box and 2-D rigid-rotor models, show that the wavefunction can be written as $\psi = A\sin(n\pi z/\ell)e^{im\phi}$. What are the allowed values of the quantum numbers n and m?

(c) Write the energy expression in terms of m, n, ℓ, a, and fundamental constants. For a nanotube of radius 5.5 Å, and length 120 Å, compute the energy spacing between the two lowest energy levels. If there are 300 π electrons in the tube, calculate, with the use of a computer, the spacing between the highest occupied and lowest empty orbitals. What are their (m, n) quantum numbers?

4.4 Ignoring vibrational energy, what is the total nuclear energy of the CO_2 molecule in the $J'' = 1$, $M'' = 0$ eigenstate, whose center of mass is moving with a velocity $v_R = 0.2$ km/sec? If the molecule absorbs a photon, what will the energy of that photon be, and what are the final quantum numbers J' and M'?

4.5 Introduce real-valued expressions for the $\ell = 1$ spherical harmonics as follows:

$$Y_x = (-Y_{11} + Y_{1-1})/\sqrt{2};$$

$$Y_y = (Y_{11} + Y_{1-1})/(i\sqrt{2});$$

$$Y_z = Y_{10}.$$

(a) Show that Y_x, Y_y, and Y_z have the same angular parts as the coordinates x, y, and z, respectively.

(b) Evaluate the following expressions, reducing the results to constants:

(i) $\langle Y_x | \ell_z | Y_z \rangle$.

(ii) $\langle Y_x | \ell_x^2 + \ell_y^2 | Y_z \rangle$.

4.6 Using the same real-valued spherical harmonics as in Problem 5, evaluate each of the following:

(a) $\langle Y_y | \ell_y | Y_y \rangle$.

(b) $\langle Y_y | \ell_z | Y_y \rangle$.

(c) $\langle Y_x | \ell_z | Y_y \rangle$.

(d) $\langle Y_x | \ell_+ \ell_- | Y_x \rangle$.

(e) $\langle Y_z | \ell_x | Y_z \rangle$.

(f) $\langle Y_z | \ell_x^2 + \ell_y^2 | Y_z \rangle$.

4.7 What is the commutator $[\hat{\ell}_+, \hat{\ell}_x]$? The final expression should contain a single operator.

4.8 Consider the wavefunction

$$\psi = N(Y_{11} + cY_{1-1}),$$

where N and c are constants and Y_{11} and Y_{1-1} are spherical harmonics.

(a) Normalize ψ. The result will contain the constant c, but N should be eliminated.

(b) Find $\langle \ell_z \rangle$, using ψ.

(c) For what value of c will ψ be an eigenfunction of $\ell_+^2 + \ell_-^2$? What is the eigenvalue?

4.9 (a) An atom moving inside a C_{60} cage is a lot like a particle in a spherical box. If the potential is zero inside the sphere $(r < a)$ and infinite outside, what is the Schrödinger equation inside? Use spherical polar coordinates.

(b) What are the angular eigenfunctions?

(c) What is the Schrödinger equation for the radial wavefunction $R(r)$?

(d) What boundary condition should be applied at $r = a$?

(e) Is $\psi(r, \theta, \phi) = Nr^{-1} \sin kr$ an acceptable solution of the Schrödinger equation for this problem? (Here, N and k are constants.) If it is, what are the allowed energies E for the bound states?

4.10 Consider a problem in which there are only two states, α and β. If the states $\langle \alpha |$ and $\langle \beta |$ are orthogonal, then $\langle \alpha | \beta \rangle = \langle \beta | \alpha \rangle = 0$. Use this idea to prove that

$$\sum_s | s \rangle \langle s | = 1.$$

This formula indicates that $\sum_s | s \rangle \langle s |$ has the same effect on any ket as does multiplication by unity. Prove also that $(| s \rangle \langle s |)^2 = | s \rangle \langle s |$ and that $(| s \rangle \langle s |)(| t \rangle \langle t |) = \delta_{st} | s \rangle \langle s |$. The form $| s \rangle \langle s |$ is called a *projection operator*, because it projects out the $| s \rangle$ component from an arbitrary state.

5

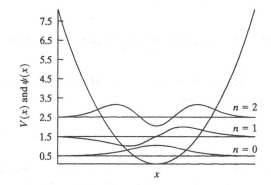

Molecular Vibrations and Time-Independent Perturbation Theory

Vibrations are an important form of molecular energy. The usual picture for vibrations is a harmonic model; this chapter discusses the energies and wavefunctions of harmonic vibrations, both for single vibrations and for the normal coordinates of a molecule. Time-independent perturbation theory is a standard and important method for improving wavefunctions and energies and is easily introduced in the context of molecular vibrations.

5.1 DIATOMIC MOLECULE VIBRATIONS

Let us begin with a problem we examined in the previous chapter, namely, the Hamiltonian of a diatomic molecule; here we study the vibrational motions that we previously ignored. This problem leads us rather directly to a consideration of the eigenvalues and eigenfunctions of the harmonic oscillator, which is one of the most important problems in quantum mechanics. Then we study the generalization of this treatment to vibrations of polyatomic molecules, a topic that is of consequence in the interpretation of infrared and Raman spectroscopy. The description of molecular vibrations provides an excellent example of the application of time-independent perturbation theory, so the last section of the chapter deals with this important approximation method for solving the Schrödinger equation.

Recall from Eq. (4.6) that the Hamiltonian of a vibrating and rotating diatomic molecule is

$$H = -\frac{\hbar^2}{2\mu} \nabla^2 + V(r), \tag{5.1}$$

where r is the internuclear distance and $V(r)$ is the potential function. Here we consider r to be variable (rather than fixed, as in Chapter 4) and take $V(r)$ to be a typical diatomic molecule potential, which usually looks like that depicted in Figure 5.1. If we confine ourselves to states with *small* displacements from equilibrium, we can expand the potential around its minimum at $r = r_e$. Since the first derivative vanishes at that point, we get

$$V(r) = V(r_e) + \frac{1}{2} V''(r_e)(r - r_e)^2 + \cdots, \tag{5.2}$$

where V'' is the second derivative of V, also known as the Hessian or force constant. In the following discussion, we consider only terms up through quadratic, which means that we replace the exact potential by a *parabola* and the oscillator motions are that of a *harmonic* oscillator.

The Schrödinger equation for the relative motion [analogous to Eq. (4.10), but for the case of three dimensions] is

$$H\chi = W\chi, \tag{5.3}$$

with the kinetic energy operator taken from Eq. (4.26) and using W as the total energy. Substituting Eq. (4.26) into Eq. (5.1) and the result into Eq. (5.3), we get

$$\left\{ \frac{-\hbar^2}{2\mu} \frac{1}{r} \frac{\partial^2}{\partial r^2} r + \frac{\hat{\mathbf{J}}^2}{2\mu r^2} + V(r_e) + \frac{1}{2} k(r - r_e)^2 \right\} \chi = W\chi, \tag{5.4}$$

where we have used $k = V''(r_e)$ and $\hat{\mathbf{J}}^2$ for the angular-momentum operator, following standard notation. The angular eigenfunctions of this equation are the spherical harmonics, just as with the rigid rotor. Since the angular and the radial motion can be separated in Eq. (5.4), we can write χ as the product

$$\chi(r, \theta, \phi) = Y_{JM_J}(\theta, \phi)\zeta(r), \tag{5.5}$$

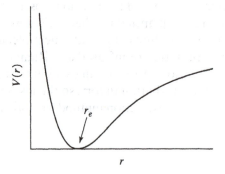

Figure 5.1   Typical diatomic molecule potential.

and the Schrödinger equation reduces to

$$\left\{-\frac{\hbar^2}{2\mu}\frac{1}{r}\frac{d^2}{dr^2}r + \frac{J(J+1)\hbar^2}{2\mu r^2} + V(r_e) + \frac{1}{2}k(r-r_e)^2\right\}\zeta = W\zeta. \quad (5.6)$$

We now assume that $r = r_e$ in the centrifugal term, as this term usually varies much more slowly with r than does the potential-energy term. If we define the vibrational energy by

$$E = W - V(r_e) - \frac{J(J+1)\hbar^2}{2\mu r_e^2}, \quad (5.7)$$

then the Schrödinger equation is

$$\left\{-\frac{\hbar^2}{2\mu}\frac{1}{r}\frac{d^2}{dr^2}r + \frac{1}{2}k(r-r_e)^2\right\}\zeta = E\zeta. \quad (5.8)$$

Now we define a new wavefunction

$$\psi = r\zeta, \quad (5.9)$$

and substitute it into the Schrödinger equation, giving

$$\left\{-\frac{\hbar^2}{2\mu}\frac{d^2}{dr^2} + \frac{1}{2}k(r-r_e)^2\right\}\psi = E\psi. \quad (5.10)$$

Finally, let us define the *displacement* from equilibrium:

$$x = r - r_e. \quad (5.11)$$

In terms of x, the Schrödinger equation is

$$\left\{-\frac{\hbar^2}{2\mu}\frac{d^2}{dx^2} + \frac{1}{2}kx^2\right\}\psi = E\psi. \quad (5.12)$$

This is precisely the Schrödinger equation for a simple *harmonic oscillator*. Note, however, that the variable x ranges from $-r_e$ to $+\infty$, rather than from $-\infty$ to ∞, as it would for a harmonic oscillator in one dimension, such as that considered in Chapter 1 (Section 1.3.4). This difference will be ignored in the discussion that follows.

5.2   RAISING AND LOWERING OPERATORS FOR THE HARMONIC OSCILLATOR

The Schrödinger equation (5.12) can be solved as a conventional differential equation; however, it is easier to solve if one rewrites H in terms of the raising and lowering operators

$$b = \left(\frac{\mu\omega}{2\hbar}\right)^{\frac{1}{2}}\left(x + \frac{ip}{\mu\omega}\right) \quad (5.13a)$$

and

$$b^+ = \left(\frac{\mu\omega}{2\hbar}\right)^{\frac{1}{2}}\left(x - \frac{ip}{\mu\omega}\right), \tag{5.13b}$$

where $\omega = (k/\mu)^{1/2}$ is the classical angular frequency [Eq. (1.8)] and $(p = -i\hbar d/dx)$ is the momentum operator. These equations may be inverted to give

$$x = \sqrt{\frac{\hbar}{2\mu\omega}}(b + b^+) \tag{5.14a}$$

and

$$p = -i\sqrt{\frac{\mu\omega\hbar}{2}}(b - b^+). \tag{5.14b}$$

If these expressions are now substituted into the Hamiltonian, one obtains

$$H = \frac{1}{2}\hbar\omega(bb^+ + b^+b). \tag{5.15}$$

Note that we have to be careful about the ordering of b and b^+, as they do not commute. Their commutator is

$$[b, b^+] = \frac{\mu\omega}{2\hbar}\left[x + \frac{ip}{\mu\omega}, x - \frac{ip}{\mu\omega}\right] \tag{5.16}$$

$$= \frac{i}{\hbar}[p, x] = 1.$$

Using this commutator, we find that

$$H = \hbar\omega\left(b^+b + \frac{1}{2}\right). \tag{5.17}$$

This form of the Hamiltonian is such that the eigenfunctions of H will also be eigenfunctions of b^+b. Thus, these eigenfunctions will satisfy the equation

$$b^+b|\lambda\rangle = \lambda|\lambda\rangle, \tag{5.18}$$

where we define λ to be the eigenvalue associated with b^+b and we label the eigenfunctions $|\lambda\rangle$ with this eigenvalue.

To solve for the eigenvalues and eigenfunctions, we use the commutator given by Eq. (5.16) to show that

$$b^+b(b^+|\lambda\rangle) = b^+(1 + b^+b)|\lambda\rangle = (\lambda + 1)b^+|\lambda\rangle. \tag{5.19}$$

This equation shows that $b^+|\lambda\rangle$ is an eigenfunction of b^+b with an eigenvalue of $\lambda + 1$. This means that b^+ acts on $|\lambda\rangle$ to *raise* the value of the eigenvalue by one. Similarly, if we consider the quantity $b|\lambda\rangle$, we find that

$$b^+bb|\lambda\rangle = (bb^+ - 1)b|\lambda\rangle = (\lambda - 1)b|\lambda\rangle. \tag{5.20}$$

This means that b operates on $|\lambda\rangle$ to *lower* the eigenvalue by one. There is, however, a lower limit to the value of λ. The limit exists because $\langle\lambda|b^+b|\lambda\rangle$, which, from Eq. (5.18), is just λ, is inherently nonnegative. This is because $\langle\lambda|b^+$ is the complex conjugate of $b|\lambda\rangle$, so $\langle\lambda|b^+b|\lambda\rangle$ is an absolute square.

To avoid applying b so many times to $|\lambda\rangle$ that it produces a state with a negative eigenvalue, we terminate the series by choosing one of the eigenvalues λ to have the value zero. The eigenfunction corresponding to $\lambda = 0$ would then satisfy $\langle 0|b^+b|0\rangle = 0$, which implies that

$$b|0\rangle = 0. \tag{5.21}$$

Since the eigenvalues above $\lambda = 0$ can differ from that by at most an integer, it is convenient to define an integer quantum number n that is identical to the eigenvalue λ. Thus, we have

$$b^+b|n\rangle = n|n\rangle, \tag{5.22}$$

which means that

$$H|n\rangle = \hbar\omega\left(b^+b + \frac{1}{2}\right)|n\rangle = \hbar\omega\left(n + \frac{1}{2}\right)|n\rangle, \tag{5.23}$$

so the energies are

$$E_n = \hbar\omega\left(n + \frac{1}{2}\right), \qquad n = 0, 1, 2, \dots. \tag{5.24}$$

These energies are *equally* spaced (see Fig. 5.2), in distinct contrast to the particle-in-box spectrum of Fig. 3.2 or the rotational spectrum of Eq. (4.42).

Note that $\langle n|b^+b|n\rangle = n$. Also, since $\langle n|b^+$ is the complex conjugate of $b|n\rangle$, we have

$$b|n\rangle = \sqrt{n}|n - 1\rangle. \tag{5.25a}$$

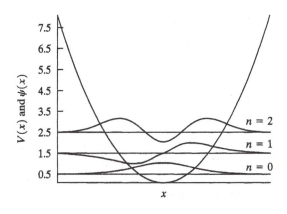

Figure 5.2 Energy levels and eigenfunctions of harmonic oscillator for $n = 0, 1, 2$. (For clarity, successive wavefunctions have been shifted up so that the local zero is at the eigenvalue.)

By a similar reasoning with $\langle n | bb^+ | n \rangle$ and using the commutator given by Eq. (5.16), we find that

$$b^+ | n \rangle = \sqrt{n + 1} | n + 1 \rangle. \tag{5.25b}$$

To determine the eigenfunctions $| n \rangle$, we write the explicit operator form of Eq. (5.21), using Eq. (5.13a), but omitting the multiplicative constant:

$$\left(x + \frac{\hbar}{\mu\omega} \frac{d}{dx} \right) | 0 \rangle = 0. \tag{5.26}$$

This equation may be rearranged to yield

$$-\frac{d | 0 \rangle}{| 0 \rangle} = \frac{\mu\omega}{\hbar} x \, dx. \tag{5.27}$$

After integrating and exponentiating, we get

$$| 0 \rangle = N e^{-\mu\omega x^2/2\hbar} = N e^{-\alpha x^2/2}, \tag{5.28}$$

where $\alpha = \mu\omega/\hbar$. The normalization constant is determined from

$$N^2 \int_{-\infty}^{\infty} e^{-\alpha x^2} \, dx = 1, \tag{5.29}$$

which gives

$$N = \sqrt[4]{\frac{\alpha}{\pi}}. \tag{5.30}$$

By the repeated application of b^+ to $| 0 \rangle$, one can use Eq. (5.25b) to show that

$$| n \rangle = \frac{(b^+)^n}{(n!)^{1/2}} | 0 \rangle. \tag{5.31}$$

Substituting Eq. (5.28) into this formula then gives

$$| n \rangle = \frac{N}{(2^n n!)^{1/2}} e^{-\alpha x^2/2} H_n(\sqrt{\alpha} x), \tag{5.32}$$

where H_n is a *Hermite* polynomial, some specific examples of which are as follows:

$$H_n(z) = \begin{cases} 1, & n = 0 \\ 2z, & n = 1 \\ 4z^2 - 2, & n = 2 \\ 8z^3 - 12z, & n = 3 \\ 16z^4 - 48z^2 + 12, & n = 4 \end{cases} \tag{5.33}$$

A few eigenfunctions are plotted in Fig. 5.2, and the corresponding probability densities are shown in Fig. 5.3. Note that the harmonic oscillator functions are even functions of x for n even and odd functions of x for n odd. The nodal structure of the eigenfunctions is very much analogous to that for the one-dimensional box wavefunctions in Fig. 3.2, but in contrast to the box

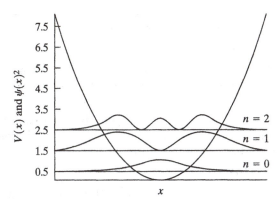

Figure 5.3 Probability densities $|\psi_n(x)|^2$ associated with the $n = 0$–2 states of the harmonic oscillator, using a plotting format that is analogous to Fig. 5.2.

wavefunctions, the harmonic oscillator wavefunctions have a nonzero probability for x where the potential energy is greater than the total energy. Such regions are forbidden in classical mechanics, but can exist in quantum mechanics, although only where the wavefunction decays rapidly (usually exponentially). This motion, called *tunnelling*, is an important example of a *quantum effect* and is responsible for such processes as proton transfer between water molecules in solution and electronic conductivity in scanning tunneling microscopy.

EXERCISE 5.1   Using the definitions of b and b^+, prove directly that the Hamiltonian operators of Eqs. (5.12) and (5.15) are identical.

Answer   First we use the expressions for b and b^+ in Eqs. (5.13) to show that

$$b^+b = \frac{\mu\omega}{2\hbar}\left(x - \frac{ip}{\mu\omega}\right)\left(x + \frac{ip}{\mu\omega}\right)$$

$$= \frac{\mu\omega}{2\hbar}\left(x^2 + \frac{p^2}{\mu^2\omega^2} + \frac{i}{\mu\omega}(xp - px)\right).$$

Since $[x, p] = i\hbar$ [from Eq. (5.16) or Eq. (2.36)], we have

$$b^+b = \frac{\mu\omega}{2\hbar}x^2 + \frac{p^2}{2\mu\hbar\omega} - \frac{1}{2}.$$

Now we substitute into Eq. (5.17) to get

$$\left(b^+b + \frac{1}{2}\right)\hbar\omega = \hbar\omega\left(\frac{\mu\omega}{2\hbar}x^2 + \frac{p^2}{2\mu\hbar\omega}\right)$$

$$= \frac{1}{2}\mu\omega^2 x^2 + \frac{p^2}{2\mu}$$

$$= \frac{1}{2}kx^2 + \frac{p^2}{2\mu},$$

which is what we wanted to show.

EXERCISE 5.2   The parity operator P is defined such that

$$P\psi(x) = p\psi(-x),$$

where $p = \pm 1$ (even or odd parity, respectively). Prove that P and the Hamiltonian in Eq. (5.12) commute. Find the parity eigenvalue p for the first five eigenfunctions of the harmonic oscillator.

Answer   The commutator is $[P, H]\psi = P(H\psi) - H(P\psi)$. But the parity of a product is the product of the parities. (That is, an even function times an odd function is odd, an odd times an odd is even, and an even times an even is even.) So

$$P(H\psi) = (PH)(P\psi) = HP\psi$$

because H is even (both x^2 and $(-i\hbar d/dx)^2$ are unchanged for $x \to -x$). Therefore,

$$[P, H]\psi = HP\psi - Hp\psi = 0, \text{ or } [P, H] = 0.$$

The oscillator eigenfunctions are given by Eq. (5.32). With the Hermite polynomials of Eq. (5.33), we see that

$$\psi_0, \psi_2, \psi_4, \psi_6, \text{ etc., are even (parity} = 1)$$

and

$$\psi_1, \psi_3, \psi_5, \psi_7, \text{ etc., are odd (parity} = -1).$$

EXERCISE 5.3   Evaluate $\langle 0|bb^+bHb^+bb^+|0\rangle$, $\langle 0|bbbHb^+b^+|0\rangle$, $\langle 0|b^+b^+b^+Hbbb|0\rangle$, and $\langle 0|bbbHb^+b^+b^+|0\rangle$, where H is the harmonic oscillator Hamiltonian.

Answer

$$\langle 0|bb^+bHb^+bb^+|0\rangle = \langle 1|H|1\rangle = 3/2\hbar\omega;$$

$$\langle 0|bbbHb^+b^+|0\rangle = \langle 3|H|2\rangle = 0;$$

$$\langle 0|b^+b^+b^+Hbbb|0\rangle = 0, \text{ because } b|0\rangle = 0;$$

$$\langle 0|bbbHb^+b^+b^+|0\rangle = 6\langle 3|H|3\rangle\hbar\omega = 6(3 + \tfrac{1}{2})\hbar\omega = 21\,\hbar\omega.$$

5.3   POLYATOMIC MOLECULE VIBRATIONS

The generalization of the theory of the previous section to molecules with three or more atoms is complex in general because of lack of separability of vibrational and rotational motion. Here we give a relatively simple treatment that circumvents the problem, though not without some sacrifice in sophistication and rigor. The final Hamiltonian is Eq. (5.44), and the solutions of the Schrödinger equation are given by Eqs. (5.48) and (5.49); the manipulations leading to these results involve some matrix algebra that beginning students might want to skip.

We begin with the Hamiltonian of a molecule containing N atoms moving in three dimensions [Eq. (2.9b)]:

$$H = \sum_{i\alpha} -\frac{\hbar^2}{2m_i}\frac{\partial^2}{\partial x_{i\alpha}^2} + V. \tag{5.34}$$

Here, the label i identifies the atoms and $\alpha = x, y, z$ the Cartesian components. The potential V is assumed to describe bound motions, so we suppose that it has a minimum at an equilibrium geometry that is labelled by the coordinates $x_{i\alpha}^e$. For motions of the atoms near equilibrium, the potential can be expanded in a Taylor series that includes only a few terms. Thus, we write

$$V(\{x_{i\alpha}\}) = V(\{x_{i\alpha}^e\}) + \sum_{i\alpha} \frac{\partial V}{\partial x_{i\alpha}}\bigg|_{x_{i\alpha}^e} (x_{i\alpha} - x_{i\alpha}^e)$$

$$+ \frac{1}{2} \sum_{i\alpha} \sum_{j\beta} \frac{\partial^2 V}{\partial x_{i\alpha} \partial x_{j\beta}}\bigg|_{x_{i\alpha}^e, x_{j\beta}^e} (x_{i\alpha} - x_{i\alpha}^e)(x_{j\beta} - x_{j\beta}^e) \qquad (5.35)$$

$$+ V_A.$$

In this expression, the first derivatives will all be zero as we expand about the minimum of the potential. V_A includes the third and higher derivative terms, so we will call it the anharmonic part of the potential. This part will be neglected in the rest of the section, but in the next section we shall examine its influence on eigenvalues with the use of perturbation theory. The Hamiltonian at this point is

$$H = \sum_{i\alpha} -\frac{\hbar^2}{2m_i} \frac{\partial^2}{\partial x_{i\alpha}^2} + \frac{1}{2} \sum_{i\alpha} \sum_{j\beta} F_{i\alpha, j\beta}(x_{i\alpha} - x_{i\alpha}^e)(x_{j\beta} - x_{j\beta}^e), \qquad (5.36)$$

where

$$F_{i\alpha, j\beta} = \frac{\partial^2 V}{\partial x_{i\alpha} \partial x_{j\beta}}\bigg|_{x_{i\alpha}^e, x_{j\beta}^e} \qquad (5.37)$$

is the matrix of force constants (the Hessian matrix) and we have omitted the constant term $V(\{x_{i\alpha}^e\})$, as this can be subtracted from the total energy W as in Eq. (5.7) to define a vibrational energy E.

The Schrödinger equation governing vibrational motion is

$$H\psi = E\psi. \qquad (5.38)$$

To solve this equation, we introduce *normal coordinates* Q_k with dimension $(\text{mass})^{1/2}$ (length), using the expression

$$x_{i\alpha} = x_{i\alpha}^e + m_i^{-\frac{1}{2}} \sum_k L_{i\alpha, k} Q_k, \qquad (5.39)$$

where the $L_{i\alpha, k}$ are coefficients that can be grouped into a matrix $\mathbf{L}$ of dimension $3N \times 3N$ (i.e., the number of atoms times the number of Cartesian coordinates for each atom). For reasons that will be apparent shortly, the matrix $\mathbf{L}$ is determined by diagonalizing the mass-weighted Hessian matrix $\mathbf{H}$, which is defined by

$$H_{i\alpha, j\beta} = \frac{F_{i\alpha, j\beta}}{\sqrt{m_i m_j}}. \qquad (5.40)$$

If the eigenvalues of this matrix are denoted λ_k, then the eigenvalues and eigenvectors satisfy

$$\mathbf{HL} = \mathbf{L} \wedge, \tag{5.41}$$

where $\wedge$ is a diagonal matrix with the λ_k's as diagonal elements. From Appendix A [Eq.(A.14)], we note that $\mathbf{L}$ is an orthogonal matrix $(\tilde{L} = L^{-1})$. This means that the inverse of Eq. (5.39) is

$$Q_k = \sum_{i\alpha} m_i^{1/2} L_{i\alpha, k}\left(x_{i\alpha} - x_{i\alpha}^e\right). \tag{5.42}$$

Let us now express the Hamiltonian in Eq. (5.36) in terms of normal coordinates. This requires a straightforward application of the chain rule to the kinetic-energy operator, using Eq. (5.42), and direct substitution of Eq. (5.39) into the potential energy. The result is

$$H = -\frac{\hbar^2}{2} \sum_{i\alpha} \sum_k \sum_{k'} L_{i\alpha, k} L_{i\alpha, k'} \frac{\partial^2}{\partial Q_k \partial Q_{k'}}$$

$$+ \frac{1}{2} \sum_{i\alpha} \sum_{j\beta} H_{i\alpha, j\beta} \sum_k \sum_{k'} L_{i\alpha, k} L_{j\beta, k'} Q_k Q_{k'}, \tag{5.43}$$

where we have replaced $\mathbf{F}$ by $\mathbf{H}$, using Eq. (5.40). This expression can be simplified by moving the sums over k and k' to the outside and invoking orthogonality of the $\mathbf{L}$ matrix in the kinetic energy and Eq. (5.41) in the potential energy. In both cases the result yields zero if $k \neq k'$, so the double sum reduces to the single-sum result

$$H = \sum_k \left\{ -\frac{\hbar^2}{2} \frac{\partial^2}{\partial Q_k^2} + \frac{1}{2} \lambda_k Q_k^2 \right\}. \tag{5.44}$$

A comparison of this equation with Eq. (5.12) indicates that Eq. (5.44) is a sum of harmonic oscillator Hamiltonians in which the reduced mass μ is unity and the force constant k is the same as λ_k. This means that the harmonic oscillator frequency for mode k is $\omega_k = (\lambda_k)^{1/2}$.

Equation (5.44) demonstrates that the role of the normal-coordinate transformation given by Eq. (5.39) is to convert the Hamiltonian of Eq. (5.36), wherein all degrees of freedom are coupled together, to the uncoupled sum in Eq. (5.44). Although the orthogonal matrix $\mathbf{L}$ that accomplishes this transformation has dimensions $3N \times 3N$, there are only $3N - 6$ vibrational coordinates for most polyatomic molecules ($3N - 5$ for linear molecules). The reason is that six of the frequencies that result from diagonalization of the $\mathbf{H}$ matrix are zero. These frequencies correspond to the three coordinates that cause the center of mass of the molecule to translate and three other coordinates that correspond to overall rotation of the molecule. For all six of these modes, the force constant λ_k is zero, and the only contribution to the energy in Eq. (5.44) comes from the kinetic-energy terms. We have already noted (in Chapter 4) that translational motions of the center of mass are completely separable from internal motions, so this contribution to the energy expression can be ignored. The corresponding separation of rotation and vibration works only in the limit of infinitesimal vibrational motions. We will assume this limit in the development that follows, which means that the sum in Eq. (5.44) is restricted to the $3N - 6$ terms for which λ_k is nonzero. However, it should be noted that because all coordinates have zero-point motions, vibration and rotation are always coupled to some extent. To treat this problem properly requires imposing the constraints of conservation of angular momentum on Eq. (5.44), which is beyond the scope of the text.

Let us now introduce raising and lowering operators into Eq. (5.44). By analogy with Eqs. (5.13), these are defined by

$$b_k = \left(\frac{\omega_k}{2\hbar}\right)^{1/2}\left(Q_k + \frac{iP_k}{\omega_k}\right) \tag{5.45a}$$

and

$$b_k^+ = \left(\frac{\omega_k}{2\hbar}\right)^{1/2}\left(Q_k - \frac{iP_k}{\omega_k}\right), \tag{5.45b}$$

and the commutation relation is

$$[b_k, b_{k'}^+] = \delta_{kk'}. \tag{5.46}$$

The Hamiltonian thus reduces to

$$H = \sum_{k=1}^{3N-6} \hbar\omega_k\left(b_k^+ b_k + \frac{1}{2}\right), \tag{5.47}$$

and the energy eigenvalues are

$$E = \sum_{k=1}^{3N-6} \hbar\omega_k\left(n_k + \frac{1}{2}\right), \tag{5.48}$$

where n_k is the quantum number associated with the kth vibrational mode. The eigenstates can be expressed as products of functions taken from Eq. (5.32) as

$$|n_1 n_2 n_3 \cdots n_{3N-6}\rangle = \prod_{k=1}^{3N-6} \left(\frac{\alpha_k}{\pi}\right)^{1/4} \frac{e^{-\alpha_k Q_k^2/2}}{(2^{n_k} n_k!)^{1/2}} H_{n_k}(\sqrt{\alpha_k}\, Q_k), \tag{5.49}$$

where $\alpha_k = \omega_k/\hbar$.

EXERCISE 5.4   The water molecule has three normal modes, whose frequencies are $\omega_1 = 3,833 \text{ cm}^{-1}$, $\omega_2 = 1,649 \text{ cm}^{-1}$, and $\omega_3 = 3,943 \text{ cm}^{-1}$. What is the energy of the (112) state? What is the energy difference between (112) and (100)?

Answer   Use $E = \sum\limits_{s,\text{ modes}} (n_s + \tfrac{1}{2})\hbar\omega_s$ to give

$$E(100) = (3/2)\omega_1 + (1/2)\omega_2 + (1/2)\omega_3 = 8546 \text{ cm}^{-1},$$

$$E(112) = 18081 \text{ cm}^{-1},$$

$$E(112) - E(100) = 9,535 \text{ cm}^{-1}.$$

5.4   TIME-INDEPENDENT PERTURBATION THEORY

The oscillator problems that we have considered in this chapter can often be represented with the Hamiltonian

$$H = H_0 + \lambda V, \tag{5.50}$$

where H_0 is a zeroth-order Hamiltonian for which solutions to the Schrödinger equation may be obtained easily and V is a time-independent perturbation that is small compared to H_0 in some sense. For oscillator problems, we are often able to choose H_0 as the harmonic oscillator Hamiltonian and V as the anharmonic part of the potential. The parameter λ eventually will be set to unity, but we will use it to keep track of the order of the perturbation. This means that terms which contain λ^k are of kth order in the perturbation expansion.

Let us denote the eigenfunctions of H_0 by ϕ_n^0 and the corresponding eigenvalues as E_n^0. This implies that

$$H_0\phi_n^0 = E_n^0\phi_n^0. \tag{5.51}$$

In the perturbation theory approximation, we assume that the exact wavefunction ψ_n may be expanded in powers of λ as

$$\psi_n = \phi_n^0 + \lambda\phi_n^{(1)} + \lambda^2\phi_n^{(2)} + \cdots. \tag{5.52}$$

The corresponding energy expansion is

$$E_n = E_n^0 + \lambda E_n^{(1)} + \lambda^2 E_n^{(2)} + \cdots. \tag{5.53}$$

Substituting these expansions into the full Schrödinger equation $(H\psi_n = E_n\psi_n)$ and equating like powers of λ gives

$$H_0\phi_n^0 = E_n^0\phi_n^0 \qquad \text{(zeroth order),} \tag{5.54a}$$

$$V\phi_n^0 + H_0\phi_n^{(1)} = E_n^{(1)}\phi_n^0 + E_n^0\phi_n^{(1)} \qquad \text{(first order),} \tag{5.54b}$$

$$V\phi_n^{(j-1)} + H_0\phi_n^{(j)} = \sum_{k=0}^{j} E_n^{(k)}\phi_n^{(j-k)} \qquad (j\text{th order, } j \geq 1). \tag{5.54c}$$

The zeroth-order equation is satisfied by assumption. To solve Eq. (5.54b), we expand $\phi_n^{(1)}$ in terms of the complete set of zeroth-order states, that is,

$$\phi_n^{(1)} = \sum_{k \neq n} C_{nk}^{(1)} \phi_k^0 \tag{5.55}$$

where C_{nk} is a coefficient. It follows from Eq. (5.55) that $\langle \phi_n^0 | \phi_n^{(1)} \rangle = 0$; in other words, the perturbation causes changes to the wavefunction that are orthogonal to the zeroth-order states. Substitution of Eq. (5.55) into Eq. (5.54b), followed by multiplication by ϕ_n^0 and integration, then leads to

$$E_n^{(1)} = \langle \phi_n^0 | V | \phi_n^0 \rangle, \tag{5.56}$$

and it follows that

$$C_{nk}^{(1)} = \frac{\langle \phi_k^0 | V | \phi_n^0 \rangle}{E_n^0 - E_k^0} \ (k \neq n). \tag{5.57}$$

One can develop expressions for higher order terms in a similar manner. For example

$$E_n^{(2)} = \sum_{k \neq n} \frac{\langle \phi_n^0 | V | \phi_k^0 \rangle \langle \phi_k^0 | V | \phi_n^0 \rangle}{E_n^0 - E_k^0}. \tag{5.58}$$

The only time that Eqs. (5.57) and (5.58) run into serious trouble is when the state n is degenerate with one of the states k, as that makes the denominators in those equations vanish, and as a result, the perturbation theory corrections to the energy diverge. In that case we have to use an alternative approach, *degenerate perturbation theory,* in which one first finds the exact eigenvalues associated with a Hamiltonian matrix contructed from the degenerate states. In this calculation, the terms in the perturbation that couple the degenerate states are included in the Hamiltonian, so that after diagonalization, the degeneracy is lifted. Perturbation theory is then applied to the diagonalized states, including those terms in the perturbation that were not included in the first step.

EXERCISE 5.5   Prove that, for a system with only two states, if V has no diagonal matrix elements, the average energy is the same in the zeroth order as in the second order.

Answer   Since V has no diagonal matrix elements, $\langle \alpha | V | \alpha \rangle = \langle \beta | V | \beta \rangle = 0$, where the two states are labeled α and β. Using Eq. (5.58), we find that

$$E_\alpha^{(2)} = E_\alpha^{(0)} + \frac{\langle \alpha | V | \beta \rangle \langle \beta | V | \alpha \rangle}{E_\alpha^0 - E_\beta^0}$$

and

$$E_\beta^{(2)} = E_\beta^{(0)} + \frac{\langle \beta | V | \alpha \rangle \langle \alpha | V | \beta \rangle}{E_\beta^0 - E_\alpha^0}.$$

Adding these and dividing by two gives

$$\frac{E_\alpha^{(2)} + E_\beta^{(2)}}{2} = \frac{E_\alpha^{(0)} + E_\beta^{(0)}}{2}.$$

This means that the average energy is the same in the zeroth, first, and second orders.

5.5   EXAMPLES

A simple example of first order perturbation theory is a harmonic oscillator in its ground state, subject to a quartic perturbation $V = \gamma x^4$, with γ a constant. In this case,

$$E_n^{(1)} = \gamma \langle n | x^4 | n \rangle. \tag{5.59}$$

If we express x in terms of the raising and lowering operators, then, for the ground state, we have

$$E_0^{(1)} = \gamma \left(\frac{\hbar}{2\mu\omega} \right)^2 \langle 0 | (b + b^+)^4 | 0 \rangle, \tag{5.60a}$$

or

$$E_0^{(1)} = 3\gamma \left(\frac{\hbar}{2\mu\omega} \right)^2. \tag{5.60b}$$

EXERCISE 5.6   Derive Eq. (5.60b) from Eq. (5.60a).

Answer   To go from the first equation to the second in this derivation, we have used several consequences of Eqs. (5.25). First, note that $b|0\rangle = 0$ and $\langle 0|b^+ = 0$, so any term with a leading b^+ or a trailing b must vanish. Second, the only nonzero products of b and b^+ must have the same number of raising and lowering operators, as the number of raising operations must match the number of lowering operations in order to convert $|0\rangle$ into $\langle 0|$. As a result, the only nonzero products coming from $(b + b^+)^4$ are bb^+bb^+ and bbb^+b^+. We can then apply Eqs. (5.25) directly to evaluate these two terms, yielding 1 and 2, respectively, for a sum of 3, which is the coefficient in Eq. (5.60b).

Alternatively, one can use the commutator relation of Eq. (5.16) to express $bb^+ = b^+b + 1$ for both of the terms bb^+bb^+ and bbb^+b^+. This is done twice for the bb^+bb^+ term, yielding $(b^+b + 1)^2$. Then each of the $b^+b + 1$ terms that result gives unity, as the leading b^+ and trailing b give a vanishing result when applied to $\langle 0|$ and $|0\rangle$ respectively. The commutator can also be applied to the bbb^+b^+ term, starting with the bb^+ in the middle to give $bb^+ + bb^+bb^+$ and then using the commutator again to reduce each of these terms to unity. The three factors of unity again sum to 3.

Note that a cubic potential would not contribute in the first order in this example because of symmetry, but would show up in the second order.

Another example is the interaction of a diatomic molecule with a static electric field $\mathbf{E}$ (the *Stark effect*). Here, we take the diatomic vibrations to be described by a harmonic oscillator and assume that the dipole moment $\boldsymbol{\mu}$ of the molecule interacts with the field $\mathbf{E}$ via a perturbation

$$V = \boldsymbol{\mu} \cdot \mathbf{E}. \tag{5.61}$$

Now we expand $\boldsymbol{\mu}$ in a Taylor series about the diatomic equilibrium position:

$$\boldsymbol{\mu} = \boldsymbol{\mu}_e + \boldsymbol{\mu}_e' \cdot \mathbf{x} + \cdots. \tag{5.62}$$

The constant term $\boldsymbol{\mu}_e$ contributes an energy $\boldsymbol{\mu}_e \cdot \mathbf{E}$ in first order and zero in higher orders. This contribution to the interaction energy is called the first-order Stark effect, and the magnitude of the energy shift resulting from it varies linearly with E. The first-order Stark-effect term will vanish for molecules with high symmetry. For these molecules the most important effect comes from the linear term. In the first order, this term is

$$E_n^{(1)} = \langle n | \boldsymbol{\mu}_e' \cdot \mathbf{E} x | n \rangle = 0, \tag{5.63}$$

because the perturbation has odd symmetry. The second-order term is

$$E_n^{(2)} = \sum_{k \neq n} \frac{\langle n | x | k \rangle \langle k | x | n \rangle}{\hbar \omega (n - k)} (\boldsymbol{\mu}_e' \cdot \mathbf{E})^2. \tag{5.64}$$

Evaluation of the coordinate matrix element [using Eq. (5.25)] leads to two important intermediate states, $k = n \pm 1$. The final result is

$$E_n^{(2)} = \frac{-(\boldsymbol{\mu}_e' \cdot \mathbf{E})^2}{2\mu\omega^2}. \tag{5.65}$$

This is an example of the second-order Stark effect, and we see that the shift in energy here is proportional to the square of the applied field.

It turns out that it is also possible to solve this problem exactly. To do so, we write down the complete Hamiltonian of the perturbed oscillator, namely,

$$H = -\frac{\hbar^2}{2\mu} \frac{d^2}{dx^2} + \frac{1}{2} \mu\omega^2 x^2 + (\boldsymbol{\mu}_e' \cdot \mathbf{E}) x, \tag{5.66}$$

where we have (for simplicity) ignored the constant term $\boldsymbol{\mu}_e \cdot \mathbf{E}$. If we now complete the square with the two potential terms, we get

$$H = -\frac{\hbar^2}{2\mu} \frac{d^2}{dx^2} + \frac{1}{2} \mu\omega^2 \left(x + \frac{\boldsymbol{\mu}_e' \cdot \mathbf{E}}{\mu\omega^2} \right)^2 - \frac{1}{2} \frac{(\boldsymbol{\mu}_e' \cdot \mathbf{E})^2}{\mu\omega^2}. \tag{5.67}$$

This form of the Hamiltonian indicates that the potential is still harmonic, with the equilibrium position centered on $x = -(\boldsymbol{\mu}_e' \cdot \mathbf{E})/\mu\omega^2$. In addition, there is a constant term on the right that represents an overall shift to each eigenvalue. The overall energy expression is thus

$$E_n = \hbar\omega \left(n + \frac{1}{2} \right) - \frac{(\boldsymbol{\mu}_e' \cdot \mathbf{E})^2}{2\mu\omega^2}. \tag{5.68}$$

This is exactly the same as $E_n = E_n^{(0)} + E_n^{(2)}$, with $E_n^{(2)}$ given by Eq. (5.65).

SUGGESTED READINGS

Although the harmonic oscillator and time-independent perturbation theory are treated in every quantum chemistry and quantum mechanics text (cited at the end of Chapters 1 and 2, respectively), polyatomic vibrations are less commonly treated. Among the textbooks that deal with this subject are the following:

P. W. Atkins and R. S. Friedman, *Molecular Quantum Mechanics*, 3rd ed. (Oxford, New York, 1997).

E. B. Wilson, Jr., J. C. Decius, and P. C. Cross, *Molecular Vibrations* (Dover, New York, 1955).

S. Califano, *Vibrational States* (Wiley, New York, 1976).

PROBLEMS

5.1 Sketch the $n = 6$ wavefunction of the harmonic oscillator and its square. How many nodes are there?

5.2 The fourth Hermite polynomial is

$$H_4(z) = 16z^4 - 48z^2 + 12.$$

Find the most probable position for the $n = 4$ harmonic oscillator. Is it closer to the origin ($x = 0$) or to the turning point (where the kinetic energy vanishes)? Compare your answer with the $n = 0$ state, and comment on the relation of the two results to the Bohr correspondence principle of Section 3.1.5 (*Hint:* In what location does the classical oscillator spend the most time?)

5.3 Which of the following are zero for a harmonic oscillator ($|0\rangle$ = ground state)?

$$\langle 0 | bbb^+b^+ | 0 \rangle;$$

$$\langle 0 | b^+bb^+b | 0 \rangle;$$

$$\langle 0 | b^+bb^+ | 0 \rangle.$$

5.4 Evaluate the following for a harmonic oscillator:

$$\langle 2 | (b + b^+)^2 | 0 \rangle.$$

5.5 Consider the H_2O molecule in its ground electronic state.

   (a) Write down the Hamiltonian governing the molecule's vibrational motions, taking the potential to be harmonic. Express your answer using raising and lowering operators.

   (b) Give expressions for the ground-state energy and wavefunction. Express the wavefunction in normal coordinates.

   (c) The first neglected terms in the Hamiltonian are the cubic anharmonicities. Consider the cubic term $V = C_{112}Q_1^2Q_2$, where the Q_i's are the normal coordinates. Express V in terms of raising and lowering operators.

   (d) Evaluate the first-order perturbation theory expression for the contribution of the cubic term in (c) to the total energy of the ground state.

   (e) With what excited states does the ground state mix if V is included in the second order? Specify the quantum numbers of the states.

5.6 Use first-order perturbation theory to determine the ground-state energy of the quartic oscillator

$$H = \frac{p^2}{2\mu} + \gamma x^4.$$

Use a harmonic oscillator to define the zeroth-order Hamiltonian. In the first order, what choice of harmonic frequency gives the lowest zeroth-plus first-order energy? What is the ground-state energy for this frequency?

5.7 Consider a two-dimensional rigid rotor that is perturbed by interacting with an applied external field. The Hamiltonian is $H = J^2/2I + \mu E \cos \phi$, where μ is the dipole moment and E the applied electric field.

(a) If $E = 0$, what are energy levels and wavefunctions of the rotor?

(b) How do these energy levels change if the perturbation is treated as a first-order problem?

(c) Now consider the ground state of the rotor. What is the second-order expression for the ground-state energy? Give explicit expressions for all sums and integrals, and evaluate at least one integral.

5.8 Equation (5.57) permits a calculation of the first-order perturbed wavefunction, and Eq. (5.58) allows a calculation of the second-order energy.

As an example, consider the four cubic anharmonicity terms for the two-dimensional harmonic oscillator. (For example, these terms arise in the description of H_2O if the bend is ignored.) The anharmonicities are

$$V_{a1} = a_1 x_1^3,$$

$$V_{a2} = a_2 x_2^3,$$

$$V_{a3} = a_3 x_1^2 x_2,$$

and

$$V_{a4} = a_4 x_1 x_2^2,$$

where a_1, a_2, a_3, a_4 are constants. Write the first-order perturbed wavefunction $\psi_{0,0}^{(1)}$, arising from the ground state $|0,0\rangle$ of the two-dimensional oscillator.

6

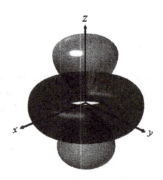

The Hydrogen Atom

The hydrogen atom is the first example of a physical system that one encounters in the study of electronic structure. The hydrogen atom orbitals and energy levels will be crucial for understanding both behavior of larger atoms, and the basis sets that exist for the description of molecules

6.1 THE SCHRÖDINGER EQUATION

A hydrogen atom consists of a nucleus of charge Ze (where $Z = 1$ for a proton, but is here allowed to have other values so that hydrogenlike atoms such as He^+ are described) and an electron of charge $-e$. The kinetic-energy operator is therefore the same as that for the two nuclei in a diatomic molecule (except for the obvious change in mass). In Chapter 4, we demonstrated how to express the kinetic-energy operator for that system in terms of center-of-mass coordinates. If this expression is combined with the coulomb potential from Eq. (1.1), the resulting Hamiltonian is

$$H = \frac{-\hbar^2}{2\mu} \nabla^2 - \frac{Ze}{4\pi\varepsilon_0 r},$$
(6.1)

where

$$\mu = \frac{m_e m_n}{m_e + m_n} \cong \frac{1,837\, m_e}{1,838\, m_e} \cong .99946\, m_e$$
(6.2)

is the reduced mass. In Eq. (6.2), we have substituted the masses for the case where m_n is a proton mass. Since this mass is about 2,000 times the electron mass, one can, to a very good approximation, assume that μ is simply the electron mass. Accordingly, we shall assume this hereafter, and we will also use m instead of m_e for the electron mass.

In Eq. (6.1), since $V(r)$ depends only on the distance r between the particles, it is appropriate to convert to polar coordinates. The result in this case is

$$H = \frac{-\hbar^2}{2m}\left\{\frac{1}{r}\frac{\partial^2}{\partial r^2}r\right\} + \frac{\hat{\ell}^2}{2mr^2} - \frac{Ze^2}{4\pi\varepsilon_0 r}. \tag{6.3}$$

Note that the dependence of the Hamiltonian on the polar angles ϕ and θ is entirely contained in $\hat{\ell}^2$. This means that H and $\hat{\ell}^2$ commute, so from our discussion in Section 2.7.3, the angular eigenfunctions of H can also be chosen to be eigenfunctions of $\hat{\ell}^2$. Also, from our discussion in Section 4.6, we can choose them as eigenfunctions of $\hat{\ell}_z$ as well, and this means that we can use the spherical harmonics $Y_{\ell m}(\theta, \phi)$ as angular eigenfunctions of H. Accordingly, we can write the complete solution as products of angular and radial eigenfunctions as

$$\psi = R(r)Y_{\ell m}(\theta, \phi), \tag{6.4}$$

where $R(r)$ is the radial wavefunction. If Eq. (6.4) is substituted into the Schrödinger equation, using the Hamiltonian in Eq. (6.3) and invoking Eq. (4.60) for the spherical harmonics, we find that

$$\left[\frac{-\hbar^2}{2m}\left\{\frac{1}{r}\frac{\partial^2}{\partial r^2}r\right\} + \frac{\hbar^2}{2mr^2}\ell(\ell + 1) - \frac{Ze^2}{4\pi\varepsilon_0 r}\right]R = ER, \tag{6.5}$$

which may be rearranged to

$$\frac{1}{r}\frac{\partial^2}{\partial r^2}rR - \frac{\ell(\ell + 1)}{r^2}R + \frac{2mZe^2}{4\pi\varepsilon_0\hbar^2 r}R = \frac{-2mE}{\hbar^2}R. \tag{6.6}$$

The first term in Eq. (6.6) is given by

$$\frac{1}{r}\frac{\partial^2}{\partial r^2}(rR) = \frac{1}{r}\frac{\partial}{\partial r}\left(R + r\frac{\partial R}{\partial r}\right)$$

$$= \frac{1}{r}\left(\frac{\partial R}{\partial r} + \frac{\partial R}{\partial r} + r\frac{\partial^2 R}{\partial r^2}\right) \tag{6.7}$$

$$= \frac{2}{r}\frac{\partial R}{\partial r} + \frac{\partial^2 R}{\partial r^2},$$

so that the radial Schrödinger equation becomes

$$\left[\frac{\partial^2}{\partial r^2} + \frac{2}{r}\frac{\partial}{\partial r} - \frac{\ell(\ell + 1)}{r^2} + \frac{2mZe^2}{4\pi\varepsilon_0\hbar^2 r}\right]R = \left(\frac{-2mE}{\hbar^2}\right)R. \tag{6.8}$$

6.2 RADIAL SOLUTIONS AND EIGENVALUES

The boundary conditions on Eq. (6.8) are that R must be well behaved at $r = 0$ (i.e., R can be zero or finite, but must not diverge) and that R must go to zero as $r \to \infty$ (corresponding to the electron being bound to the nucleus). Rather than solve this differential equation explicitly, let's guess a solution and see how it works out. One solution that is compatible with the boundary conditions is $R = e^{-\alpha r}$, with α an unknown constant. When this equation is substituted into Eq. (6.8), we find that

$$\left\{ \alpha^2 - \frac{2\alpha}{r} - \frac{\ell(\ell + 1)}{r^2} + \frac{2mZe^2}{4\pi\varepsilon_0 r\hbar^2} + \frac{2mE}{\hbar^2} \right\} e^{-\alpha r} = 0. \tag{6.9}$$

If we group together terms with like powers in r^{-n}, we find that, in order to satisfy this equation, we must have $\ell = 0$ and

$$\frac{-2mE}{\hbar^2} = \alpha^2 = \frac{m^2 Z^2 e^4}{(4\pi\varepsilon_0)^2 \hbar^4}. \tag{6.10}$$

This implies that

$$\alpha = \frac{mZe^2}{4\pi\varepsilon_0 \hbar^2} \tag{6.11}$$

and

$$E = -\frac{mZ^2 e^4}{(4\pi\varepsilon_0)^2 2\hbar^2}. \tag{6.12}$$

This solution has no nodes, so, according to our discussion of nodes in Chapter 3, E in Eq. (6.12) is the ground-state energy.

Other radial solutions may be developed by using trial solutions that are the products of $e^{-\alpha r}$ and a polynomial in r. From these, the general solution may be written as follows:

$$R_{n\ell}(r) = -\left[\frac{4Z^3(n - \ell - 1)!}{n^4 a_0^3 [(n + \ell)!]^3} \right]^{1/2} \left(\frac{2Zr}{na_0} \right)^\ell e^{-Zr/na_0} L_{n+\ell}^{2\ell+1}\left(\frac{2Zr}{na_0} \right), \tag{6.13}$$

where n is a new quantum number known as the *principal* quantum number and has the values $n = 1, 2, \ldots, \infty$. In addition, the orbital angular-momentum quantum number ℓ is bounded by $\ell = 0, 1, 2, \ldots, n - 1$, and L_k^j is a function known as *the associated Laguerre polynomial* (with k and j being indices that determine the order of the polynomial). Finally,

$$a_0 = \frac{4\pi\varepsilon_0 \hbar^2}{me^2} = .529 \text{ Å} \tag{6.14}$$

is known as the *Bohr radius*, because it turns out to be the radius of the orbit with the lowest energy in the Bohr model. Note, however, that the present derivation is not related to the Bohr model.

Examples (not normalized) of radial functions with the lowest energy are as follows:

$$\begin{cases} R_{10} = e^{\frac{-Zr}{a_0}}; \\ \\ R_{20} = \left(1 - \frac{Zr}{2a_0}\right)e^{\frac{-Zr}{2a_0}}; \\ \\ R_{21} = re^{\frac{-Zr}{2a_0}}. \end{cases} \tag{6.15}$$

EXERCISE 6.1   Prove directly that $R_{10}(r)Y_{00}(\theta, \phi)$ is an eigenfunction of H in Eq. (6.3) with $\ell = 0$.

Answer   First note that $\ell^2 Y_{00} = 0$. This means that when we evaluate $H R_{10} Y_{00}$, we get

$$H(Y_{00}R_{10}) = -\frac{\hbar^2}{2m} Y_{00} \frac{1}{r} \frac{\partial^2}{\partial r^2} re^{-Zr/a_0} - \frac{Ze^2}{4\pi\varepsilon_0 r} Y_{00} R_{10}$$

$$= -\frac{\hbar^2}{2m} Y_{00} \frac{1}{r}\left(-\frac{2Z}{a_0} e^{-Zr/a_0} + \frac{Z^2 r}{a_0^2} e^{-Zr/a_0}\right) - \frac{Ze^2}{4\pi\varepsilon_0 r} Y_{00} R_{10}$$

$$= \left[\frac{2\hbar^2 Z}{2ma_0 r} - \frac{\hbar^2 Z^2}{2ma_0^2} - \frac{Ze^2}{4\pi\varepsilon_0 r}\right] Y_{00} R_{10}.$$

Now we make the replacement $a_0 = (4\pi\varepsilon_0\hbar^2)/(me^2)$ [Eq. (6.14)]. This makes the first and third terms of the last expression (in brackets) cancel, yielding the result

$$HY_{00}R_{10} = EY_{00}R_{10},$$

with

$$E = -\frac{\hbar^2 Z^2}{2ma_0^2} = -\frac{Z^2 e^2}{2a_0 4\pi\varepsilon_0},$$

in agreement with Eq. (6.12).

6.3   ENERGY EIGENVALUES; SPECTROSCOPY OF THE H ATOM

The energy eigenvalues associated with the hydrogen atom are given by

$$E_n = \frac{-mZ^2 e^4}{2\hbar^2} \frac{1}{(4\pi\varepsilon_0)^2 n^2} \qquad (n = 1, 2, 3, 4, \dots). \tag{6.16}$$

$$= \frac{-e^2}{2a_0} \frac{Z^2}{n^2} \frac{1}{4\pi\varepsilon_0}$$

Note that these energies depend on the quantum number n, but not on ℓ. Note also that the energy is proportional to the constant $e^2/(4\pi\varepsilon_0 a_0)$. This quantity is called the *hartree* and is denoted E_h, so that

$$\frac{e^2}{4\pi\varepsilon_0 a_0} = E_h. \tag{6.17}$$

E_h has the numerical value 27.211 eV = 627.51 kcal/mol. (Energy conversion factors are given in Appendix D.)

From the expression for the energy level, it is not difficult to derive the following formula for the wavelength λ associated with a transition between states with quantum numbers n' and n'':

$$\frac{1}{\lambda} = \frac{\nu}{c} = \frac{E_{n'} - E_{n''}}{hc} = \frac{1}{4\pi\varepsilon_0} \frac{e^2 Z^2}{2a_0 hc} \left[\left(\frac{1}{n''}\right)^2 - \left(\frac{1}{n'}\right)^2 \right]$$

$$= Z^2 R_H \left[\left(\frac{1}{n''}\right)^2 - \left(\frac{1}{n'}\right)^2 \right].$$

(6.18)

Here, R_H = 109737.31 cm^{-1} = 13.605698 eV is known as the *Rydberg constant*. In energy units, $R_H = 0.5 E_h$.

Eq. (6.18) describes the famous spectroscopic line progressions [the Lyman, Balmer, and Paschen series—for example, Eq. (1.32)] that inspired the original Bohr model. Indeed, the formula is identical to that resulting from the Bohr model, an accident that is unique for the coulomb potential that binds the electron to the nucleus. By contrast, many other kinds of potentials lead to quantum energy levels that differ significantly from their Bohr (semiclassical) counterparts.

Figure 6.1 shows a schematic of the energy levels from Eq. (6.16). Note that the levels get closer together the energy increases. This contrasts with the particle-in-box (Figure 3.2), rigid-rotor (Eq. 4.42), and harmonic-oscillator (Figure 5.2) energies. The decreasing gap between energy levels with increasing energy that

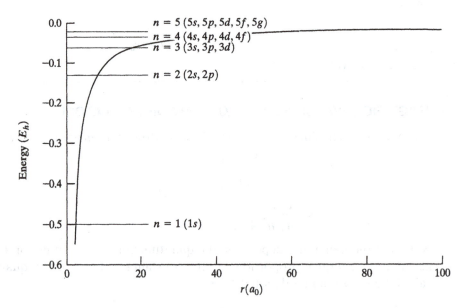

Figure 6.1 Hydrogen atom energy levels, superimposed on a plot of the coulomb potential.

falls out of Eq. (6.16) arises from the shape of the coulomb potential. In addition, the number of bound energy levels is infinite, with the energy approaching zero from below as $n \to \infty$. There also exist states of the hydrogen atom with energies above zero, but these are not bound states. Instead, they are associated with collisions between protons and electrons.

EXERCISE 6.2   Compare the excitation energies between the two lowest energy levels for the H atom with $Z = 1$ and a free electron in a cubic box the length of whose side is $\ell_0 = 1.058\text{Å}(2\,a_0)$. Comment on the difference.

Answer   For the H atom,

$$E_2 - E_1 = \left[\left(-\frac{1}{8} E_h \right) - \left(-\frac{1}{2} E_h \right) \right] = \frac{3}{8} E_h \approx \frac{3}{8} (27.21 \text{ eV}) = 10.20 \text{ eV}.$$

For the cubic box,

$$E_{2,1,1} - E_{1,1,1} = \frac{h^2}{8m\ell_0^2} \left[(2^2 + 1^2 + 1^2) - (1^2 + 1^2 + 1^2) \right]$$

$$= \frac{3}{8} \frac{h^2}{m\ell_0^2} = \frac{3}{8} \frac{4\pi^2\hbar^2}{4ma_0^2}$$

$$= \frac{3}{8} \pi^2 \left(\frac{\hbar^2}{ma_0^2} \right).$$

Using Eqs. (6.14) and (6.17), we see that the term in brackets is E_h, so $E_{2,1,1} - E_{1,1,1} = (3\pi^2/8)E_h = 3.70E_h = 100.7$ eV. The excitation energy in the box is much bigger than that of the H atom for two reasons. First, in the real H atom, the electron has a coulomb attraction to the nucleus, which is absent in the boxed particle. Second, the electron wavefunction in the H atom can extend well beyond the Bohr radius. This makes the box larger and the excitation energy smaller.

6.4   PROPERTIES OF HYDROGEN AND HYDROGENLIKE WAVEFUNCTIONS

6.4.1   Orbital Labels

The complete hydrogen or hydrogenlike wavefunction or *orbital* is given by

$$\psi_{n\ell m} = R_{n\ell}(r)Y_{\ell m}(\theta, \phi). \tag{6.19}$$

These wavefunctions are all mutually orthonormal according to the formula

$$\langle \psi_{n\ell m} | \psi_{n'\ell'm'} \rangle = \int \psi_{n\ell m}^* \psi_{n'\ell'm'} r^2 dr \sin\theta d\theta d\phi$$

$$= \int R_{n\ell}^* R_{n'\ell'} r^2 dr \int Y_{\ell m}^* Y_{\ell'm'} \sin\theta d\theta d\phi = \delta_{nn'}\delta_{\ell\ell'}\delta_{mm'}. \tag{6.20}$$

Table 6.1 lists the quantum numbers, spectroscopic names ($1s$, $2s$, $2p$, etc.), and energies associated with the lowest energy solutions.

TABLE 6.1 QUANTUM NUMBERS, ENERGIES, AND
NAMES (SPECTROSCOPIC SYMBOLS) OF LOWEST
STATES OF HYDROGEN ATOM

n	ℓ	Degeneracy	Name	Energy (E_h)
1	0	1	$1s$	$-Z^2/2$
2	0	1	$2s$	$-Z^2/8$
2	1	3	$2p$	$-Z^2/8$
3	0	1	$3s$	$-Z^2/18$
3	1	3	$3p$	$-Z^2/18$
3	2	5	$3d$	$-Z^2/18$

6.4.2 Radial Wavefunctions and Radial Probability Distributions

Figure 6.2 plots the radial wavefunctions from Eq. (6.13). Note that only s-orbitals are nonzero at the origin ($r = 0$). This is an important property in NMR spectroscopy. The number of radial nodes is $n - \ell - 1$ (not counting the origin and ∞).

The *radial probability density* $P_{n\ell}(r)$ is defined by the statement that $P_{n\ell}(r)dr$ is the probability of finding the electron between r and $r + dr$. Angular coordinates are integrated out. In view of the form of the hydrogen wavefunction in Eq. (6.19), and using the middle terms in Eq. (6.20), one can show that the radial probability distribution is related to the radial wavefunction by

$$P_{n\ell}(r) = r^2 \int \int |\psi_{n\ell m}|^2 \sin\theta d\theta d\phi = R_{n\ell}^2(r)r^2. \tag{6.21}$$

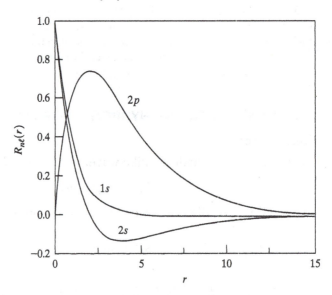

Figure 6.2 Hydrogen-atom radial functions.

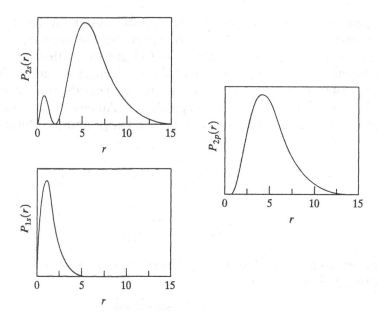

Figure 6.3 Hydrogen-atom radial probability densities.

Figure 6.3 plots the radial probability distribution for the lowest states analogous to those whose radial wavefunctions were plotted in Figure 6.2. Notice that $P_{n\ell} \geq 0$ for all r (as must be true for any probability). Notice also that there are nodes in $P_{n\ell}(r)$, at values of r for which the wavefunction ψ has a radial node.

EXERCISE 6.3   Elementary texts often discuss orbitals in terms of the "90% probability contour," the contour defined such that 90% of the probability density is inside it. Is the probability contour the same for normalized and unnormalized versions of an atomic orbital?

Answer   The 90% probability contour is defined by

$$\frac{4\pi \int_0^{R_c} |\psi_{n\ell m}|^2 r^2 dr}{4\pi \int_0^{\infty} |\psi_{n\ell m}|^2 r^2 dr} = 0.90.$$

This contour does not change if we multiply ψ by an arbitrary constant, so it is not dependent on whether ψ is normalized.

6.4.3 Angular Eigenfunctions and Nodes

In making plots of hydrogen wavefunctions, it is convenient to use angular wavefunctions that are real. These functions are easily obtained from the spherical harmonics by taking linear combinations of pairs of states having the same ℓ and $|m|$, namely, $Y_{\ell m} \pm Y_{\ell -m}$. Since the energy of the hydrogen atom does not depend on m, any linear combination of n, ℓ eigenstates with different m yields a wavefunction that is still a solution of the Schrödinger equation. Of course, the linear combination would not be an eigenfunction of other operators, such as $\hat{\ell}_z$, but this is not important for many applications. Among the solutions generated from these linear combinations are the familiar p_x and p_y orbitals (recall that $\psi_{p_z} \propto Y_{1,0}$):

$$\psi_{p_x} = -\frac{\left(Y_{1,1} - Y_{1,-1}\right)}{\sqrt{2}}; \tag{6.22a}$$

$$\psi_{p_y} = -\frac{\left(Y_{1,1} + Y_{1,-1}\right)}{\sqrt{2}i}. \tag{6.22b}$$

Substitution of the expressions for the spherical harmonics in Eq. (4.44b) into Eqs. (6.22) gives

$$\psi_{p_x} \propto \sin\theta \cos\phi$$

and

$$\psi_{p_y} \propto \sin\theta \sin\phi,$$

which indicates that $\psi_{p_x} \propto x$ and $\psi_{p_y} \propto y$. These p orbitals have one angular node, and in general, the real angular eigenfunctions have ℓ angular nodes. The d orbitals are defined as follows:

$$\psi_{d_{z^2}} = Y_{20};$$

$$\psi_{d_{xz}} = -\frac{\left(Y_{21} - Y_{2-1}\right)}{\sqrt{2}};$$

$$\psi_{d_{yz}} = -\frac{\left(Y_{21} + Y_{2-1}\right)}{i\sqrt{2}};$$

$$\psi_{d_{x^2-y^2}} = \frac{\left(Y_{22} + Y_{2-2}\right)}{\sqrt{2}};$$

$$\psi_{d_{xy}} = \frac{\left(Y_{22} - Y_{2-2}\right)}{i\sqrt{2}}.$$

6.4.4 Combined Wavefunctions

The combined hydrogen wavefunctions are summarized in Table 6.2. Figure 6.4 shows contours of wavefunction amplitudes in three-dimensional space. Two important characteristics of these and other orbitals are as follows:

TABLE 6.2 HYDROGEN-ATOM WAVEFUNCTIONS

State	Wavefunction
$1s$	$\pi^{-1/2}(Z/a_0)^{3/2}e^{-Zr/a_0}$
$2s$	$\pi^{-1/2}(Z/2a_0)^{3/2}(1 - Zr/2a_0)e^{-Zr/2a_0}$
$2p_{+1}$	$-\dfrac{1}{8}\pi^{-1/2}(Z/a_0)^{5/2}re^{-Zr/2a_0}\sin\theta e^{+i\phi}$
$2p_0$	$\pi^{-1/2}(Z/2a_0)^{5/2}re^{-Zr/2a_0}\cos\theta$
$2p_{-1}$	$\dfrac{1}{8}\pi^{-1/2}(Z/a_0)^{5/2}re^{-Zr/2a_0}\sin\theta e^{-i\phi}$
$3s$	$\pi^{-1/2}(Z/a_0)^{3/2}\dfrac{1}{3\sqrt{3}}\left(1 - \dfrac{2Zr}{3a_0} + \dfrac{2Z^2r^2}{27a_0^2}\right)e^{-Zr/3a_0}$
$3p_{+1}$	$-\pi^{-1/2}(Z/a_0)^{3/2}\dfrac{2\sqrt{2}}{27}\left(\dfrac{Zr}{a_0} - \dfrac{Z^2r^2}{6a_0^2}\right)e^{-Zr/3a_0}\sin\theta e^{i\phi}$
$3p_0$	$\pi^{-1/2}(Z/a_0)^{3/2}\dfrac{2\sqrt{2}}{27}\left(\dfrac{Zr}{a_0} - \dfrac{Z^2r^2}{6a_0^2}\right)e^{-Zr/3a_0}\cos\theta$
$3p_{-1}$	$\pi^{-1/2}(Z/a_0)^{3/2}\dfrac{2\sqrt{2}}{27}\left(\dfrac{Zr}{a_0} - \dfrac{Z^2r^2}{6a_0^2}\right)e^{-Zr/3a_0}\sin\theta e^{-i\phi}$
$3d_{+2}$	$\pi^{-1/2}(Z/a_0)^{7/2}\dfrac{1}{81\sqrt{2}}r^2e^{-Zr/3a_0}\sin^2\theta e^{2i\phi}$
$3d_{+1}$	$-\pi^{-1/2}(Z/a_0)^{7/2}\dfrac{\sqrt{2}}{81}r^2e^{-Zr/3a_0}\sin\theta\cos\theta e^{i\phi}$
$3d_0$	$\pi^{-1/2}(Z/a_0)^{7/2}\dfrac{1}{81\sqrt{6}}r^2e^{-Zr/3a_0}(3\cos^2\theta - 1)$
$3d_{-1}$	$\pi^{-1/2}(Z/a_0)^{7/2}\dfrac{\sqrt{2}}{81}r^2e^{-Zr/3a_0}\sin\theta\cos\theta e^{-i\phi}$
$3d_{-2}$	$\pi^{-1/2}(Z/a_0)^{7/2}\dfrac{1}{81\sqrt{2}}r^2e^{-Zr/3a_0}\sin^2\theta\, e^{-2i\phi}$

1. For $\ell = 0$, we get s-orbitals that are spherically symmetric. So the nodal surfaces in this case are spherical.
2. For $\ell \neq 0$, but $m = 0$, we get $Y_{\ell 0} \propto S_{\ell 0}(\theta) \propto P_\ell(\cos\theta)$. This gives the z-type orbitals $(p_z, d_{z^2}, f_{z^3}, \dots)$. For $\ell \neq 0$ and $m \neq 0$, we use the real combinations of the spherical harmonics as previously described to define orbital shapes. The total number of nodes in the combined radial and angular states is $n - 1$.

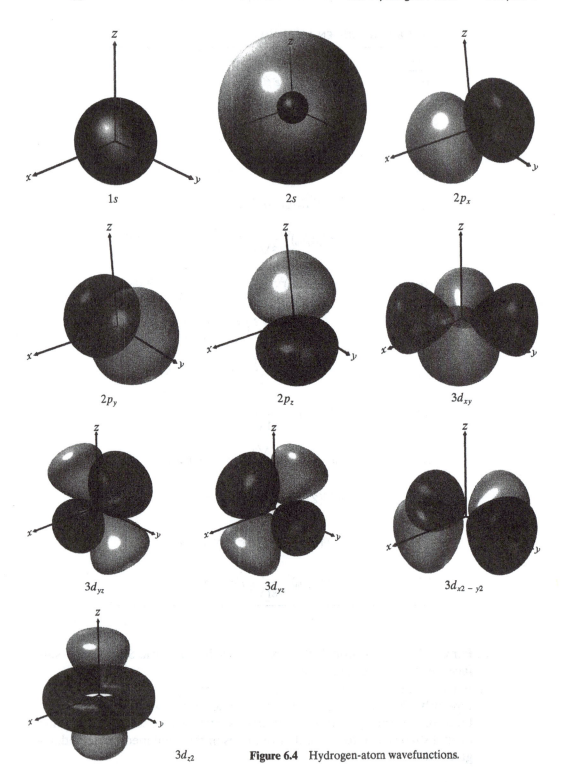

Figure 6.4 Hydrogen-atom wavefunctions.

EXERCISE 6.4   Use the data in Table 6.2 and the results $e^{i\phi} = (\cos\phi + i\sin\phi)$ [Eq. (A.4)], $x = r\sin\theta\cos\phi$, and $y = r\sin\theta\sin\phi$ to prove that

$$\psi_{d_{xy}} \propto \frac{1}{i}(\psi_{322} - \psi_{32-2}).$$

Answer

$$\psi_{322} = A_{322}R_{22}\sin^2\theta e^{2i\phi}$$

and

$$\psi_{32-2} = A_{322}R_{22}\sin^2\theta e^{-2i\phi},$$

since the normalization constants A_{322} and A_{32-2} are identical. So

$$\psi_{322} - \psi_{32-2} = AR(e^{2i\phi} - e^{-2i\phi})\sin^2\theta$$
$$= 4Aie^{-Zr/3a_0}(r\sin\theta\cos\phi)(r\sin\theta\sin\phi)$$
$$= 4Aie^{-Zr/3a_0}xy \propto \psi_{d_{xy}}.$$

6.5   ATOMIC UNITS

At this point, it is convenient to introduce atomic units to describe the electronic Hamiltonians of atoms and molecules. Atomic units are a "natural" set of units in which fundamental constants are used to define the unit charge, unit mass, and unit angular momentum. Specifically, we choose

$$\hbar = 1 \qquad \text{(action)},$$
$$m_e = 1 \qquad \text{(mass)},$$

and

$$q_e = e/(4\pi\varepsilon_0)^{1/2} = 1 \qquad \text{(charge)}.$$

Important derived units are

$$a_0 = \frac{4\pi\varepsilon_0\hbar^2}{m_e e^2} = 1 \text{ bohr} \qquad \text{(length unit)},$$

$$-2E_{1s}(\text{H atom}) = \frac{e^2}{4\pi\varepsilon_0 a_0} \cong 27.211 \text{ eV} = 1 \text{ hartree } (E_h) \qquad \text{(energy unit)},$$

and

$$(4\pi\varepsilon_0)^2\hbar^3/m_e e^4 \cong 2.42 \times 10^{-17} \text{ s} \qquad \text{(time unit)}.$$

For a hydrogen atom, the Hamiltonian, in atomic units, is

$$H = -\frac{1}{2}\nabla^2 - \frac{Z}{r}, \tag{6.23}$$

and the energy levels are given by

$$E_n = \frac{-Z^2}{2n^2}. \tag{6.24}$$

Note that all fundamental constants have disappeared in Eq. (6.24). This demonstrates one of the great advantages of using atomic units to describe electronic structure, namely, that the results will not depend on the precise values of the fundamental constants.

A summary of atomic units is given in Appendix D.

EXERCISE 6.5   (1) Show that the speed of light, c, is about 137 atomic units. (2) Show that in the Bohr model of the hydrogen atom, for the ground state, the period of revolution of the electron around the proton is 2π time units. (3) Near stars, atoms can be highly ionized. What is the $1s \rightarrow 2p$ transition energy in C^{+5}?

Answer

(1) It is easiest to start with the fine-structure constant α from Appendix D. We have

$$\alpha = \frac{e^2}{4\pi\varepsilon_0\hbar c} = \frac{1}{137.036}$$

$$= \left(\frac{e^2}{4\pi\varepsilon_0 a_0}\right) \cdot a_0 \cdot \left(\frac{1}{\hbar c}\right)$$

$$= E_h \cdot a_0 \cdot \frac{1}{\hbar} \cdot \frac{1}{c} = \frac{1}{137.036}.$$

But in atomic units, $E_h \equiv 1$, $a_0 = 1$, and $\hbar = 1$. So $c = 137.036$.

(2) In the Bohr model, the angular momentum rp is $n\hbar$. For the ground state, $n = 1$ and $r = a_0$, so $p = n\hbar/r = \hbar/a_0 = mv$, and it follows that the velocity is $\hbar/(ma_0)$. The circumference of the orbit is $2\pi a_0$, so that

$$\text{period} = \frac{\text{circumference}}{\text{velocity}} = \frac{2\pi a_0}{\dfrac{\hbar}{ma_0}} = \frac{2\pi m a_0^2}{\hbar} = 2\pi,$$

in atomic units.

(3) C^{+5} is an H-like atom, with only one electron. The energy is therefore

$$E = -\frac{1}{2}\frac{Z^2}{n^2} \quad \text{in hartrees [Eq. (6.24)].}$$

So $E_{2p} - E_{1s} = -\frac{1}{2}Z^2(1/4 - 1/1) = 3/8Z^2 = 3/8(36) = 13.5$ hartrees $= 367$ eV.

SUGGESTED READINGS

The following books contain especially detailed descriptions of the hydrogen atom:

P. W. Atkins and R. S. Friedman, *Molecular Quantum Mechanics*, 3rd ed. (Oxford, New York, 1997).

M. Karplus and R. N. Porter, *Atoms and Molecules* (Benjamin, New York, 1970).

I. N. Levine, *Quantum Chemistry*, 5th ed. (Prentice Hall, Upper Saddle River, NJ, 2000).

J. P. Lowe, *Quantum Chemistry*, 2nd ed. (Academic Press, New York, 1993).

PROBLEMS

6.1 (a) Draw the third radial wavefunction $R(r)$ for the hydrogen atom:

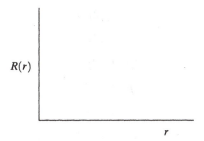

(b) Draw contours of the H-atom wavefunction $\psi_{210}(r, \theta, \phi)$:

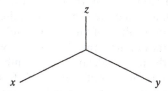

(c) Draw the third radial probability density of the hydrogen atom:

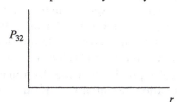

6.2 Which of the following are legitimate H-atom wavefunctions?
(a) $\psi = Nre^{-Zr}\sin\theta$.
(b) $\psi = N(1 - Zr)e^{-Zr}\cos\theta$.
(c) $\psi = N(1 - Zr)e^{-Zr}$.

6.3 What is the probability of finding an electron in a hydrogen $1s$ orbital inside the nucleus? The diameter of a proton is about 1×10^{-12} cm.

6.4 (a) Starting with the radial Schrödinger equation for the hydrogen atom and a guessed solution of the form

$$R(r) = (1 + br)e^{-ar},$$

find a and b such that this solution works.
(b) Normalize the solution in (a).
(c) Find $\langle r \rangle$ for this solution.
(d) Find the values of r where the radial probability density is maximal.

6.5 Find the maxima in $P_{n\ell}$ for the $1s$, $2s$, and $2p$ orbitals. Interpret your results in terms of nodes. Would you expect $2p$ or $2s$ to be more available for covalent bond formation? Are the maxima the same for $2p_x$, $2p_y$, $2p_+$, $2p_-$, and $2p_z$?

6.6 For the He^+ atom, which of the following do not vanish?

$$\langle 1s|\ell_z|2p_x\rangle;$$

$$\langle 2p_x|\ell_z|2p_x\rangle;$$

$$\langle 2p_x|\ell_x|2p_z\rangle;$$

$$\langle 2p_x|\ell_x|2p_x\rangle;$$

$$\langle 2p_x|\ell_+|2p_z\rangle;$$

$$\langle 3d_{z^2}|\ell^2|2p_z\rangle;$$

$$\langle 3d_{x^2-y^2}|\ell^2|3d_{x^2-y^2}\rangle.$$

6.7 (a) Show, for the $(n, \ell, m) = (1, 0, 0)$ and $(2, 1, m)$ states of the H atom, that $\langle 1/r\rangle = Z/(n^2a_0)$. Compare this result with the total energy of these states. The result is an example of a quite general one called the *virial theorem*, which relates the expectation values of kinetic (V) and potential (T) energies. The relationship is generally of the form $-k\langle T\rangle = \langle V\rangle$, where k is a constant, depending on the nature of the potential. For the coulomb potential, what is k?

(b) For the simple harmonic oscillator, evaluate $\langle T\rangle$ and $\langle V\rangle$ directly. (Use the ground state.) Does the virial theorem hold? What is k for the harmonic force?

6.8 The first line in the Lyman series for the H atom corresponds to the $n = 1 \rightarrow n = 2$ transition. If the interaction between radiation and the electron is $V = e\mathbf{E} \cdot \mathbf{r} = e(E_x x + E_y y + E_z z)$, which (n, ℓ, m) states mix with the state $(1, 0, 0)$ to give this absorption line, called Lyman α? What is its wavelength? In which region of the electromagnetic spectrum is it found? (*Hint*: Section 4.5 discusses selection rules.)

6.9 (a) In the semiclassical approximation, the electron in an electromagnetic field undergoes a perturbation potential $V = \mathbf{E} \cdot \boldsymbol{\mu} = e\mathbf{E} \cdot \mathbf{r}$, where $\mathbf{E}$ and $\mathbf{r}$ are the field and the electron coordinate, $\mathbf{r} = (x, y, z)$, and $\boldsymbol{\mu}$ is the dipole moment. If the field acts only on the z-axis, determine the energies of the $2s$, $2p_{+1}$, $2p_{-1}$, and $2p_z$ states to a first-order approximation. Then use second-order perturbation theory with only this set of four states to show which of the states change their energy. Interpret your result.

(b) For the unperturbed H atom, we used notions of degeneracy to show that $2p_{+1}$ and $2p_{-1}$ could be combined to make real wavefunctions $2p_x$ and $2p_y$. Can this still be done to the wavefunctions with first-order energies? With second-order energies? (*Hint*: Use symmetry to evaluate whether the matrix elements vanish.)

7

$$E_{\text{trial}} = \frac{\langle \phi | H | \phi \rangle}{\langle \phi | \phi \rangle} \geq E_0$$

The Helium Atom

The helium atom provides an important challenge to the generalization of quantum mechanics to many-electron atoms, as the Schrödinger equation cannot be separated and approximations must be used. Variational theory provides a powerful and systematic approach to developing these approximations, and the methods that we present here are of general application in quantum electronic structure theory.

7.1 SCHRÖDINGER EQUATION

To obtain the Hamiltonian of He, we first neglect the nuclear kinetic energy, which means that we assume that the nuclear mass is infinite. Then, using coordinates that are defined in Figure 7.1, we write

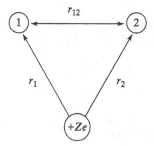

Figure 7.1 Coordinate definitions for He atom.

$$H = \frac{\hat{p}_1^2}{2m_e} + \frac{\hat{p}_2^2}{2m_e} - \frac{Ze^2}{4\pi\varepsilon_0 r_1} - \frac{Ze^2}{4\pi\varepsilon_0 r_2} + \frac{e^2}{4\pi\varepsilon_0 r_{12}}, \tag{7.1}$$

or

$$H = \frac{1}{2}\left(\hat{p}_1^2 + \hat{p}_2^2\right) - \frac{Z}{r_1} - \frac{Z}{r_2} + \frac{1}{r_{12}} \text{ (in atomic units).} \tag{7.2}$$

The terms in this equation are easily recognized to be the two electron kinetic energies, the two attractive coulomb interactions between the electrons and the nucleus, and the electron–electron repulsive coulomb interaction. Another way to write the Hamiltonian is

$$H = h_1(1) + h_2(2) + \frac{1}{r_{12}}, \tag{7.3}$$

where the first two terms on the right are the hydrogenlike Hamiltonians for each electron.

The Schrödinger equation for this problem is

$$H\Psi(\mathbf{r}_1, \mathbf{r}_2) = E\Psi(\mathbf{r}_1, \mathbf{r}_2). \tag{7.4}$$

Unfortunately, this equation is *not separable*, and it is not soluble analytically in closed form. We (really) have run out of exactly solvable systems. Fortunately, close-to-exact numerical solutions do exist, as we shall see. Before getting into that, however, it is more useful to look for approximate solutions.

7.2 INDEPENDENT-PARTICLE MODEL

The simplest approximation is simply to ignore the $1/r_{12}$ term in Eq. (7.2). This gives us the *independent-particle* Hamiltonian

$$H \cong h_1(\mathbf{r}_1) + h_2(\mathbf{r}_2), \tag{7.5}$$

whose eigenfunctions are just the products of hydrogenlike wavefunctions:

$$\psi^{(0)}(\mathbf{r}_1, \mathbf{r}_2) = \phi_{n_1\ell_1 m_1}(\mathbf{r}_1)\phi_{n_2\ell_2 m_2}(\mathbf{r}_2). \tag{7.6}$$

If we use the truncated form [Eq. (7.5)] for H, then the energy associated with these solutions for the ground state of He $(n_1 = n_2 = 1)$ is

$$E = \frac{-Z^2}{2n_1^2} - \frac{Z^2}{2n_2^2} = -\frac{4}{2} - \frac{4}{2} = -4E_h. \tag{7.7}$$

The exact ground-state energy is $-2.904E_h$ (as determined by summing the first and second ionization energies of He). The value of the energy derived from Eq. (7.6) for ψ with Eq. (7.5) for H is far too low, with most of the error coming from ignoring $1/r_{12}$ in Eq. (7.5).

A significant improvement over this result is obtained by using the complete Hamiltonian [Eq. (7.1)] to evaluate the energy for the trial function of Eq. (7.6). This is equivalent to doing *first-order perturbation theory* [see Eq. (5.56)], and the result is

$$E = \frac{\langle \phi(1)\phi(2) | H | \phi(1)\phi(2) \rangle}{\langle \phi_1\phi_2 | \phi_1\phi_2 \rangle} = \frac{-Z^2}{2} - \frac{Z^2}{2} + \left\langle \psi \left| \frac{1}{r_{12}} \right| \psi \right\rangle$$

$$= -Z^2 + \frac{5}{8}Z = -2.75E_h.$$

(7.8)

The integral involving $1/r_{12}$ is reasonably complicated. Here we just give the result $\left(\frac{5}{8}Z\right)$ and defer the evaluation of the integral to Appendix B.

Note the inequality

$$\langle \psi^{(0)} | H | \psi^{(0)} \rangle > E_{\text{tot}}.$$

(7.9)

This turns out to be a special case of the variational theorem, which we discuss next.

7.3   THE VARIATIONAL METHOD

The *variational theorem* states that if ϕ is any well-behaved function that satisfies the boundary conditions associated with the problem of interest, then the expectation value of H, calculated using ϕ, will obey the inequality

$$E_{\text{trial}} = \frac{\langle \phi | H | \phi \rangle}{\langle \phi | \phi \rangle} \geq E_0,$$

(7.10)

where E_0 is the exact ground-state energy. The ϕ's here are called *trial* wavefunctions, and the expectation value is E_{trial}.

To prove the theorem, we rearrange Eq. (7.10) to yield the inequality

$$I = \langle \phi | H - E_0 | \phi \rangle \geq 0.$$

(7.11)

Now we expand ϕ in terms of the exact eigenfunctions of H. These eigenfunctions, ψ_k, satisfy $H\psi_k = E_k\psi_k$. This expansion is therefore

$$\phi = \sum_k c_k\psi_k.$$

(7.12)

If this expression is now substituted into Eq. (7.11), we get

$$I = \sum_{k,\ell} c_k^* c_\ell \langle \psi_k | H - E_0 | \psi_\ell \rangle = \sum_{k,\ell} c_k^* c_\ell (E_k - E_0)\delta_{k\ell}$$

$$= \sum_k |c_k|^2 (E_k - E_0).$$

(7.13)

Each term in the last sum is inherently positive or zero, because the absolute square therein cannot be negative and the energy difference $(E_k - E_0)$ is also positive or zero, given that E_0 is the ground-state energy.

The variational theorem is very powerful in that it enables us to determine which trial function gives the most accurate energy. It is always the one that gives the lowest value of E_{trial}. The determination of the best trial function is called *variational optimization*. There are many different ways to do this, but most commonly, one guesses a function containing many parameters and then minimizes the energy with respect to the values of these parameters.

As an example, let us consider the He-atom trial function

$$\phi(1, 2) = u_{1s}(1)u_{1s}(2), \tag{7.14}$$

where u_{1s} is taken to be a hydrogenlike $1s$ orbital, with the "charge" Z' chosen to be an adjustable parameter. In atomic units, we thus have

$$u_{1s} = \pi^{-1/2}Z'^{3/2}e^{-Z'r}. \tag{7.15}$$

If we choose $Z' = Z = 2$, then we recover the result from the aforementioned perturbation theory. This satisfies the previously noted inequality [Eq. (7.9)], but we can generate a more accurate estimate of the energy by minimizing the trial energy with respect to the value of Z'. The trial energy may be evaluated similarly to Eq. (7.8), resulting in

$$E_{trial} = \frac{5}{8}Z' + (Z')^2 - 2ZZ'. \tag{7.16}$$

Minimization is now accomplished by setting the first derivative to zero:

$$\frac{\partial E_{trial}}{\partial Z'} = 0 = \left(\frac{5}{8} + 2Z' - 2Z\right). \tag{7.17}$$

This leads to

$$Z' = Z - \frac{5}{16} = \frac{27}{16} = 1.6875, \tag{7.18}$$

where, in the last term, we have used $Z = 2$. The resulting energy is then obtained by substituting the optimized Z' back into Eq. (7.16), giving

$$E(Z') = -\left(Z - \frac{5}{16}\right)^2 = -\left(\frac{27}{16}\right)^2 \approx -2.8377E_n. \tag{7.19}$$

This value is significantly closer to the exact value, $-2.906E_h$, than the result $Z' = 2$. Figure 7.2 shows how the trial energy varies with Z', verifying that $Z' = 27/16$ gives indeed the minimum energy for this choice of trial function. Physically, $Z' < Z$ because one electron *screens* the nuclear attraction and prevents it from acting fully on the other electron. $Z' < 2$ makes the atom bigger, reflecting this screening.

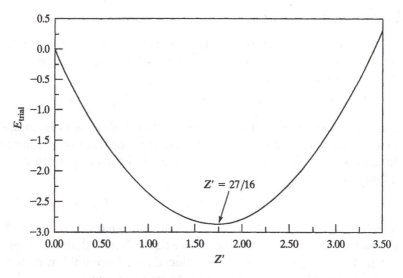

Figure 7.2   Trial energy as function of parameter Z'.

EXERCISE 7.1   Using Eqs. (7.14) and (7.15), evaluate $\langle r_1 \rangle$. Compare this with the value of $\langle r \rangle$ for the H atom, and for He$^+$. Does your calculation agree with the statement in elementary chemistry books that He has a smaller atomic radius than H?

Answer   All these wavefunctions can be written (in atomic units) as $u_1(r) = Ae^{-\xi r}$, where $\xi = 1, 27/16$, and 2 for H, He, and He$^+$, respectively, and A is a normalization constant. Then

$$\langle r \rangle = \frac{\int_0^\infty r u_1^2 r^2 dr}{\int_0^\infty u_1^2 r^2 dr} = \frac{\int_0^\infty r^3 e^{-2\xi r} dr}{\int_0^\infty r^2 e^{-2\xi r} dr}.$$

Now we evaluate the numerator using integration by parts:

$$\int_0^\infty r^3 e^{-2\xi r} dr = (r^3)\left(\frac{e^{-2\xi r}}{-2\xi}\right)\Big|_0^\infty + \int_0^\infty \frac{3}{2\xi} r^2 e^{-2\xi r} dr.$$

The term $r^3 e^{-2\xi r} \big|_0^\infty$ vanishes, giving

$$\langle r \rangle = \frac{\dfrac{3}{2\xi} \int_0^\infty r^2 e^{-2\xi r} dr}{\int_0^\infty r^2 e^{-2\xi r} dr} = \frac{3}{2\xi}.$$

So as ξ gets bigger, $\langle r \rangle$ gets smaller. In particular,

$$r(\text{H}) : r(\text{He}) : r(\text{He}^+) = 1 : 16/27 : 1/2.$$

So He is indeed smaller than H, but He$^+$ is smaller yet.

7.4 BETTER WAVEFUNCTIONS

We can determine the ground-state energy of He more accurately if we use a more flexible trial wavefunction. One straightforward generalization of what we have already done is to assume that

$$\phi = \pi^{-1}Z'^{3/2}Z''^{3/2}\left(e^{-Z'r_1}e^{-Z''r_2} + e^{-Z'r_2}e^{-Z''r_1}\right), \tag{7.20}$$

which is just the symmetrized product of two hydrogen $1s$ functions with different effective charges. To minimize the energy for this two-parameter wavefunction, we set derivatives with respect to Z' and with respect to Z'' equal to zero. This leads to

$$Z' = 1.19, \quad Z'' = 2.18, \quad \text{and} \quad E_{\text{trial}} = -2.8757,$$

which is even closer to the correct result. Note that the values of the exponents are such that one electron is much more tightly bound than the other. This demonstrates how electron repulsion is reduced by allowing the wavefunction to take account of correlation between the electron motions.

The following even better two-parameter wavefunction was developed by Hylleraas:

$$\phi = N^{-1}e^{-Z'(r_1+r_2)}(1 + br_{12}). \tag{7.21}$$

Here, one multiplies a wavefunction similar to Eq. (7.14) by a term $1 + br_{12}$ that depends explicitly on the distance r_{12} between the electrons. Variational optimization of E_{trial} with respect to Z' and b gives

$$Z' = 1.849, \quad b = 0.364, \quad \text{and} \quad E_{\text{trial}} = -2.892.$$

Since $b > 0$, ϕ will be larger when the electrons are farther apart from one another. Equation (7.21) thus describes radial correlation, but it also describes angular correlation in which two electrons with the same r_1 and r_2 have a higher probability of being at opposite sides of the nucleus.

Still more accurate results can be obtained by using even more parameters, such as in the function

$$\phi = N^{-1}e^{-Z'(r_1+r_2)}\ \left(1 + \text{sum of terms involving } r_1, r_2, \text{ and } r_{12}\right). \tag{7.22}$$

This type of wavefunction was studied by Hylleraas in the 1930s and then Pekeris and Schwartz more recently. The best result obtained to date, $E_{\text{trial}} = -2.9037243756$, matches experimental results to within the limits of error after small corrections for relativity and quantum electrodynamics are included.

SUGGESTED READINGS

Variational theory is covered in more detail in many books, including the following texts:

I. N. Levine, *Quantum Chemistry*, 5th ed. (Prentice Hall, Upper Saddle River, NJ, 2000).

J. P. Lowe, *Quantum Chemistry*, 2nd ed. (Academic Press, New York, 1993).

High-level treatments of the He atom and similar atoms are presented in the following two journal articles:

C. L. Pekeris, *Phys. Rev.* 115, 1216 (1959).

C. Schwartz, *Phys. Rev.* 128, 1146 (1962).

PROBLEMS

7.1 Show that if a trial function ϕ_1 gives a certain energy, then a new function $\phi_1 + c\phi_2$, where c is a variational parameter, cannot give a higher trial energy if c is optimized. For simplicity, assume that $\langle \phi_1 | \phi_2 \rangle = 0$, that ϕ_1 and ϕ_2 are normalized, and that c is real.

7.2 In this problem, we use variational theory to solve the particle-in-box problem. Assume a trial function

$$\phi(x) = x(a - x),$$

where a is the width of the box.

(a) Does this function satisfy the boundary conditions associated with the particle in a box?

(b) Normalize ϕ.

(c) Evaluate the trial energy. Does it obey the variational theorem? What is the fractional error in the ground-state energy?

(d) How might this wavefunction be improved to give a more accurate value of the energy?

7.3 Consider the following Gaussian trial function for the He atom:

$$\phi = e^{-\alpha(r_1^2 + r_2^2)}.$$

In this function, α is a variational parameter.

(a) Write down the complete expression for determining the trial energy in a variational calculation. Be sure to indicate the volume element, integration limits, and the Hamiltonian.

(b) Suppose that the resulting expression for the trial energy is

$$E_{\text{trial}} = \frac{11}{4}\alpha - \frac{11\sqrt{\alpha}}{2}.$$

What is the optimum value of α? The optimum energy?

(c) Which is the better trial wavefunction, the Gaussian one presented or the product of $1s$ functions,

$$\phi = e^{-Z'(r_1 + r_2)}?$$

Why?

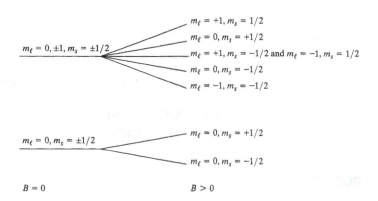

8

Electron Spin and the Pauli Principle

Electrons are indistinguishable from one another, which means that there is a fundamental symmetry that the wavefunction must obey in describing the simultaneous behavior of many electrons. A generalization of this symmetry requirement leads to the Pauli principle, the last postulate in our presentation of quantum mechanics. Electron spin is an angular momentum, and its formal operator properties can be understood as those of an angular momentum. The Pauli principle produces typical quantum mechanical behavior, such as exchange among different electrons. The coupling of spin magnetic moments with applied magnetic fields produces the Zeeman effect, electron paramagnetic resonance, and nuclear magnetic resonance, and these in turn provide a general tool for probing nuclear and electronic structure that can be understood with the tools of electronic structure theory that we have developed thus far.

8.1 ELECTRON SPIN

The wavefunction obtained in Chapter 7 for the ground state of He is symmetric with respect to interchanging the spatial coordinates. If this property were generalizable, we would have a ground-state wavefunction for Li that could be represented as a product of three hydrogen $1s$ functions. However, such a generalization turns out to be seriously in error, because it ignores the consequences of electron spin on the symmetry properties of the wavefunction. As mentioned in Chapter 1, the Dirac approach deduces spin as a natural consequence of developing relativistic quantum mechanics. That electrons have an internal spinlike angular momentum was suggested in 1922 by Uhlenbeck and Goudsmit, who used the idea to discuss the fine structure of spectral lines.

In the present nonrelativistic treatment, there is no exact classical analogue to spin, so we will deal with it by asserting the following postulate:

Electrons have an intrinsic angular momentum s *of magnitude* $\hbar/2$.

The properties of the spin eigenfunctions are analogous to those of the spherical harmonics, namely,

$$\begin{cases} \hat{S}^2 |sm_s\rangle = \hbar^2 s(s + 1)|sm_s\rangle = \hbar^2 \left(\frac{1}{2}\right)\left(\frac{3}{2}\right)|sm_s\rangle & (8.1a) \\[2em] \hat{S}_z |sm_s\rangle = \hbar m_s |sm_s\rangle = \pm \frac{\hbar}{2}|sm_s\rangle & (8.1b) \end{cases}$$

and

$$[\hat{S}_x, \hat{S}_y] = i\hbar S_z \quad \text{and} \quad [\hat{S}^2, \hat{S}_z] = 0. \tag{8.2}$$

The quantum numbers s and m_s have the values 1/2 and ±1/2. The two spin eigenfunctions are often labelled by the symbols α and β, where

$$|\alpha\rangle = \left|\frac{1}{2} \frac{1}{2}\right\rangle \tag{8.3a}$$

and

$$|\beta\rangle = \left|\frac{1}{2} -\frac{1}{2}\right\rangle. \tag{8.3b}$$

One can also define raising and lowering operators $\hat{S}_\pm$ whose properties are analogous to $\hat{\ell}_\pm$, thus giving

$$\hat{S}_+|\beta\rangle = \hbar|\alpha\rangle, \tag{8.4a}$$

$$\hat{S}_+|\alpha\rangle = 0, \tag{8.4b}$$

$$\hat{S}_-|\alpha\rangle = \hbar|\beta\rangle, \tag{8.4c}$$

and

$$\hat{S}_-|\beta\rangle = 0, \tag{8.4d}$$

and from these we can derive $\left(\text{with } \hat{S}_{\pm} = \hat{S}_x \pm i\hat{S}_y\right)$

$$\hat{S}_x|\alpha\rangle = \frac{1}{2}\hbar|\beta\rangle, \tag{8.5a}$$

$$\hat{S}_y|\alpha\rangle = \frac{i}{2}\hbar|\beta\rangle, \tag{8.5b}$$

$$\hat{S}_x|\beta\rangle = \frac{1}{2}\hbar|\alpha\rangle, \tag{8.5c}$$

and

$$\hat{S}_y|\beta\rangle = \frac{-i}{2}\hbar|\alpha\rangle. \tag{8.5d}$$

8.2 THE PAULI PRINCIPLE

The Pauli principle refers to the properties of the wavefunction with respect to the interchange of identical particles. It is convenient to express this principle using an operator $\hat{P}_{12}$, which permutes the particles 1 and 2 as follows:

$$\hat{P}_{12}\psi(x_1, x_2) = \psi(x_2, x_1). \tag{8.6}$$

Note in this equation that the coordinates x_1 and x_2 are both the *space* and *spin* coordinates of the particles.

The importance of symmetry in quantum mechanics derives from the fact that the Hamiltonian operator is *invariant* to the permutation of any two identical particles. From this, it follows that $\hat{H}$ and $\hat{P}_{12}$ must commute; that is,

$$[\hat{H}, \hat{P}_{12}] = 0. \tag{8.7}$$

From Chapter 2, we know that if $\hat{H}$ and $\hat{P}_{12}$ commute, they may have *simultaneous* eigenfunctions. This means that the solutions of the Schrödinger equation may be chosen to be eigenfunctions of $\hat{P}_{12}$. But what *are* the possible eigenvalues of $\hat{P}_{12}$? Well, if

$$\hat{P}_{12}\psi(x_1, x_2) = p_{12}\psi(x_1, x_2), \tag{8.8}$$

where p_{12} is an eigenvalue, then applying $\hat{P}_{12}$ twice gives

$$\hat{P}_{12}^2\psi(x_1, x_2) = p_{12}^2\psi(x_1, x_2) = \psi(x_1, x_2). \tag{8.9}$$

The last form arises from Eq. (8.6) because permuting the same two particles twice returns them to their original location. Equation (8.9) implies that the eigenvalues should be

$$p_{12} = \pm 1, \tag{8.10}$$

which means that wavefunctions can be either symmetric or antisymmetric with respect to the permutation of any two identical particles. Note that both of these eigenvalues are physically possible, since the observable quantity $|\psi|^2$ is in both cases invariant to the permutation of identical particles; that is,

$$|\hat{P}_{12}\psi|^2 = |\psi|^2. \tag{8.11}$$

Now we ask the question, "Are there any restrictions on the value of p_{12}?" Nothing in the theory developed so far forces us to choose one value over the other, but experimental observation reveals that the allowed values of p_{12} are constrained according to the *Pauli principle*, given in the following postulate:

> **POSTULATE 6:** There exists an intrinsic angular momentum called *spin* for each elementary particle. For integer spins, the particles are called *bosons*, and $\hat{P}_{12}\psi = \psi$. For half-integer spin, the particles are called *fermions*, and $\hat{P}_{12}\psi = -\psi$.

Particles with half-integer spins ($s = 1/2$) include electrons, protons, and ^{3}He nuclei. Particles with integer spins include photons ($s = 1$), ^{4}He nuclei (spin 0), and deuterium nuclei (spin 1). Note that the Pauli principle refers to the symmetry of the wavefunction with respect to the interchange of both space and spin coordinates of the particles. This means that our treatment of the He atom in Chapter 7 was incomplete (though, fortunately, not incorrect), so it is now appropriate to consider again what the proper wavefunctions are.

8.3   He-ATOM WAVEFUNCTIONS, INCLUDING SPIN

The Hamiltonian operator (Eq. 7.3) for the He atom does not include spin, which means that its eigenfunctions will be products of spatial and spin parts. If we use the approximation that the spatial part is represented by $1s$ orbitals for each electron, we can write the following four possible wavefunctions for the ground state of He:

$$\Phi_1 = \phi_{1s}(r_1)\alpha_1\phi_{1s}(r_2)\beta_2; \tag{8.12a}$$

$$\Phi_2 = \phi_{1s}(r_1)\beta_1\phi_{1s}(r_2)\alpha_2; \tag{8.12b}$$

$$\Phi_3 = \phi_{1s}(r_1)\alpha_1\phi_{1s}(r_2)\alpha_2; \tag{8.12c}$$

$$\Phi_4 = \phi_{1s}(r_1)\beta_1\phi_{1s}(r_2)\beta_2. \tag{8.12d}$$

However, these wavefunctions do not satisfy the Pauli principle. To find ones that do, we take advantage of the fact that

$$\psi = (1 - \hat{P}_{12})\Phi_{1,2,3,\,\text{or}\,4} \tag{8.13}$$

will always be antisymmetric with respect to the interchange of electrons 1 and 2. (To prove this, apply $\hat{P}_{12}$ to this wavefunction, using the fact that $p_{12}^2 = 1$.) If Eqs. (8.12) are now substituted into Eq. (8.13), we obtain

$$(1 - \hat{P}_{12})\Phi_3 = (1 - \hat{P}_{12})\Phi_4 = 0, \tag{8.14a}$$

$$\hat{P}_{12}\Phi_1 = \Phi_2, \tag{8.14b}$$

and

$$\hat{P}_{12}\Phi_2 = \Phi_1. \tag{8.14c}$$

This means that the only wavefunction for the ground state of He that is consistent with the Pauli principle is

$$\psi = N(1 - \hat{P}_{12})\Phi_1 = N\phi_{1s}(1)\phi_{1s}(2)(\alpha_1\beta_2 - \beta_1\alpha_2), \tag{8.15}$$

where N is a normalization constant. This wavefunction has the same spatial part as that considered in Chapter 7, so if we were to calculate the energy (by evaluating the expectation value of H), we would obtain the same result (all the spin parts integrate out, since H does not operate on spin coordinates). This means that all of the effort we expended in Chapter 7 to solve the Schrödinger equation as accurately as possible was not wasted. However, spin does influence the energy of other states of He, as we shall see later.

To normalize Eq. (8.15) we evaluate

$$\langle\psi|\psi\rangle = N^2\langle\phi_{1s}(1)|\phi_{1s}(1)\rangle\langle\phi_{1s}(2)|\phi_{1s}(2)\rangle\langle\alpha_1\beta_2 - \beta_1\alpha_2|\alpha_1\beta_2 - \beta_1\alpha_2\rangle = 1. \tag{8.16}$$

The spatial wavefunctions are already normalized, as are the spin functions, so we get

$$N^2\{\langle\alpha_1\beta_2 - \beta_1\alpha_2|\alpha_1\beta_2 - \beta_1\alpha_2\rangle\} = 2N^2 = 1, \tag{8.17}$$

or

$$N = \frac{1}{\sqrt{2}}. \tag{8.18}$$

It is useful to consider the properties of the spin function that we have just examined with respect to the *total* spin operators $\hat{S}^2$ and $\hat{S}_z$. These operators are related to the single-electron operators $\hat{S}_1$ and $\hat{S}_2$ by the vector-addition formula,

$$\mathbf{S} = \mathbf{S}_1 + \mathbf{S}_2. \tag{8.19}$$

(The "hats" on the spin operators are omitted hereafter.) The z-component of this formula is

$$S_z = S_{1z} + S_{2z}, \tag{8.20}$$

so applying Eq. (8.20) the spin function $(\alpha_1\beta_2 - \beta_1\alpha_2)$ yields

$$S_z\psi = (S_{1z} + S_{2z})(\alpha_1\beta_2 - \beta_1\alpha_2) =$$

$$\frac{\hbar}{2}\alpha_1\beta_2 - \frac{\hbar}{2}\alpha_1\beta_2 + \frac{\hbar}{2}\beta_1\alpha_2 - \frac{\hbar}{2}\beta_1\alpha_2 = \hbar \cdot 0 \cdot \psi = 0, \tag{8.21}$$

since this function always has an α spin wavefunction for one electron paired with a β spin wavefunction for the other electron. Now consider the operator $\mathbf{S}^2$

that can also be derived from Eq. (8.19). Using Eq. (4.72) and applying that operator to the spin function gives

$$\mathbf{S}^2\psi = (\mathbf{S}_1 + \mathbf{S}_2)^2\psi = (\mathbf{S}_1^2 + \mathbf{S}_2^2 + 2\mathbf{S}_1 \cdot \mathbf{S}_2)\psi$$
$$= (\mathbf{S}_1^2 + \mathbf{S}_2^2 + 2S_{z1}S_{z2} + S_{+1}S_{+2} + S_{-1}S_{+2})\psi \tag{8.22}$$
$$= \hbar^2\left\{\frac{3}{4} + \frac{3}{4} - \frac{2}{4} - 1\right\}\psi = 0.$$

To derive this result, we used the formula

$$(S_{+1}S_{-2} + S_{-1}S_{+2})(\alpha_1\beta_2 - \beta_1\alpha_2) = [-\alpha_1\beta_2 + \beta_1\alpha_2]\hbar = (-\hbar)(\alpha_1\beta_2 - \beta_1\alpha_2). \tag{8.23}$$

We conclude that the wavefunction

$$\psi = \frac{1}{2}(\phi_{1s}(1)\phi_{1s}(2))(\alpha_1\beta_2 - \beta_1\alpha_2) \tag{8.24}$$

is an eigenfunction of S_z and $\mathbf{S}^2$, with an eigenvalue of zero in both cases. We will call this a *singlet state*, because the spin multiplicity (the number of independent states associated with the same value of $\mathbf{S}^2$) is unity. For a more general spin state wherein

$$\mathbf{S} = \mathbf{S}_1 + \mathbf{S}_2 + \mathbf{S}_3 + \cdots \tag{8.25}$$

and the total spin eigenfunctions satisfy the equation

$$\hat{S}^2|SM_S\rangle = \hbar^2 S(S + 1)|SM_S\rangle, \tag{8.26}$$

the spin multiplicity is $(2S + 1)$, and the states are labelled as follows:

$$\begin{cases} S = 0 & \text{(singlet)} \\ S = \dfrac{1}{2} & \text{(doublet)} \\ S = 1 & \text{(triplet)} \\ S = \dfrac{3}{2} & \text{(quartet)} \end{cases} \tag{8.27}$$

Since the nonrelativistic Hamiltonian is independent of spin, it will commute with $\mathbf{S}^2$ and S_z. Thus,

$$[H, S_z] = [H, \mathbf{S}^2] = [\mathbf{S}^2, S_z] = 0, \tag{8.28}$$

which means that simultaneous eigenfunctions may be defined. It is therefore *convenient* to label the states by their spin multiplicity. This labelling is useful even when the wavefunction we have developed is not an exact eigenfunction of H, as is the case with Eq. (8.24). Indeed, since this wavefunction is already an eigenfunction of $\mathbf{S}^2$ and S_z, and since the space and spin parts are multiplicative, improvements in the space part using, say, variational methods can be introduced without changing the spin part in any way. This is fortunate, for it means that the steps we outlined in Chapter 7 are still correct.

EXERCISE 8.1 What are $[S_x, \mathbf{S}^2]$ and $[S_y, \mathbf{S}^2]$? Is your result consistent with the fact that, in free space, no preferred axis system is defined?

Answer $\mathbf{S}^2 = S_x^2 + S_y^2 + S_z^2$, so that

$$[S_x, \mathbf{S}^2] = [S_x, S_x^2] + [S_x, S_y^2] + [S_x, S_z^2]$$

$$= (S_y[S_x, S_y] + [S_x, S_y]S_y) + (S_z[S_x, S_z] + [S_x, S_z]S_z)$$

$$= i\hbar(S_y S_z + S_z S_y) - i\hbar(S_z S_y + S_y S_z)$$

$$= 0.$$

One can similarly show that $[S_y, \mathbf{S}^2] = 0$, and the same procedure leads to $[S_z, \mathbf{S}^2] = 0$. These results are consistent with the fact that in free space there is no distinction among x-, y-, and z-axes. We choose the z-axis for the definition of spin components only by arbitrary convention.

There is one additional complication. If we allow for relativistic effects like spin–orbit coupling, then the complete Hamiltonian depends on spin; that is,

$$H_{\text{actual}} \cong H + H_{\text{relativistic}}(x, p, S), \tag{8.29}$$

and therefore, $[H_{\text{actual}}, \mathbf{S}^2] \neq 0$. This means that the correct eigenfunctions no longer have a well-defined spin multiplicity, which in fact gives rise to processes like phosphorescence, intersystem crossing, etc., where the spin state of the atom changes. For light atoms like He, the effect on energies is very small, so we neglect it. However, in Chapter 9 we return to this topic.

8.4   EXCITED STATE OF He

The first excited state of He can be approximated by promoting one of the $1s$ electrons to a $2s$ orbital. The configuration associated with this state is denoted He($1s, 2s$), and a reasonable approximation to the spatial part of the wavefunction is simply the product $\phi_{1s}(1)\phi_{2s}(2)$. To allow for electron spin, we form product wavefunctions analogous to those in Eq. (8.12) as follows:

$$\Phi_1 = \phi_{1s}(1)\phi_{2s}(2)\alpha_1\alpha_2; \tag{8.30a}$$

$$\Phi_2 = \phi_{1s}(1)\phi_{2s}(2)\beta_1\beta_2; \tag{8.30b}$$

$$\Phi_3 = \phi_{1s}(1)\phi_{2s}(2)\alpha_1\beta_2; \tag{8.30c}$$

$$\Phi_4 = \phi_{1s}(1)\phi_{2s}(2)\beta_1\alpha_2. \tag{8.30d}$$

None of these functions obeys the Pauli principle. However, we can generate wavefunctions that do by using equation (8.13):

$$\frac{1}{\sqrt{2}}(1 - P_{12})\Phi = \psi'.$$ (8.31)

This equation generates the following wavefunctions:

$$\psi_1' = \frac{1}{\sqrt{2}}(\alpha_1 \alpha_2)(\phi_{1s}(1)\phi_{2s}(2) - \phi_{2s}(1)\phi_{1s}(2));$$ (8.32a)

$$\psi_2' = \frac{1}{\sqrt{2}}(\beta_1 \beta_2)(\phi_{1s}(1)\phi_{2s}(2) - \phi_{2s}(1)\phi_{1s}(2));$$ (8.32b)

$$\psi_3' = \frac{1}{\sqrt{2}}(\alpha_1\beta_2\phi_{1s}(1)\phi_{2s}(2) - \beta_1\alpha_2\phi_{2s}(1)\phi_{1s}(2));$$ (8.32c)

$$\psi_4' = \frac{1}{\sqrt{2}}(\beta_1\alpha_2\phi_{1s}(1)\phi_{2s}(2) - \alpha_1\beta_2\phi_{2s}(1)\phi_{1s}(2)).$$ (8.32d)

All of these functions obey the Pauli principle; however, it turns out that this is not sufficient to give us eigenfunctions that are acceptable in all respects. Consider, for example, the operator $\mathbf{S}^2$ of Eq. (8.22):

$$\mathbf{S}^2 = \mathbf{S}_1^2 + \mathbf{S}_2^2 + 2S_{z1}S_{z2} + S_{+1}S_{-2} + S_{-1}S_{+2}.$$ (8.33)

If this operator is now applied to the products of two spin functions that appear in Eqs. (8.32), we obtain

$$\mathbf{S}^2(\alpha_1\alpha_2) = \hbar^2\left(\frac{3}{4} + \frac{3}{4} + \frac{2}{4}\right)(\alpha_1\alpha_2) = \hbar^2 2(\alpha_1\alpha_2),$$ (8.34a)

$$\mathbf{S}^2(\beta_1\beta_2) = \hbar^2\left(\frac{3}{4} + \frac{3}{4} + \frac{2}{4}\right)(\beta_1\beta_2) = \hbar^2 2(\beta_1\beta_2),$$ (8.34b)

$$\mathbf{S}^2(\alpha_1\beta_2) = \hbar^2\left(\frac{3}{4} + \frac{3}{4} - \frac{2}{4}\right)(\alpha_1\beta_2) + \hbar^2(\beta_1\alpha_2) = \hbar^2(\alpha_1\beta_2 + \beta_1\alpha_2),$$ (8.34c)

and

$$\mathbf{S}^2(\beta_1\alpha_2) = \hbar^2\left(\frac{3}{4} + \frac{3}{4} - \frac{2}{4}\right)(\beta_1\alpha_2) + \hbar^2(\alpha_1\beta_2) = \hbar^2(\alpha_1\beta_2 + \beta_1\alpha_2).$$ (8.34d)

Clearly, the third and fourth of these products are not eigenfunctions of $\mathbf{S}^2$. However, if we form the linear combinations $\alpha_1\beta_2 \pm \beta_1\alpha_2$, we find that

$$\mathbf{S}^2(\alpha_1\beta_2 + \beta_1\alpha_2) = 2\hbar^2(\alpha_1\beta_2 + \beta_1\alpha_2)$$ (8.35a)

and

$$\mathbf{S}^2(\alpha_1\beta_2 - \beta_1\alpha_2) = 0. \tag{8.35b}$$

This means that the *correct* eigenfunctions of $\mathbf{S}^2$ and S_z are

$$S = 1 \quad \begin{cases} \alpha_1\alpha_2 \\ \beta_1\beta_2 \\ \dfrac{1}{\sqrt{2}}(\alpha_1\beta_2 + \beta_1\alpha_2) \end{cases} \tag{8.36a}$$

and

$$S = 0 \quad \frac{1}{\sqrt{2}}(\alpha_1\beta_2 - \beta_1\alpha_2). \tag{8.36b}$$

The $S = 0$ function is just the *singlet* spin state analogous to what we found for the ground state of He, while the $S = 1$ state is of *triplet* multiplicity. Note that for the triplet, the electrons can have $m_s = 1/2$ for both electrons, $-1/2$ for both, or a symmetric mixture of $+1/2$ and $-1/2$.

If we now form the linear combinations $\psi_3' \pm \psi_4'$ of Eqs. (8.32), we find that the space and spin parts may be factored. The spin parts are the singlet and triplet wavefunctions just defined. Furthermore, the space part of the triplet is just the same as for ψ_1' and ψ_2'. This means that we can group the excited states of He into the singlet state

$$\psi_{\text{singlet}} = \frac{1}{2}(\alpha_1\beta_2 - \beta_1\alpha_2)(\phi_{1s}(1)\phi_{2s}(2) + \phi_{2s}(1)\phi_{1s}(2)) \tag{8.37}$$

and the three triplets

$$\psi_{\text{triplet}} = \frac{1}{\sqrt{2}}[\phi_{1s}(1)\phi_{2s}(2) - \phi_{2s}(1)\phi_{1s}(2)] \cdot \begin{cases} \alpha_1\alpha_2 \\ \beta_1\beta_2 \\ \dfrac{1}{\sqrt{2}}(\alpha_1\beta_2 + \beta_1\alpha_2) \end{cases}. \tag{8.38}$$

Both the singlet and triplet states are consistent with the Pauli principle, but the spatial part of the singlet state is symmetric with respect to interchanging the electrons, while the spatial part of the triplet state is antisymmetric. These differences are important in determining the relative energies of the singlet and triplet states.

EXAMPLE 8.2   Define the operator $S_- = S_{1-} + S_{2-}$, and show that, like ℓ_- of Eqs. (4.78) and (4.83), it can produce the $|SM_S\rangle$ states, starting from $|SM_S = S\rangle$. Consider $S = 0$ and $S = 1$.

Answer   We apply S_- to the four possible spin states associated with $S = 0$ and $S = 1$:

$$\frac{S_-(\alpha_1\beta_2 - \beta_1\alpha_2)}{\sqrt{2}} = \frac{(S_{1-} + S_{2-})(\alpha_1\beta_2 - \beta_1\alpha_2)}{\sqrt{2}} = \frac{\hbar(\beta_1\beta_2 - \beta_1\beta_2)}{\sqrt{2}} = 0;$$

$$\frac{S_-(\alpha_1\beta_2 + \beta_1\alpha_2)}{\sqrt{2}} = \frac{(S_{1-} + S_{2-})(\alpha_1\beta_2 + \beta_1\alpha_2)}{\sqrt{2}} = \frac{\hbar(\beta_1\beta_2 + \beta_1\beta_2)}{\sqrt{2}} = \hbar\sqrt{2}(\beta_1\beta_2);$$

$$S_-(\alpha_1\alpha_2) = (S_{1-} + S_{2-})(\alpha_1\alpha_2) = \alpha_1\beta_2 + \beta_1\alpha_2;$$

$$S_-(\beta_1\beta_2) = 0.$$

All of these results are consistent with the equation

$$S_-|SM_S\rangle = \hbar\sqrt{S(S + 1) - M(M - 1)}\,|SM_S - 1\rangle,$$

where $|0 - 1\rangle$ and $|1 - 2\rangle$ are both impossible and therefore vanish.

8.5   ENERGIES OF He(1s2s) STATES

Now let us calculate the energy using the $M_s = 1$ triplet state defined in Eq. (8.38). The general formula is

$$E = \frac{1}{2}\langle[\phi_{1s}(1)\phi_{2s}(2) - \phi_{2s}(1)\phi_{1s}(2)]\alpha_1\alpha_2|H|\alpha_1\alpha_2[\phi_{1s}(1)\phi_{2s}(2) - \phi_{2s}(1)\phi_{1s}(2)]\rangle,$$

$$(8.39)$$

and since the Hamiltonian depends only on spatial coordinates, the space and spin parts of this expression may be separated to give

$$E = \frac{1}{2}\langle\alpha_1\alpha_2|\alpha_1\alpha_2\rangle\{\langle\phi_{1s}(1)\phi_{2s}(2)|H|\phi_{1s}(1)\phi_{2s}(2)\rangle + \langle\phi_{2s}(1)\phi_{1s}(2)|H|\phi_{2s}(1)\phi_{1s}(2)\rangle$$

$$- \langle\phi_{1s}(1)\phi_{2s}(2)|H|\phi_{2s}(1)\phi_{1s}(2)\rangle - \langle\phi_{2s}(1)\phi_{1s}(2)|H|\phi_{1s}(1)\phi_{2s}(2)\rangle\} \quad (8.40)$$

$$= \langle\phi_{1s}(1)\phi_{2s}(2)|H|\phi_{1s}(1)\phi_{2s}(2)\rangle - \langle\phi_{1s}(1)\phi_{2s}(2)|H|\phi_{2s}(1)\phi_{1s}(2)\rangle,$$

where we have interchanged some of the coordinate labels 1 and 2 in condensing the first form to the second. Because we are integrating over these coordinates, they can be relabelled as we wish.

Now we substitute the Hamiltonian from Eq. (7.3), viz.,

$$\hat{H} = h_1 + h_2 + \frac{1}{r_{12}}, \tag{8.41}$$

to give

$$E = \left\langle 1s(1)2s(2) \left| h_1 + h_2 + \frac{1}{r_{12}} \right| 1s(1)2s(2) \right\rangle$$
$$- \left\langle 1s(1)2s(2) \left| h_1 + h_2 + \frac{1}{r_{12}} \right| 2s(1)1s(2) \right\rangle, \tag{8.42}$$

or

$$E = \varepsilon_{1s} + \varepsilon_{2s} + J_{1s2s} - K_{1s2s}, \tag{8.43}$$

where

$$J_{1s2s} = \left\langle 1s(1)2s(2) \left| \frac{1}{r_{12}} \right| 1s(1)2s(2) \right\rangle \tag{8.44a}$$

and

$$K_{1s2s} = \left\langle 1s(1)2s(2) \left| \frac{1}{r_{12}} \right| 2s(1)1s(2) \right\rangle. \tag{8.44a}$$

The resulting expression consists of the sum of one-electron energies for the $1s$ and $2s$ orbitals, plus two terms J and K that arise from repulsion between the electrons. The J term, often called a *coulomb integral*, represents the repulsion between the $1s$ and $2s$ charge clouds and is always positive. The K term, called an *exchange integral*, arises from the required antisymmetry of the wavefunction. It does not appear if the simple product wavefunction $1s(1)\alpha_1 2s(2)\alpha_2$ is used to evaluate the energy; however, all three components of the triplet state in Eq. (8.38) give the same energy expression (8.43). The exchange integral represents the repulsion between the *overlap density* $1s2s$ and itself. The sign of such integrals is not guaranteed, but it is usually positive. Since K appears with a negative sign in Eq. (8.43), a positive exchange integral *lowers* the energy of the triplet state relative to what it would have been for the distinguishable electron wavefunction. Physically, this situation arises because the spatial part of the triplet wavefunction is antisymmetric with respect to interchanging the electrons. That is, the wavefunction vanishes when the two electrons have the same coordinates. This *exchange hole* in the wavefunction lessens the repulsion and lowers the total energy.

If we use the *singlet* wavefunction Eq. (8.37) to calculate the energy, we find that

$$E_{(singlet)} = \varepsilon_{1s} + \varepsilon_{2s} + J_{1s2s} + K_{1s2s}. \tag{8.45}$$

This expression is identical to that for the triplet energy, except that the exchange integral appears with the opposite sign, thus *raising* the energy relative to the triplet. The reason is that the spatial part of the singlet wavefunction is symmetric with respect to interchanging the electrons. This causes the electron density to be larger when the two electrons are close together than it would be for an antisymmetric wavefunction, so their repulsion is greater. In fact, the exchange term causes the repulsion to be larger than for a distinguishable electron wavefunction.

EXERCISE 8.3   Evaluate the ground-state energy of He, using notation similar to Eq. (8.45).

Answer

$$E = \langle \psi | H | \psi \rangle = \frac{1}{2} \langle (\alpha_1\beta_2 - \beta_1\alpha_2)(1s(1)1s(2)) | H | \langle (\alpha_1\beta_2 - \beta_1\alpha_2)(1s(1)1s(2)) \rangle$$

$$= \left\langle 1s(1)1s(2) \left| h_1 + h_2 + \frac{1}{r_{12}} \right| 1s(1)1s(2) \right\rangle$$

$$= \varepsilon_{1s} + \varepsilon_{1s} + \left\langle 1s(1)1s(2) \left| \frac{1}{r_{12}} \right| 1s(1)1s(2) \right\rangle$$

$$= 2\varepsilon_{1s} + J_{1s1s}.$$

8.6 INTERACTION OF ELECTRON SPIN WITH MAGNETIC FIELDS

Whenever a charged particle has either spin or orbital angular momentum, it has a magnetic moment that will interact with externally applied magnetic fields. For an electron orbiting a nucleus (say, in a hydrogen atom), it is easy to determine the magnetic moment from classical physics, which tells us that if a current I circulates around an area A, the magnitude of the magnetic moment equals the product of the current and the area (i.e., $M = IA$). [This is the correct expression in SI units. Had we used cgs units, an extra factor of $1/c$, where c is the speed of light, would be needed.] If we have an electron (charge $-e$) with velocity v revolving about a nucleus in a circular orbit of radius r, then the area is $A = \pi r^2$, and the current is $I = \text{charge}/(\text{orbital period}) = -e(2\pi r/v)^{-1}$, where $-e$ is the charge on the electron (1.6×10^{-19} C). Since the magnitude of the angular momentum in a circular orbit is $\ell = mvr$, we conclude that $I = -e\ell/(2\pi mr^2)$. Thus, $M = -e\ell/(2m) = \gamma_e\ell$, where γ_e is a negative constant called the *magnetogyric ratio* of the electron. This formula tells us that the magnitude of M is proportional to the angular momentum ℓ. The directions of the vectors associated with M and ℓ are the same, so we can write

$$\mathbf{M}(\text{orbital motion}) = \gamma_e\ell. \tag{8.46}$$

It turns out that this expression is also correct in the quantum description of magnetic moments, in which case $\mathbf{M}$ and ℓ are operators.

In considering spin angular momentum, the relationship between the magnetic momentum and the spin angular momentum **s** is less apparent than in Eq. (8.46), and in fact, the correct result can be obtained only from relativistic quantum electrodynamic theory. We shall not derive the formula here, but the final result is easy enough to write down, namely,

$$\mathbf{M}(\text{spin}) = 2.0023193044\gamma_e\mathbf{s}. \tag{8.47}$$

What Eq. (8.47) tells us is that the orbital and spin contributions to the magnetic moment have the same form, but spin has an anomalous coefficient that arises from relativistic and quantum electrodynamic effects.

The Hamiltonian associated with the interaction of a magnetic moment **M** with an externally applied magnetic field **B** is known as the *Zeeman* Hamiltonian and is given by

$$\hat{H} = -\mathbf{M} \cdot \mathbf{B}. \tag{8.48}$$

If we combine Eqs. (8.46) and (8.47) with this expression we get

$$\hat{H} = -\gamma_e \mathbf{B} \cdot (\boldsymbol{\ell} + 2.0023\mathbf{s}). \tag{8.49}$$

Taking **B** to be along the z-axis, we see that the expectation value of this Hamiltonian for a hydrogen atom is

$$E(\text{Zeeman}) = \langle \ell m_\ell s m_s | \hat{H} | \ell m_\ell s m_s \rangle = -\gamma_e \hbar B(m_\ell + 2.0023 m_s)$$
$$= \beta_e B(m_\ell + 2.0023 m_s), \tag{8.50}$$

where $\beta_e = e\hbar/(2m)$ is a positive constant known as the *Bohr magneton*. (See Appendix D for its value.)

Figure 8.1 shows how this expression for the Zeeman energy varies with field strength for two states of the hydrogen atom, namely, the ground state (in which $m_\ell = 0$) and a p state (wherein $m_\ell = 0, \pm 1$). The figure also shows that the magnetic field lifts the degeneracy of the states, and the spacing between the states increases linearly with the magnitude of the field. For a field of 1 tesla (T) = 10^4 gauss, this splitting for the ground state is 9.27×10^{-24} J, corresponding to an energy of 5.58×10^{-3} kJ/mol = 0.47 cm^{-1}, or a frequency of 1.4×10^{10} s^{-1}. Equation (8.50) gives only the first-order perturbation theory expression for the change in energy of the atom due to the Zeeman interaction.

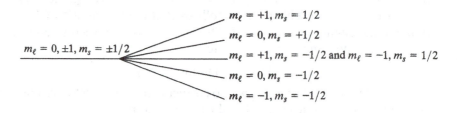

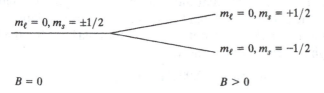

$B = 0$ $B > 0$

Figure 8.1   Energy levels of selected states of the hydrogen atom in a magnetic field. Because g is not exactly 2.0000, the $m_\ell = +1$, $m_s = -1/2$ and $m_\ell = -1$, $m_s = 1/2$ states are not precisely degenerate.

However, for typical laboratory fields, with a strength of 1T, the splitting that is noted is very small compared to, say, the Coulomb interaction between the electron and nucleus, so first-order perturbation theory should be quite accurate.

8.7 EPR AND NMR

The splitting between the states shown in Fig. 8.1 is what one measures in an electron paramagnetic resonance (EPR) spectroscopy experiment. To do this, one applies a microwave field to the atom, in addition to the constant magnetic field **B**. When $\hbar$ times the frequency of the microwave field matches the splitting between states, transitions occur between the states, causing power to be absorbed and emitted.

In molecules, the orbital contribution to the magnetic moment of the electrons is almost completely quenched (as the electrons are no longer able to move in simple circular orbits), but the spin magnetic moment is largely unperturbed. As a result, the Zeeman Hamiltonian can be described by the expression

$$\hat{H} = -g\gamma_e \mathbf{B} \cdot \mathbf{s}, \tag{8.51}$$

where g is a constant that is close, but not quite equal, to its free-electron value of 2.0023. The deviation from the free-electron g-value provides a sensitive measure of the local magnetic environment in which the electron is located and is an important component of an EPR experiment. Equation (8.51) applies best to molecules in gases and liquids, where they are able to rotate freely. For paramagnetic atoms and molecules in solids, Eq. (8.51) is replaced by $\hat{H} = -\gamma \mathbf{B} \cdot g \cdot \mathbf{s}$, where g is now a 3×3 matrix (tensor) whose components depend on the orientation of the molecule relative to the field. EPR measurements are also sensitive to the magnetic coupling between different electron spins (spin–spin interaction) and between electrons and nuclei (hyperfine interaction).

In nuclear magnetic resonance (NMR) spectroscopy, one observes the analogous interaction between the magnetic moment of the nuclei and an externally applied magnetic field. Here, the Hamiltonian is given by

$$\hat{H} = -\gamma_N \mathbf{B} \cdot \mathbf{I}, \tag{8.52}$$

where **I** is the nuclear spin operator and γ_N is the nuclear magnetogyric ratio (which is now a positive constant). This ratio can be related to a nuclear magneton β_N via the equation $g_N \beta_N = \gamma_N \hbar$, with g_N being a nuclear "g" factor. Like its electron counterpart, the nuclear magneton is defined by $\beta_N = e\hbar/(2m_p)$, where m_p is the mass of the proton. There is no simple theory of nuclear g factors, so they are treated as empirical parameters. For a proton, $g_N = 5.586$.

An important feature of nuclear Zeeman interactions is that they are much smaller than their electronic Zeeman counterparts. This can be determined from the ratio of the nuclear magneton to the Bohr magneton, which is just the inverse of the proton–electron mass ratio. Using the values in Appendix D, we find that the ratio is 0.0006, which means that the splitting of nuclear spin states will be more than a thousand times smaller than that of electron spin states. This means that radio frequencies, rather than microwave frequencies, can be used in NMR spectrometers.

As with EPR, the energy splitting in NMR depends on the local magnetic environment near each nucleus. To incorporate this feature, one typically rewrites Eq. (8.52) as

$$\hat{H} = -(1 - \sigma)\gamma_N \mathbf{B} \cdot \mathbf{I}, \tag{8.53}$$

where the chemical shift σ expresses the deviation between the Zeeman interaction associated with some reference nucleus and that for the nucleus of interest.

EXERCISE 8.4   A nucleus with spin $m\hbar$ generates a magnetic field at a distance R. This field falls off as R^{-3}, with the form for the z-component being

$$\mathbf{B} = -\frac{m\mu_0\beta_N}{4\pi}\frac{(1 - 3\cos^2\theta)}{R^3},$$

where θ is the angle between the z-axis and the orientation of the spin. Show that the average value of B vanishes. This result is important experimentally, since it shows that dipolar interactions average to zero in a freely tumbling liquid, but not in a solid.

Answer

$$\langle B \rangle = \int B(\theta)\sin\theta d\theta \propto \int_0^\pi (1 - 3\cos^2\theta)\sin\theta d\theta$$

$$= -\cos\theta\Big|_0^\pi + 3 \cdot \frac{1}{3}\cos^3\theta\Big|_0^\pi = 0.$$

This result is physically reasonable: Any angular interaction might be expected to average to zero.

SUGGESTED READINGS

I. N. Levine, *Quantum Chemistry*, 5th ed. (Prentice Hall, Upper Saddle River, NJ, 2000).

J. P. Lowe, *Quantum Chemistry*, 2nd ed. (Academic Press, New York, 1993).

A. Carrington and A. D. McLachlan, *Introduction to Magnetic Resonance* (Harper & Row, New York, 1967).

C. P. Slichter, *Principles of Magnetic Resonance*, 3rd ed. (Springer, Berlin, 1990).

PROBLEMS

8.1 Which of the following are valid wavefunctions for He? Ignore normalization.

(a) $1s(1)2s(2)(\alpha_1\beta_2 - \beta_1\alpha_2)$.

(b) $\left[1s(1)2s(1) - 2s(1)1s(1)\right]\alpha_1\alpha_2$.

(c) $\left[1s(1)2s(2) - 2s(1)1s(2)\right](\alpha_1\beta_2 - \beta_1\alpha_2)$.

(d) $1s(1)2s(2)\alpha_1\beta_2 - 2s(1)1s(2)\beta_1\alpha_2 + 1s(1)2s(2)\beta_1\alpha_2 - 2s(1)1s(2)\alpha_1\beta_2$.

8.2 Use the following diagram to plot the He $1s2s$ triplet wavefunction against the angle θ for $r_1 = r_2$:

8.3 Write down the complete expression for the coulomb integral that describes repulsion between a $2s$ function and a $3s$ function. Express your answer using polar coordinates. Be sure to give the integration volume and limits on integrals. You do not have to evaluate the integral explicitly.

8.4 If there is a Coulomb integral of the form

$$\langle 1s(1)2p_z(2)\,|\,1/r_{12}\,|\,1s(1)2p_z(2)\rangle,$$

what should the corresponding exchange integral be? Are either of these integrals zero because of symmetry?

8.5 Consider the wavefunction $\psi = N(\alpha + c\beta)$, where α and β are spin functions and N and c are constants.

(a) Normalize ψ. The result will contain the constant c.

(b) Find $\langle s_z \rangle$ for the wavefunction ψ.

(c) If a spin system has the Hamiltonian $H = As_z + Bs_x$, where A and B are constants, what is the value of the constant c that makes ψ a solution to the equation $H\psi = E\psi$? Also, what is the energy E for this solution?

8.6 Evaluate $\langle \alpha|\, s_x^4 + s_y^4\,|\alpha\rangle$.

8.7 Let the spin angular momentum of an electron be denoted by **s**. One way to describe the properties of the **s** operators is to represent them by using Pauli matrices. The operators are defined as

$$s_x = \frac{1}{2}\hbar\sigma_x, \; s_y = \frac{1}{2}\hbar\sigma_y, \quad \text{and} \quad s_z = \frac{1}{2}\hbar\sigma_z,$$

where

$$\sigma_x = \begin{pmatrix} 0 & 1 \\ 1 & 0 \end{pmatrix}, \sigma_y = \begin{pmatrix} 0 & -i \\ i & 0 \end{pmatrix}, \quad \text{and} \quad \sigma_z = \begin{pmatrix} 1 & 0 \\ 0 & -1 \end{pmatrix}.$$

(a) Show that the **s** operators satisfy the usual commutation relations for angular momentum, such as $[s_x, s_y] = i\hbar s_z$.

(b) Find s^2, and show that the diagonal elements of s^2 and s_z correspond to angular momentum eigenvalues with $s = 1/2$.

(c) Reduce the operator $s_x^2 s_y^2 s_z^2$ to a single spin operator.

8.8 One very simple model for magnetic behavior is called the *Ising model*. It assumes that electron spins exist on separated sites and that the only interactions among them are between nearest neighbors. In that case, the Hamiltonian is

$$H_{\text{Ising}} = H = J \sum_{i,j}{}' s_Z^{(i)} s_Z^{(j)},$$

where J is an exchange-type integral, $s_Z^{(i)}$ is the Z-component of the spin on site i, and the sum runs only over nearest neighbors.

For the case of two sites on a line, there are four possible eigenstates of H. (For instance, one is $|\alpha_1\beta_2\rangle$.) Draw the energy-level diagram of these four eigenstates for the two cases of $J < 0$ (a ferromagnet) and $J > 0$ (an antiferromagnet). For which of these is the total spin greater in the ground state?

8.9 Calculate the magnetic field needed for proton NMR if the radiofrequency field is 500 MHz?

8.10 If the chemical shift for a proton NMR line is 1 part per million (ppm), what is the change in magnetic field associated with this shift? Assume that the measurements are being done in a 500-Mhz spectrometer.

8.11 The Fermi contact interaction is a coupling between the nuclear and electron spins that arises because electrons in s-orbitals have a finite probability of being inside the nucleus. The interaction Hamiltonian in this case has the form

$$H(\text{fermi}) = A\mathbf{s} \cdot \mathbf{I},$$

where A is a constant and $\mathbf{s}$ and $\mathbf{I}$ are the spin operators for the electron and nucleus, respectively.

Consider a hydrogen atom in its ground state. The constant A in this case is 0.04762 cm^{-1}. What is the expectation value of $H(\text{fermi})$ for the four possible combinations of electronic and nuclear spin states (i.e., $m_s = \pm 1/2$ and $m_N = \pm 1/2$)? What resonance wavelength is associated with EPR transitions (wherein the electron spin changes quantum numbers while the nuclear spin is unchanged)?

9

$$\psi(\text{Be}) = \left(\frac{1}{4!}\right)^{1/2} \begin{vmatrix} \phi_{1s}(1)\alpha_1 & \phi_{1s}(1)\beta_1 & \phi_{2s}(1)\alpha_1 & \phi_{2s}(1)\beta_1 \\ \phi_{1s}(2)\alpha_2 & \phi_{1s}(2)\beta_2 & \cdots & \cdots \\ \phi_{1s}(3)\alpha_3 & \phi_{1s}(3)\beta_3 & \cdots & \cdots \\ \phi_{1s}(4)\alpha_4 & \phi_{1s}(4)\beta_4 & \cdots & \phi_{2s}(4)\beta_4 \end{vmatrix}$$

Many-Electron Atoms

In the nonrelativistic limit, the hydrogen atom can be solved exactly. This is not true, however, of any system with more than two particles, including all molecules and all atoms with two or more electrons. The formulation of antisymmetry in terms of Slater determinants and the mean field (Hartree–Fock) method are first introduced for atoms and will be necessary for an understanding of molecular behavior as well. The electronic structure of many-electron atoms can also be described in terms of orbital and spin angular momenta, and this in turn leads to a description of atoms that is useful in understanding their magnetic and electrical properties.

9.1 MANY-ELECTRON HAMILTONIAN AND SCHRÖDINGER EQUATION

For an atom with n electrons, the electronic Hamiltonian is

$$\hat{H} = -\frac{1}{2}\sum_i \nabla_i^2 - \sum_i \frac{Z}{r_i} + \sum_i \sum_{j<i} \frac{1}{r_{ij}} \quad \text{(in atomic units)}, \qquad (9.1)$$

where the first term is the sum of kinetic-energy operators for all the electrons, the second describes the coulomb attraction between the electrons and the nucleus, and the third describes interelectronic repulsion. The Schrödinger equation is

$$\hat{H}\psi(r_1, r_2, \ldots, r_n) = E\psi(r_1, r_2, \ldots, r_n). \qquad (9.2)$$

As with the He atom, the method separation of variables cannot be used to solve this equation, because of the interelectronic repulsion term. The primary focus of this chapter will be on developing the simplest approximate solutions based on wavefunctions that are products of orbitals. To get started with this task, we need to learn how to write down such products and still satisfy the Pauli principle.

9.2 SLATER DETERMINANTS

The wavefunctions developed in Chapter 8 were worked out the hard way, starting from products of distinguishable electron orbitals, followed by antisymmetrization and then finding linear combinations of the results that have the desired spin properties. Even the antisymmetrization becomes very tedious in treating many-electron atoms, so it is convenient to introduce an alternative way to construct wavefunctions that automatically satisfy the Pauli principle. This approach is based on a basic property of all determinants which states that if any two rows or columns of a determinant are interchanged, the sign of the determinant changes. The following simple example illustrates this property for a 2×2 determinant in which the rows are interchanged:

$$\begin{vmatrix} 1 & 2 \\ 3 & 4 \end{vmatrix} = -2; \tag{9.3a}$$

$$\begin{vmatrix} 3 & 4 \\ 1 & 2 \end{vmatrix} = +2. \tag{9.3b}$$

So if a wavefunction has the form of a *determinant* of orbitals, with each electron occupying a different row (or column) of the determinant, then it *automatically* satisfies the Pauli principle. Consider the $m_s = 1$ triplet state of He($1s2s$). From Eq. (8.38), a 2×2 determinant can be developed as follows:

$$\psi_1 = \frac{\alpha_1 \alpha_2}{\sqrt{2}} \left(\phi_{1s}(1)\phi_{2s}(2) - \phi_{2s}(1)\phi_{1s}(2) \right) = \frac{1}{\sqrt{2}} \begin{vmatrix} \phi_{1s}(1)\alpha_1 & \phi_{2s}(1)\alpha_1 \\ \phi_{1s}(2)\alpha_2 & \phi_{2s}(2)\alpha_2 \end{vmatrix}. \tag{9.4}$$

This form can be generalized to describe wavefunctions with any number of electrons. For example, a wavefunction for the Be ground state, in which the electron *configuration* (the list of orbital occupations) is $1s^2 2s^2$, is

$$\psi(\text{Be}) = \left(\frac{1}{4!}\right)^{1/2} \begin{vmatrix} \phi_{1s}(1)\alpha_1 & \phi_{1s}(1)\beta_1 & \phi_{2s}(1)\alpha_1 & \phi_{2s}(1)\beta_1 \\ \phi_{1s}(2)\alpha_2 & \phi_{1s}(2)\beta_2 & \cdots & \cdots \\ \phi_{1s}(3)\alpha_3 & \phi_{1s}(3)\beta_3 & \cdots & \cdots \\ \phi_{1s}(4)\alpha_4 & \phi_{1s}(4)\beta_4 & \cdots & \phi_{2s}(4)\beta_4 \end{vmatrix}. \tag{9.5}$$

The $(4!)^{-1/2}$ prefactor is for normalization and arises because an $n \times n$ determinant has $n!$ terms when it is expanded, each of which integrates to give unity if the orbitals are orthonormal. Each row in Eq. (9.5) consists of a list of all the occupied orbitals for a particular electron. In the case of the Be ground state, the electrons are occupied according to the *Aufbau* principle, with one α and one β spin electron for each spatial orbital. The *Aufbau* (German for *building-up*) principle requires that electrons occupy spin orbitals (the product of a space orbital ϕ and a spin function α or β) such that the lowest energy spin orbital consistent with the Pauli principle is chosen.

These determinantal wavefunctions, called Slater determinants, provide a powerful way to generate antisymmetrized products of orbitals; however, they are not really the correct wavefunctions of many-electron atoms, for two reasons. First, as we have seen for the ground state of He, writing a wavefunction as an antisymmetrized product of one-electron orbitals is only an approximation; it ignores correlation between the electron motions. Second, an individual Slater determinant is not generally an eigenfunction of $\mathbf{S}^2$ (although it is for closed shells). This is easily seen for the $M_S = 0$ triplet excited state of He, which must be written as the sum of two Slater determinants:

$$\psi_0 = \frac{1}{2}(\alpha_1\beta_2 + \beta_1\alpha_2)(\phi_{1s}(1)\phi_{2s}(2) - \phi_{2s}(1)\phi_{1s}(2))$$

$$= \frac{1}{\sqrt{2}}\left\{\frac{1}{\sqrt{2}}\begin{vmatrix}\phi_{1s}(1)\alpha_1 & \phi_{2s}(1)\beta_1 \\ \phi_{1s}(2)\alpha_2 & \phi_{2s}(2)\beta_2\end{vmatrix} + \frac{1}{\sqrt{2}}\begin{vmatrix}\phi_{1s}(1)\beta_1 & \phi_{2s}(1)\alpha_1 \\ \phi_{1s}(2)\beta_2 & \phi_{2s}(2)\alpha_2\end{vmatrix}\right\}. \tag{9.6}$$

In spite of these limitations, wavefunctions written in terms of Slater determinants are extremely useful, first because they provide a simple physical picture of electron configurations and second because they provide the starting point for the *Hartee–Fock* approximation, which is one of the most useful methods for determining electronic structure in all of quantum chemistry.

Several abbreviations are in common use for writing Slater determinants. One example is the replacement of the α and β spin functions by unbarred and barred spatial wavefunctions; for example, $1s\alpha2s\beta$ becomes $1s\overline{2s}$. In this notation, the Be-atom Slater determinant that we just wrote becomes

$$\frac{1}{\sqrt{4!}}\begin{vmatrix}1s(1) & \overline{1s}(1) & 2s(1) & \overline{2s}(1)\\ 1s(2) & & & \vdots \\ 1s(3) & & & \overline{2s}(3)\\ 1s(4) & \cdots & 2s(4) & \overline{2s}(4)\end{vmatrix}. \tag{9.7}$$

Further simplifications are introduced by using the shorthand notation

$$\left|1s(1) \quad \overline{1s}(1) \quad 2s(1) \quad \overline{2s}(1)\right| \tag{9.8a}$$

or

$$\left|1s \quad \overline{1s} \quad 2s \quad \overline{2s}\right| \tag{9.8b}$$

or

$$\left|1s_\alpha \quad 1s_\beta \quad 2s_\alpha \quad 2s_\beta\right|. \tag{9.8c}$$

EXERCISE 9.1 For odd-electron atoms, a single unpaired electron can be chosen with arbitrary m_s. Write two possible Slater determinants for the ground state of the Li atom. Is either an eigenstate of $\hat{S}_z$? Of $\hat{S}^2$?

Answer $|1s_\alpha 1s_\beta 2s_\alpha|$ and $|1s_\alpha 1s_\beta 2s_\beta|$ are the two Slater determinants. Each is an eigenfunction of $\hat{S}_z = \hat{s}_{1z} + \hat{s}_{2z} + \hat{s}_{3z}$, with respective eigenvalues $\hbar/2$ and $-\hbar/2$. Neither is an eigenfunction of $\mathbf{S}^2$.

9.3 HARTREE METHOD

Now we face a general problem: How do we determine the *orbitals* that go into the determinants? The variational theorem gives us a criterion that needs to be satisfied to achieve this optimization, but how is that criterion implemented? We shall show how this is done in two steps, first through Hartree theory, which is a relatively simple theory that ignores the Pauli principle, and then through Hartree–Fock theory, which considers the correctly antisymmetrized wavefunction.

If we ignore the Pauli principle, then the wavefunction can be written as a product of spatial orbitals, as in

$$\psi = u_1(1)u_2(2)\cdots, \tag{9.9}$$

where

$$u_i = Y_{\ell m}(\theta_i, \phi_i)g(r_i). \tag{9.10}$$

Now let us consider the effective Hamiltonian associated with each electron, which is the total Hamiltonian averaged over the other electrons. Electron 1, for example, will have a kinetic energy and a coulomb attraction to the nucleus and will be repelled by the other electrons. The average repulsion experienced by electron 1 due to the charge distribution associated with electron j is given by the integral of the product of $1/r_{1j}$ and the charge density of electron j. This charge density is simply the electron's charge e, multiplied by the square of the spatial wavefunction u_j. In atomic units, $e = 1$ and this becomes

$$\rho(j) = |u_j(j)|^2, \tag{9.11}$$

so the average repulsion induced in electron 1 by electron j is

$$v_{1j} = \int |u_j(j)|^2 \frac{1}{r_{1j}} d\tau_j. \tag{9.12}$$

This repulsion is generally written as

$$v_{1j} = J_j(1), \tag{9.13}$$

where J_j is called a *coulomb operator*. It is an operator because it depends on the coordinates of electron 1; only the coordinates of electron j are integrated over in Eq. (9.12).

The overall effective one-electron Hamiltonian for the first electron is then

$$\hat{h}_1^{\text{eff}} = \frac{-\nabla_1^2}{2} - \frac{Z}{r_1} + \sum_{j \neq 1} \int \frac{|u_j(r_j)|^2}{r_{j1}} \, d\tau_j$$

$$= \frac{-\nabla_1^2}{2} - \frac{Z}{r_1} + \sum_{j \neq 1} J_j(1).$$

(9.14)

Once we have this Hamiltonian, the variationally best wavefunction for electron 1 can be obtained by solving the one-electron Schrödinger equation

$$\hat{h}_1^{\text{eff}}(r_1)u_1(r_1) = \varepsilon_1 u_1(r_1).$$

(9.15)

In order to solve this equation, we need to have the wavefunctions u_j of all the electrons except electron 1. However, to get those functions, we need to solve equations equivalent to Eq. (9.15) for the other orbitals u_j. How do we do this? The answer is that we solve by *iteration*: We make a guess for all the u_j's and plug them into the one-electron Hamiltonians to define Schrödinger equations like Eq. (9.15) for each orbital. These equations are then solved, and the results are cycled back to redefine the one-electron Hamiltonians. The iteration is continued until it becomes self-consistent, which means that the orbitals and energies that we get from one step of the iteration to the next are the same to within some tolerance. This procedure is called the *self-consistent field (SCF)* approximation, and its implementation using a wavefunction written as a product of spatial orbitals is called *Hartree theory*. Once self-consistency has been achieved, the total electronic energy is obtained by calculating the expectation value of the full many-electron Hamiltonian as follows:

$$E_{\text{total}} = \langle \psi | \hat{H} | \psi \rangle$$

$$= \left\langle \psi \left| - \sum_i \frac{\nabla_i^2}{2} - \sum_i \frac{Z}{r_i} + \sum_i \sum_{j<i} \frac{1}{r_{ij}} \right| \psi \right\rangle$$

$$= \left\langle \psi \left| \sum_i \left(-\frac{\nabla_i^2}{2} - \frac{Z}{r_i} + \sum_{j \neq i} J_j(r_i) \right) + \left[\sum_i \left[\sum_{j<i} \frac{1}{r_{ij}} - \sum_{j \neq i} J_j(r_i) \right] \right] \right| \psi \right\rangle$$

$$= \sum_i \varepsilon_i + \sum_i \left\{ \sum_{j<i} \left\langle u_i u_j \left| \frac{1}{r_{ij}} \right| u_i u_j \right\rangle - \sum_{j \neq i} \left\langle u_i u_j \left| \frac{1}{r_{ij}} \right| u_i u_j \right\rangle \right\}.$$

(9.16)

Notice that in deriving Eq. (9.16), we have used expressions like the following to reduce integrals over the coordinates of all the electrons to integrals over the coordinates of just one electron:

$$\langle u_1(1)u_2(2)u_3(3) | f(1) | u_1'(1)u_2'(2)u_3'(3) \rangle$$
$$= \langle u_1(1) | f(1) | u_1'(1) \rangle \langle u_2(2) | u_2'(2) \rangle \langle u_3 | u_3' \rangle$$
$$= \langle u_1(1) | f(1) | u_1'(1) \rangle \delta_{22'} \delta_{33'}.$$

(9.17)

This expression assumes an orthonormal set of orbital functions for each electron. It is the variational minimization of Eq. (9.16) that leads to Eq. (9.15).

If the two sums in Eq. (9.16) are combined, we get

$$E = \sum_i \varepsilon_i - \sum_i \sum_{j>i} J_{ij}, \tag{9.18}$$

where J_{ij} is the same coulomb integral as defined in Eq. (8.44a), that is,

$$J_{ij} = \left\langle u_i(1)u_j(2) \left| \frac{1}{r_{12}} \right| u_i(1)u_j(2) \right\rangle. \tag{9.19}$$

Note that the i, j's label the spin orbitals, not the space orbitals. (See Exercise 9.2.) Note also that the total energy is not just the sum of orbital energies; instead, it is necessary to subtract off the sum of coulomb integrals because the sum of orbital energies double counts the coulomb interactions (i.e., the sum of ε_1 and ε_2 includes first the repulsion experienced by electron 1 due to 2 and then the repulsion experienced by electron 2 due to 1; however, the full Hamiltonian includes only one repulsion term).

EXERCISE 9.2 For the ground state of the He atom, the configuration is He($1s^2$). What is the orbital energy ε_1 and the total energy E_{TOT} in the Hartree approximation? Express the two energies in terms of the integrals

$$\left\langle 1s \left| -\frac{\hbar^2}{2m} \nabla^2 \right| 1s \right\rangle \equiv T_{ss}$$

and

$$\left\langle 1s \left| -\frac{Z}{r} \right| 1s \right\rangle \equiv V_s$$

and J_{11}, where r is the distance from the nucleus.

Answer   From Eq. (9.16),

$$\varepsilon_1 = \left\langle 1s \left| -\frac{1}{2} \nabla_i^2 - \frac{Z}{r_i} + J_1(r_i) \right| 1s \right\rangle$$

$$= T_{ss} + V_s + J_{11}.$$

This is the energy for each electron, since the spatial orbitals are the same. The total energy is therefore

$$E_{TOT} = \sum_i \varepsilon_i - \sum_i \sum_{j>i} J_{ij}$$

$$= 2\varepsilon_1 - J_{11} = 2T_{ss} + 2V_s + J_{11}.$$

The total energy is the sum of the two kinetic energies, the two nuclear attractions, and the average repulsion,

$$J_{11} = \left\langle 1s(1)1s(2) \left| \frac{1}{r_{12}} \right| 1s(1)1s(2) \right\rangle.$$

The sums in Eq. (9.18) run over the two $1s$ electrons. So the first sum yields twice the energy ε_1. The second sum reduces to just one term that describes the repulsion between the two electrons.

9.4 HARTREE–FOCK METHOD

The SCF or Hartree theory that we have just developed is actually an old theory often used to describe problems in classical mechanics, such as the motions of planets, in which the Pauli principle is not relevant. The theory that allows for the Pauli principle while still letting each electron have its own orbital is known as *Hartree–Fock theory*. In this theory, the assumed form of the wavefunction is a Slater determinant. If we suppose that the electronic state being described is a closed-shell singlet state in which there are α and β spins for each spatial orbital, then the Slater determinant can be written

$$\psi = \left| u_1(1)\overline{u_1}(1)u_2(1)\overline{u_2}(1)u_3(1) \ldots \right|, \qquad (9.20)$$

where the u_i's are the spatial orbitals analogous to what we used in Hartree theory ($i = 1, 2, \ldots, N$, with N being the number of distinct spatial orbitals, i.e., half the number of electrons, n, for a closed-shell example). We have denoted the α and β spin functions using the "barred" and "unbarred" abbreviations explained earlier.

At this point, Hartree–Fock theory uses the variational theorem to optimize the spatial orbitals u_i, with the trial energy evaluated by means of the Slater determinant and the full electronic Hamiltonian. The derivation is too long to give here, but the result is closely related to what we already have learned from Hartree theory, namely, that there is an effective one-electron Schrödinger equation for the orbitals. In Hartree–Fock theory, this equation for a closed-shell singlet is

$$\varepsilon_i u_i = \left[\frac{-\nabla_i^2}{2} - \frac{Z}{r_i} + \sum_{j=1}^{N} \left[2J_j(i) - K_j(i) \right] \right] u_i. \qquad (9.21)$$

The operator in brackets on the right-hand side of this equation is called the *Fock* operator. It consists of a kinetic-energy operator, a nuclear coulomb potential, and a coulomb operator $J_j(i)$ that we already know from Hartree theory. $K_j(i)$ is a new term called the *exchange operator*. When $K_j(1)$ operates on an orbital $u_\ell(1)$, it gives

$$K_j(1)u_\ell(1) = \left(\int u_j^*(2)u_\ell(2) \frac{1}{r_{12}} \, d\tau_2 \right) u_j(1). \qquad (9.22)$$

In this expression, it is important to note that the function u_ℓ ends up *inside* the integral. This means that K_j is an *integral* operator. Note also that the exchange operator enters into the one-electron Hamiltonian with a negative sign. In many circumstances this will lead to a lowering of the energy, much as the exchange integral in Eq. (8.43) signified a lowering of the triple-state energy as a result of the exchange "hole" that arises from the Pauli principle.

The set of equations (9.21), one for each distinct spatial orbital, are called the *Hartree–Fock* equations. Just as in the Hartree theory, these equations must be solved by iteration until self-consistency is achieved. Once it is, the total electronic energy may be evaluated (for a closed-shell system) using the formula

$$E_{tot} = 2 \sum_{i=1}^{N} \varepsilon_i - \sum_{i,j=1}^{N} (2J_{ij} - K_{ij}), \tag{9.23}$$

where J and K are the same types of coulomb and exchange integrals as in Eq. (8.44), namely,

$$J_{ij} = \langle u_i | J_j(i) | u_i \rangle = \iint d\tau_1 d\tau_2 \frac{|u_i|^2(1)|u_j|^2(2)}{r_{12}} \tag{9.24}$$

and

$$K_{ij} = \langle u_i | K_j(i) | u_i \rangle = \iint d\tau_1 d\tau_2 \frac{u_i^*(1)u_j(1)u_j^*(2)u_i(2)}{r_{12}}. \tag{9.25}$$

Just as with Hartree theory, the $(2J_{ij} - K_{ij})$ correction term arises because of double counting.

EXERCISE 9.3   Repeat example 9.2, but now for the Hartree–Fock, rather than the Hartree, approximation. That is, for the He$(1s^2)$ configuration, write the Slater determinant wavefunction, evaluate ε_{1s}, and compute E_{TOT}.

Answer   The He$(1s^2)$ configuration yields

$$\psi_1 = \frac{1}{\sqrt{2}} \begin{vmatrix} u_{1s}(1)\alpha_1 & u_{1s}(1)\beta_1 \\ u_{1s}(2)\alpha_2 & u_{1s}(2)\beta_2 \end{vmatrix} = u_{1s}(1)u_{1s}(2)\frac{(\alpha_1\beta_2 - \beta_1\alpha_2)}{\sqrt{2}},$$

$$\langle \psi_{HF} | H | \psi_{HF} \rangle = \langle u_{1s}(1)u_{1s}(2) | H | u_{1s}(1)u_{1s}(2) \rangle \frac{1}{2} \langle \alpha_1\beta_2 - \beta_1\alpha_2 | \alpha_1\beta_2 - \beta_1\alpha_2 \rangle$$

$$= \langle u_{1s}(1)u_{1s}(2) | H | u_{1s}(1)u_{1s}(2) \rangle,$$

and

$$\varepsilon_{1s} = \left\langle u_{1s}(i) \left| -\frac{1}{2} \nabla_i^2 - \frac{Z}{r_i} + 2J_1(i) - K_1(i) \right| u_{1s}(i) \right\rangle.$$

Here, $J_1(i) = K_1(i)$ (the exchange of identical orbitals leads to the same operator), so that

$$\varepsilon_{1s} = T_{ss} + V_s + J_{11}$$

and

$$E_{TOT} = 2T_{ss} + 2V_s + J_{11}.$$

These are the same answers (using the same notation) as for Exercise 9.2, indicating that the additional exchange term has no effect for two electrons of opposite spin in identical spatial orbitals.

EXERCISE 9.4   Replace the schematic notation

$$J_{1s,1s} = \left\langle 1s(1)1s(2) \left| \frac{1}{r_{12}} \right| 1s(1)1s(2) \right\rangle$$

by actual integrals. That is, write out the correct integral expressions for these parameters.

Answer   The u_{1s} functions, normalized, are

$$u_{1s} = \sqrt{\frac{1}{\pi a_0^3}}\, e^{-r/a_0} \qquad \text{(see Table 6.2)};$$

therefore,

$$J_{1s,1s} = \left(\frac{1}{\pi a_0^3}\right)^2 \iiint dr_1 d\theta_1 d\phi_1 r_1^2 \sin\theta_1 \iiint dr_2 d\theta_2 d\phi_2 r_2^2 \sin\theta_2 e^{-2r_1/a_0} e^{-2r_2/a_0}.$$

Appendix B shows how this integral is evaluated.

9.5   KOOPMANS'S THEOREM

Koopmans's theorem states that, if adding or subtracting an electron doesn't change the energy levels, then the ionization energy is $|\varepsilon_H|$, where ε_H is the highest occupied orbital energy, and the electron affinity is $|\varepsilon_L|$, where ε_L is the orbital energy of the lowest unoccupied orbital. The theorem can be proved rigorously for an H–F state; essentially, the corrections cancel at this level of theory. Koopmans's theorem is of real use in determining rough spectra, because it means that the orbital energies obtained from a Hartree–Fock calculation for the ground state of the atom can be used to predict the ionization potential and electron affinity. Of course, the theorem is not exactly correct and will fail when electronic *relaxation* occurs, wherein the orbitals other than the highest occupied one are different for the neutral atom and its positive or negative ion. In fact, for many spectral assignments in atoms and small molecules, the results obtained from Koopmans's theorem are not very useful; it is generally much better for the ionization energies than the electron affinities.

9.6   ELECTRON CORRELATION

The wavefunction obtained from a Hartree–Fock calculation is the best possible wavefunction that can be expressed as a single Slater determinant. This wavefunction describes electron–electron repulsions in an average sense. It does not include electron correlation, so the correlation energy may be defined as the difference between the exact total energy of the atom and the H–F energy; that is,

$$E_{\text{corr}} \equiv E_{\text{exact}} - E_{\text{HF}}, \tag{9.26}$$

where E_{exact} can be obtained from experiment or, in many cases, from more accurate calculations.

Detailed calculations show that the correlation energy E_{corr} is usually only a small fraction of the total energy of the electrons: $|E_{corr}/E_{exact}| < .01$. This means that H–F theory accounts for a large fraction of the total energy in the atom. However, E_{corr} is often comparable in size to the strengths of chemical bonds, which means that H–F theory often fails to describe the bond strengths of molecules realistically. In fact, sometimes H–F theory predicts molecules to be unbound that are, in fact, bound.

9.7   CONSTANTS OF THE MOTION

So far, we have been concerned primarily with determining energy eigenfunctions, without paying attention to what other operators might have simultaneous eigenfunctions with H. However, we know from Chapter 2 [Section 2.7.3] that any operator which commutes with H may have simultaneous eigenfunctions. The eigenvalues associated with these operators are important, as they define the *constants of the motion* that may be used to label the states of the atom. To prove this statement, we simply calculate the derivative of the expectation value of the operator. If $\hat{A}$ is a time-independent operator and $[\hat{A}, \hat{H}] = 0$, then

$$\frac{\partial}{\partial t} \langle \psi | \hat{A} | \psi \rangle = \langle \dot{\psi} | \hat{A} | \psi \rangle + \langle \psi | \hat{A} | \dot{\psi} \rangle$$

$$= -\frac{i}{\hbar} \langle \hat{H}\psi | \hat{A} | \psi \rangle + \frac{i}{\hbar} \langle \psi | \hat{A} | \hat{H}\psi \rangle \tag{9.27}$$

$$= -\frac{i}{\hbar} \{ \langle \psi | \hat{H}\hat{A} - \hat{A}\hat{H} | \psi \rangle \}$$

$$= 0.$$

[We have used the time-dependent Schrödinger equation (2.10) in the second step.] Thus, if $[\hat{A}, \hat{H}] = 0$ and $\hat{A}\psi = a\psi$, then a is called a *good quantum number* because it is time independent.

Sometimes, we have operators that "almost" commute with the Hamiltonian. Typically, this situation arises when

$$\hat{H} = \hat{H}_0 + \lambda \hat{H}_1 \tag{9.28}$$

and

$$[\hat{A}, \hat{H}_0] = 0, \tag{9.29}$$

but

$$[\hat{A}, \hat{H}_1] \neq 0. \tag{9.30}$$

As long as λ is small, a is a *nearly good quantum number*, which means that it varies slowly with time.

9.8 ANGULAR-MOMENTUM OPERATORS FOR MANY-ELECTRON ATOMS

Let us consider $\hat{H}$ to be the exact nonrelativistic Hamiltonian of a many-electron atom. Then since $\hat{H}$ does not depend on spin, any operator $\hat{\Omega}$ that depends only on spin will commute with it; that is,

$$[\hat{H}, \hat{\Omega}(\mathbf{S})] = 0. \tag{9.31}$$

For the excited states of the He atom, we learned in Chapter 8 how to define eigenfunctions of the total spin operators $\mathbf{S}^2$ and S_z (omitting the caret on operators hereafter in the chapter). Such eigenfunctions can be defined for atoms with any number of electrons, and, as with the He atom, they are usually sums of several Slater determinants.

Now consider the one-electron orbital angular-momentum operators ℓ^2 and ℓ_z, say, for electron 1. Just as with the hydrogen atom, these commute with the one-electron part of the many electron Hamiltonian:

$$\left[\left(H - \sum_{i>j} \frac{1}{r_{ij}}\right), \ell_1^2\right] = 0; \tag{9.32}$$

$$\left[\left(H - \sum_{i>j} \frac{1}{r_{ij}}\right), \ell_{1z}\right] = 0. \tag{9.33}$$

This commutativity means that the exact eigenfunctions of the one-electron part of the Hamiltonian can be written in terms of antisymmetrized products of hydrogenic orbitals. We have already used this property in Chapter 8, and in fact, it also applies to any Hamiltonian that can be written as a sum of one-electron terms;

$$H = \sum_i h_1(i), \tag{9.34}$$

such as the Fock operators in Hartree–Fock theory [where there is also an additive constant in Eq. (9.34)].

However, the one-electron angular-momentum operators do not commute with the interelectronic repulsions:

$$\left[\ell_{1z}, \frac{1}{r_{12}}\right] \neq 0. \tag{9.35}$$

This noncommutativity can be demonstrated by explicit evaluation using the equations

$$\ell_{1z} = -i\hbar \frac{\partial}{\partial \phi_1} \tag{9.36}$$

and

$$r_{12} = \sqrt{r_1^2 + r_2^2 - 2r_1 r_2 \cos \chi}, \tag{9.37}$$

where χ is the angle between the vectors $\mathbf{r}_1$ and $\mathbf{r}_2$. This is related to the polar angles $\theta_1, \phi_1, \theta_2,$ and ϕ_2 associated with the vectors $\mathbf{r}_1$ and $\mathbf{r}_2$ via the formula

$$\cos \chi = \cos \theta_1 \cos \theta_2 + \sin \theta_1 \sin \theta_2 \cos (\phi_1 - \phi_2). \qquad (9.38)$$

So, since $\cos \chi$ depends on the difference $(\phi_1 - \phi_2)$, its derivative with respect to ϕ_1 will not be zero, and the commutator is nonzero. This means that the individual electron quantum numbers ℓ_1 and ℓ_2 are not rigorous constants of the motion for a many-electron atom. However, the *sum* of angular momentum operators,

$$L_z = \ell_{1z} + \ell_{2z}, \qquad (9.39)$$

does commute with the Hamiltonian. The reason is that the commutator of L_z with $1/r_{12}$ involves the sum of derivatives $(\partial/\partial\phi_1) + (\partial/\partial\phi_2)$ acting on $\cos \chi$. Because $\cos \chi$ depends on $\phi_1 - \phi_2$, the derivative vanishes, and

$$\left[L_z, \frac{1}{r_{12}} \right] = 0. \qquad (9.40)$$

Another operator that commutes with $1/r_{12}$ is $\mathbf{L}^2 = L_x^2 + L_y^2 + L_z^2$, where the vector $\mathbf{L}$ is related to the one-electron operators by

$$\mathbf{L} = \boldsymbol{\ell}_1 + \boldsymbol{\ell}_2. \qquad (9.41)$$

We therefore have two operators, L_z and $\mathbf{L}^2$, that commute with H:

$$0 = \left[\mathbf{L}^2, H \right] = \left[L_z, H \right]. \qquad (9.42)$$

The operators L_x, L_y, and L_z obey the same commutation relations that the one-electron angular-momentum operators do [see Eqs. (4.70)–(4.71)], such as $\left[L_x, L_y \right] = i\hbar L_z$, and $\left[\mathbf{L}^2, L_z \right] = 0$. As a result, the eigenfunctions of these operators behave a lot like the spherical harmonics. Thus, we have

$$\mathbf{L}^2 | LM_L \rangle = \hbar^2 L(L + 1) | LM_L \rangle \qquad (9.43a)$$

and

$$L_z | LM_L \rangle = \hbar M_L | LM_L \rangle. \qquad (9.43b)$$

However, there are constraints on the allowed values of L and M due to conservation of angular momentum. In particular, since the vectors $\boldsymbol{\ell}_1$ and $\boldsymbol{\ell}_2$ add to give $\mathbf{L}$, L can be no larger than $\ell_1 + \ell_2$ and no smaller than $|\ell_1 - \ell_2|$. Similarly, the z-projection quantum numbers m_1 and m_2 must add algebraically to give M. Hence, we have

$$L = \ell_1 + \ell_2, \ell_1 + \ell_2 - 1, \cdots |\ell_1 - \ell_2| \qquad (9.44a)$$

and

$$M = m_1 + m_2 + \cdots. \qquad (9.44b)$$

The total *spin* operators $\mathbf{S}^2$ and S_z may also be used to determine eigenfunctions $| SM_S \rangle$ with quantum numbers S and M_S. From this fact, we conclude that the exact eigenfunctions of H can be characterized using the quantum numbers L, M_L, S, and M_S, as well as E.

It is also possible to label the states of an atom using the total angular momentum

$$\mathbf{J} = \mathbf{L} + \mathbf{S}. \tag{9.45}$$

The eigenfunctions of $\mathbf{J}^2$ and J_z have eigenvalues J and M_J, respectively, that are very much like the other angular-momentum eigenfunctions we have studied, with the restrictions that $L + S \geq J \geq |L - S|$ and $M_J = M_L + M_S$. Note, however, that although $[\mathbf{J}^2, \mathbf{L}^2] = 0$ and $[\mathbf{J}^2, \mathbf{S}^2] = 0$, $[\mathbf{J}^2, L_z]$ and $[\mathbf{J}^2, S_z]$ are nonzero, so it is not possible to label the states of atoms simultaneously with the six angular-momentum quantum numbers J, M_J, L, M_L, S, and M_S. Instead, we can use the four quantum numbers L, M_L, S, and M_S as previously noted, or we can use J, M_J, L, and S. Since all of these operators commute with H, there is nothing at this point to favor one choice over the other. However, the introduction of relativity into the description of atoms changes the situation.

9.9   RELATIVISTIC EFFECTS

The proper way to describe atoms using relativistic quantum mechanics is via the Dirac equation. However, for most applications, it is sufficient (and much easier) to use the Schrödinger equation with a Hamiltonian that has been supplemented by relativistic terms. The dominant relativistic term is one-electron spin-orbit coupling, for which the many-electron Hamiltonian is

$$H_{so} = \sum_i \xi_i(r_i) \ell_i \cdot s_i, \tag{9.46}$$

with the parameter

$$\xi \approx \frac{1}{r}\frac{\partial V}{\partial r} \cdot \frac{1}{2m^2c^2}. \tag{9.47}$$

For a one electron atom, $\xi \approx Z^4$, which means that relativistic contributions to the electronic energy are much larger for atoms with a greater nuclear charge (i.e., heavier atoms) than for those with a lesser charge (the light ones). This makes sense physically, as the electrons are more strongly attracted to the nucleus the greater the charge, so they move faster, creating larger magnetic fields.

EXERCISE 9.5   What configurations will mix with the configuration $|1s\,\overline{1s}\,2s\,\overline{2s}\,2p_z|$ by the term H_{so}?

Answer   In expanding the Slater determinant, there will be 120 terms (that is, 5!). In each term, there will be a product like $u_{1s}(1)\alpha_1 u_{1s}(2)\beta_2 u_{2s}(3)\alpha_3 u_{2s}(4)\beta_4 u_{2p_z}(5)\alpha_5$. Also,

$$H_{so} = \sum_i \xi_i \ell_i \cdot s_i = \sum_i \xi_i \left[\ell_{iz} s_{iz} + \frac{1}{2}(\ell_{i+} s_{i-} + \ell_{i-} s_{i+}) \right].$$

Operating on an s orbital with ℓ_z, ℓ_+, and ℓ_- gives zero. So the only effect of H_{so} is on $u_{2p_z}\alpha$:

$$H_{so}u_{2p_z}\alpha = \xi\left(\ell_z s_z + \frac{1}{2}\ell_+ s_- + \frac{1}{2}\ell_- s_+\right)u_{2p_z}\alpha$$

$$= 0 + \frac{\xi}{2}u_{2p_+}\beta + 0 = \frac{\xi}{2}u_{2p_+}\beta.$$

So the determinant $|1s\,\overline{1s}\,2s\,\overline{2s}\,\overline{2p_+}|$ will be mixed by H_{so}.

The inclusion of spin–orbit coupling in the Hamiltonian changes the angular-momentum operators that commute with it. In particular, the operators $\mathbf{L}^2, L_z, \mathbf{S}^2$, and S_z do not commute with H_{so}. However,

$$[H_{so}, J_z] = [H_{so}, \mathbf{J}^2] = 0, \tag{9.48}$$

which means that only J, M_J, and E are rigorously good quantum numbers. For any atom in which

$$\langle H_{so}\rangle \ll \left\langle\frac{1}{r_{12}}\right\rangle \tag{9.49}$$

(i.e., H_{so} is small compared to the interelectronic repulsions), L and S are "near-ly" good quantum numbers, so it is useful to label states with these in addition to J and M_J. In this limit, the so-called Russell–Saunders term symbols are used to identify states. These are written as

$$^{2S+1}L_J, \tag{9.50}$$

where $2S + 1$ is the spin multiplicity and the M quantum number is omitted un-less a magnetic field is applied that would lift the degeneracy of the states. Ex-amples of symbols for the Russell–Saunders term for selected configurations of several atoms are as follows:

$$
\begin{array}{ll}
\text{H}(1s) & ^2S_{1/2}; \\
\text{Li}(1s^2 2s) & ^2S_{1/2}; \\
\text{B}(1s^2 2s^2 2p) & ^2P_{3/2}, {}^2P_{1/2}; \\
\text{F}(1s^2 2s^2 2p^5) & ^2P_{3/2}, {}^2P_{1/2}; \\
\end{array}
$$

$$
\text{C}(1s^2 2s^2 2p3p) \quad
\begin{cases}
^3D_3 \quad {}^3D_2 \quad {}^3D_1 \\
^3P_2 \quad {}^3P_1 \quad {}^3P_0 \;. \\
^3S_1
\end{cases}
$$

$$
\begin{cases}
^1D_2 \\
^1P_1 \\
^1S_0
\end{cases}
$$

Notationally, the L value is replaced by the letter S for $L = 0$, P for $L = 1$, D for $L = 2, \ldots$, a scheme that is familiar from the H atom. Note that the closed subshells can be ignored in working out the term symbols. In addition, if a shell is more than half full, it can be thought of as a less-than-half-full configuration of "holes" (missing electrons) such that the number of electrons plus the number of holes equals the subshell degeneracy. Thus, the fluorine atom is treated as having one hole in the $2p$ subshell, which means that $L = 1$, $S = 1/2$, and $J = 3/2$ and $1/2$, and the terms are $^2P_{3/2}$ and $^2P_{1/2}$. The example of the carbon atom is a straightforward application for the case of two electrons with $\ell_1 = \ell_2 = 1$. Notice that we have assumed that the carbon atom is in an excited configuration $(2p3p)$ rather than the ground configuration $(2p^2)$. The reason for this is that additional rules apply in determining the allowed terms for partially filled subshells that have two or more electrons. These further rules are discussed shortly.

The energies associated with the terms that are generated from a given electronic configuration are all degenerate for a Hamiltonian that neglects electron–electron repulsion. If such repulsion is included—say, using Hartree–Fock theory—then different terms (labelled by L and S) have different energies. Different J's for a given L and S are still degenerate. However, the J degeneracy is lifted when H_{so} is included (i.e., the terms are split into *levels*). The M_J degeneracy remains at this point, but would be lifted if an external magnetic field were applied (i.e., the levels are split into *states*). Figure 9.1 shows the progress from configurations (governed by $H_{separable}$) to terms (governed by H_{nonrel}) to levels (governed by H_{rel}) for the $1s^22s2p$ configuration of Be.

When two or more electrons are in the *same* subshell, the Pauli principle excludes some terms that would otherwise arise if the preceding rules were used. The best way to describe this situation is by an example. Table 9.1 describes the microstates (orbital and spin quantum members for each electron) associated with the ground configuration of the carbon atom. Only the open shell $(2p)$ needs to be considered, so there are two electrons, divided among the three p orbitals. Application of the rules given earlier would lead to the conclusion that the allowed terms are $^{1,3}D$, $^{1,3}P$, and $^{1,3}S$. However, in order to have a D term, the $M_L = 2$ state would have to arise from putting two electrons in $m_\ell = 1$ orbitals. This would be possible only for paired electrons—that is, for the singlet term ^1D. By contrast, the $M_L = 1$ state of the P term can be constructed from $m_\ell = 0$ and

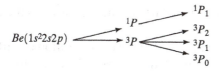

Figure 9.1   Evolution from configuration to terms to levels.

$m_\ell = 1$, which means that the electrons can be in different orbitals and both sin-glet and triplet states are possible in principle. However, the singlet state with this configuration is already used as one member of ^1D, so only ^3P is possible. The S term arises from putting both electrons in $M_S = 0$, so again, only the singlet term is allowed. We conclude that the allowed terms of carbon are $^1D_2$, $^3P_2$, $^3P_1$, $^3P_0$, and $^1S_0$.

TABLE 9.1 MICROSTATES ASSOCIATED WITH A $(p)^2$ CONFIGURATION.

m_{ℓ_1}	m_{ℓ_2}	m_{s_1}	m_{s_2}	M_L	M_S	Arrangement
1	1	1/2	−1/2	2	0	^1D
1	0	1/2	−1/2	1	0	^1D
1	0	−1/2	1/2	1	0	^3P
1	0	1/2	1/2	1	1	^3P
1	0	−1/2	−1/2	1	−1	^3P
1	−1	1/2	−1/2	0	0	^3P
1	−1	−1/2	1/2	0	0	^1D
1	−1	1/2	1/2	0	1	^3P
1	−1	−1/2	−1/2	0	−1	^3P
0	0	1/2	−1/2	0	0	^1S
0	−1	1/2	−1/2	−1	0	^1D
0	−1	−1/2	1/2	−1	0	^3P
0	−1	1/2	1/2	−1	1	^3P
0	−1	−1/2	−1/2	−1	−1	^3P
−1	−1	1/2	−1/2	−2	0	^1D

The following *Hund rules* are a set of simple rules that determine the low-est energy term and level within a given configuration:

1. For a given configuration, the term with the largest S has the lowest ener-gy (as this maximizes exchange stabilization).

2. For a given spin multiplicity, the term with the largest L has the lowest en-ergy (as this minimizes spatial overlap).

3. Allowing for spin–orbit coupling, the ordering of levels is determined by the following two conditions:

   a. If a configuration is more than half full, the highest J has the lowest en-ergy (inverted).
   b. If a configuration is less than half full, the lowest J has the lowest energy.

When these rules are applied to the $(1s^2 2s^2 2p^2)$ configuration of carbon, one finds that the lowest level is $^3P_0$. Note that Hund's rules are meant to be applied only to the ground configuration of an atom, as the higher configurations are gener-ally perturbed by *configuration interaction*, which often leads to a violation of the rules.

For the heavier atoms (such as many of the states of I, Hg, Pb, etc.), the spin–orbit interaction is sufficiently large that

$$\langle H_{so}\rangle > \left\langle \frac{1}{r_{ij}}\right\rangle. \tag{9.51}$$

In this case, the symbols for Russell–Saunders terms provide a poor labelling of the atomic states. Instead, it is preferable to use so-called j–j coupling, wherein first the ℓ and s quantum numbers for each electron are combined to give a total angular momentum j for each electron, and then the j's for each electron are combined to give the total angular-momentum quantum numbers J and M_J. Note that J and M_J are still good quantum numbers, but now it is the j's for each electron that are best thought of as "nearly" good quantum numbers.

SUGGESTED READINGS

I. N. Levine, *Quantum Chemistry*, 5th ed. (Prentice Hall, Upper Saddle River, NJ, 2000).

J. P. Lowe, *Quantum Chemistry*, 2nd ed. (Academic Press, New York, 1993).

J. Simons and J. Nichols, *Quantum Mechanics in Chemistry* (Oxford, New York, 1997).

PROBLEMS

9.1 What is the volume element $d\tau$ for the Li atom?

9.2 Write down the wavefunction of the ground state of the B atom, expressed as a Slater determinant.

9.3 Numerical Hartree–Fock calculations on the $^3S_1$ and $^1S_0$ states of He($1s2s$) give total energies of -1.752 and -1.525 hartree, respectively. What is the value of the exchange integral $K_{1s,2s}$ for He?

9.4 Koopmans' theorem says that the electron affinity and the ionization energy can be estimated, respectively, as the negative orbital energies ε of the lowest empty and highest occupied Hartree–Fock orbitals. Does this mean that the difference between these orbital energies is a good estimate of the lowest optical excitation energy (moving an electron from the highest occupied to the lowest empty orbital)? Why or why not?

9.5 Evaluate the following:

$$\langle\alpha_1\beta_2|S_{1+}S_{2-} + S_{1-}S_{2+}|\beta_1\alpha_2\rangle;$$

$$\langle\alpha_1\alpha_2|\mathbf{S}^2|\alpha_1\alpha_2\rangle, \quad \text{where} \quad \mathbf{S} = \mathbf{S}_1 + \mathbf{S}_2.$$

9.6 The process of intersystem crossing in simple organic molecules usually corresponds to the molecule moving between states of differing spin multiplicity without any external perturbation. If $H_{so} = 0$, can intersystem crossing occur?

9.7 Classify the following quantum numbers as good, nearly good, or neither:
 (a) Spin multiplicity in $^3D_2$.
 (b) Orbital angular momentum in $^5D_3$.

   (c) Total angular momentum J in $^2P_{3/2}$.

   (d) Total energy in the ground state of Be.

   (e) Spin multiplicity in the ground state of $U^{+2.}$

9.8 For the configuration $He^*(1s2p)$, write the symbols for all of the terms. Then write the actual wavefunctions as linear combinations of Slater determinants.

9.9 True (T) or false (F)?

   (a) Levels in an atom are split into states by spin–orbit coupling.

   (b) Russell–Saunders coupling applies when the electrostatic interaction between different terms is smaller than the spin–orbit interaction between different levels.

9.10 **(a)** What terms and levels arise from the following configurations?

$$1s2p \quad 2p3p \quad 3p3d.$$

   If Hund's rule applied to these terms and levels, what would be the ground level?

   (b) Using the selection rules $\Delta S = 0$, $\Delta L = \pm 1$, and $\Delta J = 0, \pm 1$, predict which transitions are allowed from the ground level in (a). Consider only transitions between terms within each configuration given.

9.11 Using Hund's rules, determine the lowest term and level of each spin multiplicity, based on the following list (assume that the shell is more than half filled):

$$^3P_2, {}^3P_1, {}^3P_0, {}^1D_2, {}^3S_0, {}^1F_3.$$

9.12 The lowest allowed term associated with the $2p^3$ configuration can be determined by applying Hund's rules and a little common sense (rather than brute force). What is that term?

9.13 Which of the following describe possible states of $C(1s^2 2s^2 2p^2)$?

   (a) $^1S_1$.

   (b) $^3P_2$.

   (c) $^3D_2$.

   (d) $^1F_2$.

9.14 Consider an atom whose electronic configuration is $3d^2$.

   (a) What are the possible terms and levels, ignoring the Pauli principle?

   (b) What are terms and levels from (a) that are allowed if the Pauli principle is taken into account?

   (c) What are the lowest energy term and the lowest energy level?

   (d) Write down the complete electronic wavefunction (angular and spin parts only) for the state in (b) with the highest L, J, and M_J.

10

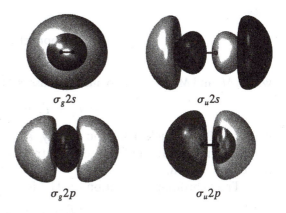

$\sigma_g 2s$

$\sigma_u 2s$

$\sigma_g 2p$

$\sigma_u 2p$

Homonuclear Diatomic Molecules

Just as the *Aufbau* principle permitted us to understand many-electron atoms in terms of adding electrons to the orbitals generated by the simplest one-electron atom (hydrogen), so the hydrogen molecule ion H_2^+ provides a set of delocalized molecular orbitals that can be used to construct approximations to the wavefunctions of homonuclear diatomics. Molecular orbital and valence bond functions take different approaches to understanding electronic structure in molecules and are the simplest prototype descriptions of electronic structure.

10.1 HYDROGEN MOLECULAR ION: BORN–OPPENHEIMER APPROXIMATION

The simplest molecule is H_2^+ which consists of two protons and one electron as pictured in Figure 10.1. The complete Hamiltonian of this molecule is

$$\hat{H} = -\frac{1}{2M_a}\nabla_a^2 - \frac{1}{2M_b}\nabla_b^2 - \frac{1}{2}\nabla_e^2 - \frac{1}{r_a} - \frac{1}{r_b} + \frac{1}{R_{ab}} \text{ (in atomic units)}, \quad (10.1)$$

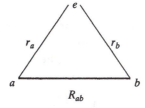

Figure 10.1 Coordinate definitions for H_2^+ molecule.

145

where M_a and M_b are the two nuclear masses. One way to write this Hamiltonian is

$$\hat{H} = T_n + T_e + V_{en} + V_{nn}, \tag{10.2}$$

where T_n is the nuclear kinetic energy, T_e is the electron kinetic energy, V_{en} is the coulomb attraction between electron and nuclei, and V_{nn} is the nuclear-nuclear repulsion.

The Schrödinger equation for H_2^+ is

$$\hat{H}\Psi = W\Psi, \tag{10.3}$$

where we use the symbol W for the total electron–nuclear energy. Just as for the He atom, this equation refers to the motions of three particles, and it is not separable. However, there is an excellent solution based on the Born–Oppenheimer approximation, which takes advantage of the huge difference between electron and nuclear masses to approximate the wavefunction as a product of a function of electron coordinates and a function of nuclear coordinates. The electron wavefunction is determined by solving the Schrödinger equation for fixed nuclear positions and thus depends parametrically on nuclear coordinates. If we use the symbol **r** for electronic coordinates and **R** for nuclear coordinates, then the complete wavefunction is

$$\Psi(\mathbf{r}, \mathbf{R}) = \psi(\mathbf{r}; \mathbf{R})\chi(\mathbf{R}), \tag{10.4}$$

where ψ is the electronic wavefunction and χ is the nuclear wavefunction. If Eq. (10.4) is substituted into the Schrödinger equation, we find that

$$\left(T_n + T_e + V_{en} + V_{nn}\right)\psi\chi = W\psi\chi. \tag{10.5}$$

If this is now divided by $\psi\chi$ and we take advantage of the fact that T_e operates on ψ, but not on χ, then we get

$$\frac{T_n\psi\chi}{\psi\chi} + \frac{T_e\psi}{\psi} + V_{en} + V_{nn} = W. \tag{10.6}$$

T_n, on the other hand, operates on both ψ and χ. If no approximations are made, then one expects to have terms like

$$T_n\psi\chi = \sum_{\alpha=a,\,b} \frac{-\hbar^2}{2M_\alpha}\left[\psi\nabla_\alpha^2\chi + \chi\nabla_\alpha^2\psi + 2\nabla_\alpha\psi \cdot \nabla_\alpha\chi\right]. \tag{10.7}$$

In the simplest Born–Oppenheimer approximation, the second and third terms in this expression, namely, those involving derivatives of the electronic wavefunction with respect to the nuclear coordinates, are neglected. This assumption enables us to separate Eq. (10.6) approximately into electronic terms and nuclear terms, yielding

$$\frac{(T_n + V_{nn})\chi}{\chi} + \frac{(T_e + V_{en})\psi}{\psi} = W. \tag{10.8}$$

Equation (10.8) is not really separable, as V_{en} depends on nuclear as well as electronic coordinates. We assume, however, that the electronic part may be equated to a "constant," $E_{e\ell}$, that depends parametrically on nuclear coordinates. This gives

$$\frac{(T_e + V_{en})\psi}{\psi} = E_{e\ell}, \tag{10.9}$$

or

$$(T_e + V_{en})\psi = E_{e\ell}\psi, \tag{10.10}$$

which defines an *electronic Schrödinger equation* in which nuclear coordinates appear only as parameters. The corresponding *nuclear Schrödinger equation* is

$$(T_n + V_{nn} + E_{e\ell})\chi = W\chi. \tag{10.11}$$

In this equation, the effective potential governing nuclear motions is the sum of V_{nn} and $E_{e\ell}$. We often group these into a single function

$$V_n = V_{nn} + E_{e\ell}, \tag{10.12}$$

which is called the *potential-energy surface.* Equation (10.12) governs the vibrational motions of molecules and also determines intermolecular forces. In most of what follows in this chapter, we shall concern ourselves with solving the electronic Schrödinger equation. The solution of the nuclear Schrödinger equation for determining vibrational motions was considered in Chapter 5 and will be used in Chapter 15 to discuss spectra.

EXERCISE 10.1   Use the form of T_n to show that a slightly more accurate form for the total potential energy is

$$V_n = V_{nn} + E_{e\ell} + \langle \psi | T_n | \psi \rangle.$$

Answer   From Eq. (10.7), $T_n \chi\psi = \chi T_n \psi + \psi T_n \chi$, plus a neglected term that involves $\nabla_\alpha \chi \, \nabla_\alpha \psi$. We have, then,

$$\langle \psi | T_e + V_{en} + T_n | \psi \rangle \chi + V_{nn}\chi = W\chi.$$

Assuming effective separability, we obtain

$$(T_e + V_{en})\psi = E_{e\ell}\psi$$

and

$$(T_n + V_{nn} + E_{e\ell} + \langle \psi | T_n | \psi \rangle)\chi = W\chi.$$

The correction term $\langle \psi | T_n | \psi \rangle$, called the *diagonal nonadiabatic correction*, is usually small.

10.2   LCAO–MO TREATMENT OF H_2^+

From Eq. (10.10), the electronic Schrödinger equation for H_2^+ is

$$\left\{ -\frac{1}{2}\nabla^2 - \frac{1}{r_a} - \frac{1}{r_b} \right\}\psi = E_{e\ell}\psi \text{ (in atomic units).} \qquad (10.13)$$

This equation, which involves just a single particle moving in a potential determined by coulomb attraction to the two nuclei, is exactly separable in ellipsoidal coordinates. However, the technical complexity associated with its exact solution is not very illuminating, so we develop instead an approximate solution based on using linear combinations of the atomic orbitals (LCAO) on each nucleus to represent the molecular orbitals (MO). The atomic orbitals associated with each nucleus are just the hydrogenlike orbitals. To get started, we assume that the only orbitals of interest are the $1s$ functions on each atom, which we respectively call $1s_a$ and $1s_b$. Thus, the LCAO–MO description of ψ is

$$\psi = C_a 1s_a + C_b 1s_b, \qquad (10.14)$$

where C_a and C_b are unknown coefficients that we want to use in a variational determination of the energy. To do this, we substitute the expression for ψ into the Schrödinger equation, yielding

$$H_{e\ell}\psi = H_{e\ell}\{C_a 1s_a + C_b 1s_b\} = E_{e\ell}\{C_a 1s_a + C_b 1s_b\}. \qquad (10.15)$$

If we multiply this equation by $1s_a^*$ and integrate, we obtain

$$H_{aa}C_a + H_{ab}C_b = E\{C_a + S_{ab}C_b\}, \qquad (10.16)$$

where

$$H_{aa} = \langle 1s_a | H_{e\ell} | 1s_a \rangle, \qquad (10.17a)$$

$$H_{ab} = \langle 1s_a | H_{e\ell} | 1s_b \rangle, \qquad (10.17b)$$

and

$$S_{ab} = \langle 1s_a | 1s_b \rangle. \qquad (10.17c)$$

Note that we have assumed that the orbitals are normalized, but that orbitals on different nuclei are not orthogonal. A second equation may be developed by multiplying by $1s_b^*$ and integrating to give

$$H_{ba}C_a + H_{bb}C_b = E\{S_{ba}C_a + C_b\}. \qquad (10.18)$$

Equations (10.16) and (10.18) may be rearranged into the matrix-vector form

$$\begin{pmatrix} (H_{aa} - E) & (H_{ab} - ES_{ab}) \\ (H_{ba} - ES_{ba}) & (H_{bb} - E) \end{pmatrix} \begin{pmatrix} C_a \\ C_b \end{pmatrix} = 0. \qquad (10.19)$$

The two equations are linear, homogeneous equations. From our discussion in Appendix A.3, they have solutions only if the determinant of the coefficient matrix vanishes—that is, if

$$\begin{vmatrix} H_{aa} - E & H_{ab} - ES_{ab} \\ H_{ba} - ES_{ba} & H_{bb} - E \end{vmatrix} = 0. \tag{10.20}$$

This equation, which is called the *secular* equation, may be solved by explicitly calculating the determinant. If we use the fact that the nuclei are identical, so that $H_{aa} = H_{bb}$, $H_{ab} = H_{ba}$, and $S_{ab} = S_{ba}$, we get

$$\left(H_{aa} - E\right)^2 - \left(H_{ab} - ES_{ab}\right)^2 = 0, \tag{10.21}$$

for which the solution (actually, two) is

$$E_{\pm} = \frac{H_{aa} \pm H_{ab}}{1 \pm S_{ab}}. \tag{10.22}$$

To complete this solution, we need to evaluate the overlap integral S_{ab} and the Hamiltonian matrix elements H_{aa} and H_{ab}. These integrals are highly specialized, so we omit the evaluation here. The results, in atomic units, are

$$S_{ab} = e^{-R}\left\{\frac{R^2}{3} + R + 1\right\}, \tag{10.23a}$$

$$H_{aa} = -\frac{1}{2} - \frac{1}{R} + \left(1 + \frac{1}{R}\right)e^{-2R}, \tag{10.23b}$$

and

$$H_{ab} = -\frac{1}{2}S_{ab} - (R + 1)e^{-R}. \tag{10.23c}$$

If these equations are substituted into the energy expression (positive sign in Eq. 10.22), we get

$$E_{+} = -\frac{1}{2} + \frac{(R + 1)e^{-2R} - R(R + 1)e^{-R} - 1}{R\left[\left(1 + R + \frac{1}{3}R^2\right)e^{-R} + 1\right]}. \tag{10.24}$$

If this equation is then substituted into Eq. (10.12), we obtain the potential curve $V_{n+} = E_{+} + 1/R$.

Figure 10.2 shows E_{+} and V_{n+} as a function of R. Evidently, both curves approach the energy of the hydrogen atom, $-1/2\ E_h$, in the $R \to \infty$ limit (the *separated-atom limit*). In addition, E_{+} must equal the He^+ ground-state energy, and $V_{n+} \to \infty$ as $R \to 0$ (the *united-atom limit*). [One subtle point to note is that Eq. (10.24) has the value $-1.5E_h$ at $R = 0$, rather than the correct value $-2.0E_h$. This happens because the $R = 0$ limit of Eq. (10.14) is the hydrogen-atom function e^{-r}

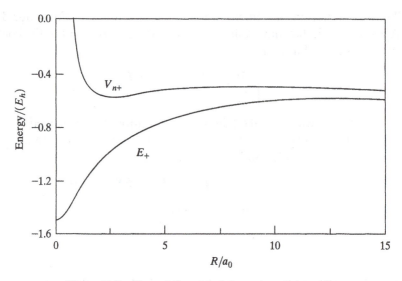

Figure 10.2 V_{n+} and E_+ versus internuclear distance R.

instead of the helium wavefunction e^{-2r}.] Between these two limits, E_+ varies monotonically in such a way that V_{n+} has a minimum, at $R = R_e = 2.490a_0$. The energy at this minimum is $V_{n+} = -0.5648E_h$, and the *difference* between this energy and the separated-atom energy is the *dissociation energy* D_e, so $D_e = 0.0648E_h$ in this case.

   The results just described are all qualitatively correct, but quantitatively inaccurate. From experiment, it is known that the correct $R_e = 2.00a_0$ and $D_e = 0.1025E_h$. Improved results may be obtained by applying variational theory. For example, if the energy is minimized with respect to the charge parameter Z' that appears in the hydrogen-atom orbitals, we get $R_e = 2.02a_0$ and $D_e = 0.0864E_h$. There are many other ways to improve the trial wavefunction, as will be discussed in detail later in this chapter and in Chapter 11. The important point for the present discussion is that the LCAO–MO approach predicts that the ground state of H_2^+ is bound. It is not difficult to show that the other solution of Eq. 10.22, E_-, does not give a minimum in V_{n-} and therefore corresponds to an unbound state of H_2^+.

   The wavefunction associated with the energy E_+ is easily obtained by substituting back into Eq. (10.19). This gives $C_a = C_b$, which implies that

$$\psi_+ = N_+(1s_a + 1s_b), \tag{10.25}$$

where N_+ is a normalization constant.

   The other solution is

$$\psi_- = N_-(1s_a - 1s_b). \tag{10.26}$$

Figure 10.3 shows the dependence of ψ_+ and ψ_- along the diatomic axis, which we denote as z. The "bonding" state ψ_+ shows enhanced electron density between the

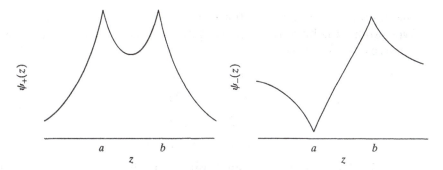

Figure 10.3 ψ_+ and ψ_- versus z coordinate.

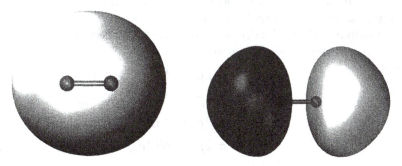

Figure 10.4   Three-dimensional contour plots of ψ_+ and ψ_-. (These are actually the true σ_g 1s and σ_u 1s orbitals of H_2^+.) The position of the nuclei is indicated using two filled black circles connected by a black line.

two nuclei, and thus, it is reasonable that this state gives a bound diatomic. By contrast, the "antibonding" state ψ_- has a node midway between the nuclei. Another format for plotting the orbitals is presented in Figure 10.4, where we plot contours of a constant wavefunction in three-dimensional space, using different shadings for positive and negative parts of the function.

10.3   OTHER H₂⁺ STATES

The states ψ_+ and ψ_- just determined are usually labelled σ_g 1s and σ_u 1s, where the σ label refers to orbitals that are invariant with respect to rotation about the molecular axis. The labels g and u respectively refer to even and odd (*gerade* and *ungerade*) symmetry with respect to the inversion operation (whereby $x \rightarrow -x$, $y \rightarrow -y$, and $z \rightarrow -z$). These labels are described in much more detail in Chapter 13. It is not difficult to construct orbitals corresponding to excited states by using excited-state hydrogenlike orbitals, provided that we keep track of the symmetry properties of the wavefunctions being constructed. For example, other σ-type orbitals can be constructed using $2s$ and $2p_z$ functions as follows (we take the z-axis as the intermolecular axis):

$$\sigma_g 2s = N(2s_a + 2s_b); \tag{10.27a}$$

$$\sigma_u 2s = N(2s_a - 2s_b); \tag{10.27b}$$

$$\sigma_g 2p = N(2p_{za} - 2p_{zb}); \tag{10.27c}$$

$$\sigma_u 2p = N(2p_{za} + 2p_{zb}). \tag{10.27d}$$

The sign change in the $2p$ orbitals relative to the $1s$ or $2s$ orbitals is due to the nodal structure of the p orbitals. Figure 10.5 shows three-dimensional plots of the orbitals.

Another class of solutions to the H_2^+ problem is the class of π-orbitals. These have angular functions that vary as $e^{\pm i\phi}$, or equivalently, $\cos \phi$ or $\sin \phi$ in the azimuthal angle ϕ. The simplest examples are obtained from linear combinations of $2p_x$ or $2p_y$ orbitals as follows:

$$\pi_u 2p_x = N(2p_{xa} + 2p_{xb}); \tag{10.28a}$$

$$\pi_g 2p_x = N(2p_{xa} - 2p_{xb}); \tag{10.28b}$$

$$\pi_u 2p_y = N(2p_{ya} + 2p_{yb}); \tag{10.28c}$$

$$\pi_g 2p_y = N(2p_{ya} - 2p_{yb}). \tag{10.28d}$$

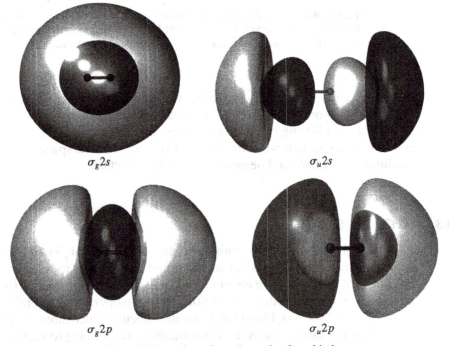

$\sigma_g 2s$ $\sigma_u 2s$

$\sigma_g 2p$ $\sigma_u 2p$

Figure 10.5 $\sigma_g 2s, \sigma_u 2s, \sigma_g 2p,$ and $\sigma_u 2p$ orbitals.

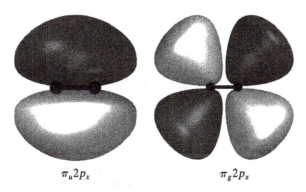

$$\pi_u 2p_x \qquad\qquad\qquad\qquad \pi_g 2p_x$$

Figure 10.6 Lowest energy π orbitals of H_2^+.

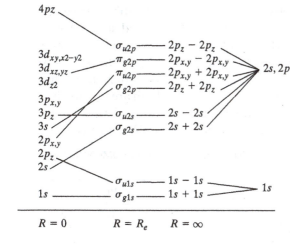

Figure 10.7 Correlation of H_2^+ orbital energies between united-atom and separated-atom limits.

Figure 10.6 shows these orbitals. Note that the real $2p_x$ orbitals all have a node in the yz-plane. In addition, the $2p_x$ antibonding orbitals have a nodal plane perpendicular to the molecular axis. One can similarly use the $3d$ orbitals to construct orbitals with σ, π, or δ symmetry.

Figure 10.7 shows qualitatively how the orbital energies vary with R from the united- $(R = 0)$ to the separated- $(R \rightarrow \infty)$ atom limits.

10.4   ELECTRONIC STRUCTURE OF HOMONUCLEAR DIATOMICS

We can now apply the *Aufbau* principle to the H_2^+ molecular orbitals to approximate the electronic structure of other homonuclear diatomic molecules. For example, the electronic configuration of H_2 is $(\sigma_g 1s)^2$, and that of He_2 would be $(\sigma_g 1s)^2(\sigma_u 1s)^2$. The only major point of ambiguity here is that, because of the curve crossings in the orbital correlation picture, the ordering of the orbitals is not unique. The correct ordering for the lighter homonuclear diatomics is given in Figure 10.8. If this ordering is applied to the homonuclear diatomics H_2^+ through Ne_2, we obtain the configurations given in Table 10.1. Included in this table are

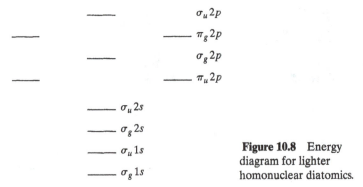

Figure 10.8   Energy diagram for lighter homonuclear diatomics.

the number of bonding electrons, n_B; the number of antibonding electrons, n_A; the bond order $(\mathrm{BO} = 1/2(n_B - n_A))$; and the predicted spin S, based on Hund's first rule. It turns out that all the predictions in the table are correct, including the prediction that B_2 and O_2 should be triplets. In addition, the bond dissociation energies are approximately correlated with the bond order, although the bond energy of Li_2 is much weaker than H_2 even though both are nominally single bonds. The table predicts that He_2 and Be_2 should be unbound, which is approximately correct; in reality, they are bound by weak van der Waals forces.

TABLE 10.1   ELECTRONIC CONFIGURATIONS FOR HOMONUCLEAR DIATOMIC MOLECULES

Diatomic	Configuration	n_B	n_A	BO	S
H_2^+	$\sigma_g 1s$	1	0	1/2	1/2
H_2	$(\sigma_g 1s)^2$	2	0	1	0
He_2^+	$(\sigma_g 1s)^2(\sigma_u 1s)$	2	1	1/2	1/2
He_2	$(\sigma_g 1s)^2(\sigma_u 1s)^2$	2	2	0	0
Li_2	$[He_2](\sigma_g 2s)^2$	4	2	1	0
Be_2	$[He_2](\sigma_g 2s)^2(\sigma_u 2s)^2$	4	4	0	0
B_2	$[Be_2](\pi_u 2p)^2$	6	4	1	1
C_2	$[Be_2](\pi_u 2p)^4$	8	4	2	0
N_2	$[C_2](\sigma_g 2p)^2$	10	4	3	0
O_2	$[N_2](\pi_g 2p)^2$	10	6	2	1
F_2	$[N_2](\pi_g 2p)^4$	10	8	1	0
Ne_2	$[F_2](\sigma_u 2p)^2$	10	10	0	0

EXERCISE 10.2   The vibrational frequencies of O_2^+ and of O_2 are 1,905 and 1,580 cm^{-1}, respectively, while those of N_2^+ and N_2 are 2,207 and 2,359 cm^{-1}, and those of C_2^+ and C_2 are 1,350 and 1,855 cm^{-1}. Rationalize these behaviors using the *Aufbau* model of diatomic binding.

Answer   Vibrational frequencies vary directly with bond order, so the results are consistent with bond orders (see Table 10.1) of 2.5 and 2 for O_2^+ and O_2, respectively, 2.5 and 3 for N_2^+ and N_2, and 1.5 and 2 for C_2^+ and C_2.

10.5 ELECTRONIC STRUCTURE OF H$_2$: MOLECULAR ORBITAL AND VALENCE BOND WAVEFUNCTIONS

In this section, we consider the electronic wavefunction of H$_2$, including electron repulsion effects that were not discussed in Section 10.4. First we write the complete Hamiltonian, with reference to the coordinates in Figure 10.8. The electronic Hamiltonian is

$$\hat{H} = -\frac{1}{2}\nabla_1^2 - \frac{1}{2}\nabla_2^2 - \frac{1}{r_{a1}} - \frac{1}{r_{a2}} - \frac{1}{r_{b1}} - \frac{1}{r_{b2}} + \frac{1}{r_{12}} \text{ (in atomic units), } (10.29)$$

and the LCAO–MO wavefunction, based on the $(\sigma_g 1s)^2$ configuration, is

$$\psi_{\text{MO}} = \frac{1}{\sqrt{2}}\begin{vmatrix} \sigma_g(1)\alpha_1 & \sigma_g(1)\beta_1 \\ \sigma_g(2)\alpha_2 & \sigma_g(2)\beta_2 \end{vmatrix} = \frac{1}{\sqrt{2}}\sigma_g(1)\sigma_g(2)(\alpha_1\beta_2 - \beta_1\alpha_2), \quad (10.30)$$

where, from Eq. (10.25),

$$\sigma_g(1) = N(1s_a(1) + 1s_b(1)). \tag{10.31}$$

Notice that the spatial part of the wavefunction ψ_{MO} is

$$\begin{aligned} \sigma_g(1)\sigma_g(2) &= N^2(1s_a(1) + 1s_b(1))(1s_a(2) + 1s_b(2)) \\ &= N^2(1s_a(1)1s_b(2) + 1s_b(1)1s_a(2) + 1s_a(1)1s_a(2) + 1s_b(1)1s_b(2)) \\ &= \psi_{\text{covalent}} + \psi_{\text{ionic}}, \end{aligned} \tag{10.32}$$

where

$$N = \frac{1}{\sqrt{2 + 2S}}$$

and $S = \langle 1s_a | 1s_b \rangle$ is the overlap between orbitals on different nuclei. In Eq. (10.32), ψ_{covalent} includes terms like $1s_a(1)1s_b(2)$, in which one electron is on one nucleus

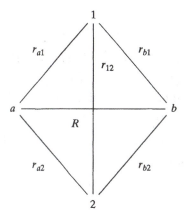

Figure 10.9 Coordinate definitions for H$_2$ molecule.

while the other is on the other nucleus, and ψ_{ionic} includes terms like $1s_a(1)1s_a(2)$, in which both electrons are on the same nucleus. ψ_{ionic} is appropriate for describing ionic bonding that would exist if H_2 were best thought of as H^+H^-. However, the bonding in H_2 is closer to being covalent, so the equal weighting of $\psi_{covalent}$ and ψ_{ionic} in defining ψ_{MO} is a source of error in the molecular orbital wavefunction.

One approach to fixing the wavefunction is simply to omit the ionic part. This is the approach of *valence bond theory*, and the resulting wavefunction for H_2 is the Heitler–London function,

$$\psi_{VB} = N'\left(1s_a(1)1s_b(2) + 1s_b(1)1s_a(2)\right)\frac{1}{\sqrt{2}}(\alpha_1\beta_2 - \beta_1\alpha_2). \quad (10.33)$$

Note that the normalization factor here is different from that in the molecular orbital wavefunction. In fact, the normalization integral is

$$1 = \langle\psi_{VB}|\psi_{VB}\rangle$$
$$= 2(N')^2\{\langle 1s_a(1)1s_b(2)\,|\,1s_a(1)1s_b(2)\rangle + \langle 1s_a(1)1s_b(2)\,|\,1s_b(1)1s_a(2)\rangle\} \quad (10.34)$$
$$= 2(N')^2(1 + S^2).$$

The properly normalized wavefunction is thus

$$\psi_{VB} = \frac{1}{\sqrt{2 + 2S^2}}\left(1s_a(1)1s_b(2) + 1s_b(1)1s_a(2)\right)\frac{1}{\sqrt{2}}(\alpha_1\beta_2 - \beta_1\alpha_2). \quad (10.35)$$

Now let us calculate the energy of H_2 using ψ_{VB}. To do this, it is convenient to group the terms in the Hamiltonian so that we have

$$H = h_a(1) + h_b(2) + \frac{1}{r_{12}} - \frac{1}{r_{b1}} - \frac{1}{r_{a2}}, \quad (10.36)$$

where h_a is the hydrogen Hamiltonian associated with nucleus a and h_b is the corresponding Hamiltonian for nucleus b. The expectation value of H is then

$$E_{VB} = \langle\psi_{VB}|H|\psi_{VB}\rangle = \frac{1}{2 + 2S^2}\{H_{ab,\,ab} + H_{ba,\,ba} + H_{ba,\,ab} + H_{ab,\,ba}\}, \quad (10.37)$$

where

$$H_{ab,\,ab} = \langle 1s_a(1)1s_b(2)\,|\,H\,|\,1s_a(1)1s_b(2)\rangle, \quad (10.38a)$$

$$H_{ab,\,ba} = \langle 1s_a(1)1s_b(2)\,|\,H\,|\,1s_b(1)1s_a(2)\rangle, \quad (10.38b)$$

and the other terms are related to these by trivial permutations. Indeed, because H is invariant to interchanging the labels a and b, we have

$$H_{ba,\,ba} = H_{ab,\,ab} \quad \text{and} \quad H_{ba,\,ab} = H_{ab,\,ba}, \quad (10.39)$$

so only Eqs. (10.38) need be considered. Upon substitution of Eq. (9.37) into those equations, we obtain

$$H_{ab,\,ab} = -\frac{1}{2} - \frac{1}{2} + Q, \quad (10.40)$$

where the $-1/2$'s come from the $1s_a(1)$ and $1s_b(2)$ orbital energies and

$$Q = \left\langle 1s_a(1)1s_b(2) \left| \frac{1}{r_{12}} - \frac{1}{r_{b1}} - \frac{1}{r_{a2}} \right| 1s_a(1)1s_b(2) \right\rangle \tag{10.41}$$

is a kind of coulomb integral that represents the difference between the force of repulsion of electron 1 on nucleus a exerted on electron 2 on nucleus b and the combined forces of attraction of electron 1 to nucleus b and electron 2 to nucleus a. The other Hamiltonian matrix element is

$$H_{ab,ba} = -S^2 + K, \tag{10.42}$$

where S is the $\langle a | b \rangle$ overlap as before and

$$K = \left\langle 1s_a(1)1s_b(2) \left| \frac{1}{r_{12}} - \frac{1}{r_{b1}} - \frac{1}{r_{a2}} \right| 1s_b(1)1s_a(2) \right\rangle. \tag{10.43}$$

K is an exchange integral (the so-called Heitler–London exchange integral) that involves repulsion and attraction associated with the *overlap charge density* $1s_a 1s_b$.

EXERCISE 10.3   Rewrite Q of Eq. (10.41) and K of Eq. (10.43) as sums involving one-electron and two-electron integrals. Note that the two-electron parts of J and K are essentially identical to the J and K integrals defined for the excited He atom in Eqs. (8.44), except that here the orbitals are $1s_a$ and $1s_b$, and in excited helium they are $1s$ and $2s$.

Answer

$$Q = \left\langle 1s_a(1)1s_b(2) \left| \frac{1}{r_{12}} - \frac{1}{r_{b1}} - \frac{1}{r_{a2}} \right| 1s_a(1)1s_b(2) \right\rangle$$

$$= \left\langle 1s_a(1)1s_b(2) \left| \frac{1}{r_{12}} \right| 1s_a(1)1s_b(2) \right\rangle - \left\langle 1s_a(1) \left| \frac{1}{r_{b1}} \right| 1s_a(1) \right\rangle - \left\langle 1s_b(2) \left| \frac{1}{r_{a2}} \right| 1s_b(2) \right\rangle$$

and

$$K = \left\langle 1s_a(1)1s_b(2) \left| \frac{1}{r_{12}} - \frac{1}{r_{b1}} - \frac{1}{r_{a2}} \right| 1s_b(1)1s_a(2) \right\rangle$$

$$= \left\langle 1s_a(1)1s_b(2) \left| \frac{1}{r_{12}} \right| 1s_b(1)1s_a(2) \right\rangle - \left\langle 1s_a(1) \left| \frac{1}{r_{b1}} \right| 1s_b(1) \right\rangle \langle 1s_b(2) | 1s_a(2) \rangle$$

$$- \left\langle 1s_b(2) \left| \frac{1}{r_{a2}} \right| 1s_a(2) \right\rangle \langle 1s_a(1) | 1s_b(1) \rangle.$$

Note that the only true two-electron integrals are

$$Q = \left\langle 1s_a(1)1s_b(2) \left| \frac{1}{r_{12}} \right| 1s_a(1)1s_b(2) \right\rangle$$

and

$$K = \left\langle 1s_a(1)1s_b(2) \left| \frac{1}{r_{12}} \right| 1s_b(1)1s_a(2) \right\rangle.$$

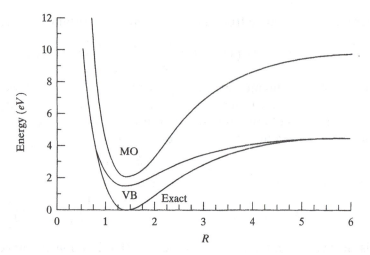

Figure 10.10 V_n versus H_2 internuclear distance.

Figure 10.10 shows the MO and VB energies, including the $1/R$ repulsions for H_2, as a function of the internuclear distance, along with the exact result. The figure has several notable features. First, both MO and VB predict that H_2 has a minimum energy that is below the energy of two separated hydrogen atoms $(-1E_h)$. If we define the difference between the minimum and the separated atoms to be the dissociation energy, then D_e = 2.68 eV for MO and 3.20 eV for VB if hydrogen $1s$ orbitals are used to evaluate the energy. If the $1s$ orbitals are screened (i.e., if they are allowed to have an effective charge Z'), then these energies become 3.49 eV (MO) and 3.78 eV (VB). The experimental dissociation energy is 4.75 eV, so we see that the relatively simple MO and VB wavefunctions that we have used can account for 73% or 80% of the dissociation energy. The VB result is more accurate than MO, as we had anticipated.

Another feature of the figure is that the VB curve dissociates to the correct energy while the MO curve does not. It is easy to see why this happens from the forms of the wavefunctions, since the VB curve represents two H atoms in this limit while MO is a superposition of two H's and $H^+ + H^-$. Because $H^+ + H^-$ has a much higher energy than H + H, the MO energy is much higher. One of the major virtues of the VB description is that it describes the dissociation of covalent bonds correctly.

10.6   IMPROVEMENTS TO MO AND VB RESULTS FOR H_2

Although the MO and VB calculations correctly predict chemical bonds in H_2, it is clear from Figure 10.10 that the accuracy of either result is not very good. Improvements can be obtained by improving the orbitals used to construct the wavefunction or by improving the two-electron wavefunctions. Here we consider simple examples of each possibility.

First, one can improve the orbitals in the MO or VB wavefunction by including "polarization functions." In this approach, we replace the $1s$ functions described earlier with

$$1s \rightarrow N(e^{-Z'r} + \lambda_1 z e^{-Z''r}), \tag{10.44}$$

where the second function is a $2p_z$ orbital centered on the same atom and λ_1 is a variational parameter. The p_z function allows the wavefunction to distort away from spherical symmetry, thereby taking into account the polarization of the orbital on each center due to the presence of the other center. If this type of polarization function is used in the VB wavefunction of H_2, the dissociation energy increases from 3.78 eV to 4.04 eV.

A more flexible form for the wavefunction can be used in which the relative amounts of ψ_{covalent} and ψ_{ionic} are variationally optimized. The idea here is to replace Eq. (10.32) with

$$\psi = \psi_{\text{covalent}} + \lambda_2 \psi_{\text{ionic}}, \tag{10.45}$$

where λ_2 is a parameter to be optimized. This is an attempt to generate a more flexible wavefunction than either the MO or VB wavefunction, and it gives $D_e = 4.02$ eV, using screened $1s$ functions as the orbital basis. It turns out that this is only a small improvement over the VB result (3.78 eV), as the contribution of the ionic terms is small ($\lambda_2 \approx 0.2$).

A general approach to improving the wavefunction is to replace a single Slater determinant with a sum of Slater determinants with coefficients that are variationally optimized. Equation (10.45) is a simple form of such a function known as a *configuration-interaction* expansion. In the more general case, the determinants would be constructed using excited atomic orbitals $2s$, $2p$, etc. We will discuss this approach in more detail in Chapter 11.

Another way of generating more accurate wavefunctions involves using functions that depend explicitly on the interelectronic coordinate r_{12}. This type of function is analogous to the Hylleraas wavefunction for He, and it can produce extremely accurate results, although the calculations are difficult. In 1936, in an application to H_2, James and Coolidge used a 13-term wavefunction that included r_{12}, as well as polarization functions and excited configurations. They obtained a dissociation energy $D_e = 4.7198$ eV. In the 1960's, Kolos and Wolniewicz used a similar expansion with 2,000 terms to give a value that matches experimental results to seven figures.

While the Hylleraas-type wavefunctions yield highly accurate results with relatively compact wavefunctions, the approach does not lend itself to the automated description of electronic states of molecules with more electrons and is not often used. Instead, the most commonly used method for describing the electronic structure of molecules is the MO method, and improvements thereto are constructed using MO wavefunctions. We therefore return to MO-based methods in the next chapter in a more general discussion of molecular electronic structure.

SUGGESTED READINGS

The classical study of the electronic structure of the hydrogen molecule is W. Kolos and L. Wolniewicz, *J. Chem. Phys.* 49, 404 (1968). Other texts with extensive discussions of homonuclear diatomic molecules include the following:

I. N. Levine, *Quantum Chemistry*, 5th ed. (Prentice Hall, Upper Saddle River, NJ, 2000).

J. P. Lowe, *Quantum Chemistry*, 2nd ed. (Academic Press, New York, 1993).

PROBLEMS

10.1 True or False?
   (a) In the united-atom limit, the $\sigma_u 1s$ orbital of H_2^+ turns into a $2p_z$ function of He^+.
   (b) If there is a minimum in the potential-energy function V as a function of the internuclear separation, then the electronic energy E_{el} must also have a minimum at the same place.

10.2 Write down the complete electronic-plus-nuclear Hamiltonian of He_2^+. Use atomic units.

10.3 Sketch the following:
   (a) The $\pi_u 3p$ orbital of H_2^+.
   (b) E_{el} versus R for the $\sigma_u 2s$ state of H_2^+.

10.4 Calculate the ground state of H_2^+, using extended Hückel theory. In this theory, one employs the usual LCAO expansion to define the wavefunction, but the Hamiltonian matrix elements are approximated as

$$H_{1s_a, 1s_a} = H_{1s_b, 1s_b} = -\varepsilon_{1s}$$

and

$$H_{1s_a, 1s_b} = \frac{1}{2} K S_{1s_a, 1s_b} \left(H_{1s_a, 1s_a} + H_{1s_b, 1s_b} \right),$$

where ε_{1s} is the $1s$ orbital energy, $S_{1s_a, 1s_b}$ is the overlap integral, and K is a universal "fudge factor" with the value 1.75. Let

$$S_{1s_a, 1s_b} = \left(1 + R + \frac{1}{3} R^2 \right) e^{-R},$$

and use a computer or calculator to determine the dependence of the ground-state energy on R. Make a plot of the result, and also plot the sum of the electronic and nuclear repulsion energies. What is the H_2^+ equilibrium internuclear distance and dissociation energy, according to this model?

10.5 The normalized σ_g orbital of H_2^+ is given by

$$\psi = N(1s_a + 1s_b),$$

where $N = [2(1 + S)]^{-1/2}$ and the overlap is

$$S = \left(1 + R + R^2/3 \right) e^{-R} \qquad \text{(in atomic units).}$$

   (a) Use a calculator to evaluate $|\psi|^2$, employing the line connecting the nuclei for $R = 1.8$ at the coordinates $x = 0$, $y = 0$, and $z = -4, -3, -2, -1.5, -1.0, -0.9, -0.5, 0, 0.5, 0.9, 1.0, 1.5, 2, 3,$ and 4 where z measures the distance along the molecule's axis and $z = 0$ is at the midpoint between the nuclei.

(b) What is the difference in density,

$$\Delta\rho = |\psi|^2 - \frac{1}{2}\left(1s_a^2 + 1s_b^2\right),$$

at the same points as in Part (a)? Where is $\Delta\rho$ the most positive?

10.6 Using experimental data given in this chapter, compute ΔE for the reaction $2H_2^+ \rightarrow 2H^+ + H_2$. Comment on the result: Is the H_2^+ molecule stable to this metathesis-type reaction?

10.7 Write down a Slater determinant for the ground state of He_2.

10.8 Express the valence bond wavefunction of H_2 as a sum of two Slater determinants.

10.9 State two advantages and one disadvantage of VB wavefunctions over MO wavefunctions.

10.10 The simplest MO approximation for the excited state of H_2 is a singlet, with one electron promoted from $\sigma_g 1s$ to $\sigma_u 1s$. Call this state ψ_x, and call the ground state $(\sigma_g 1s \bar{\sigma}_g 1s)\psi_0$. Evaluate $\langle\psi_0|\psi_x\rangle = S_{0x}$ and $\langle\psi_0|H|\psi_x\rangle = H_{0x}$. Comment on your results.

10.11 The lowest singlet excited state of H_2, called ψ_x, is prepared by promoting one electron from σ_g to σ_u. Can this state be constructed as a single Slater determinant? If so, do so. If not, express the state as a combination of Slater determinants.

11

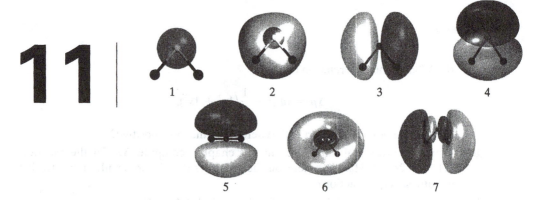

1 2 3 4

5 6 7

Ab Initio and Density Functional Methods

By quantum chemistry, we usually mean the description of the electronic structure of molecules. The most important methods for determining that structure are based on either a Hartree–Fock model or a density functional approach. In this chapter, these two dominant themes are introduced, as are several methods for improving the general, but quantitatively inadequate, description that simple density functional theory and the Hartree–Fock description provide.

11.1 LCAO–MO–SCF THEORY FOR MOLECULES

In this section, we combine the LCAO–MO theory of Chapter 10 with the Hartree–Fock (SCF) theory from Chapter 9. The resultant LCAO–MO–SCF theory, which is the basic tool of many *ab initio* molecular orbital methods, can be found in numerous computer codes, and is an important tool of contemporary chemical science.

The complete (nonrelativistic) electronic Hamiltonian is

$$\hat{H}_{e\ell} = \sum_{i=1}^{N} \left\{ -\frac{1}{2}\nabla_i^2 - \sum_{\alpha=1}^{M} \frac{Z_\alpha}{r_{i\alpha}} + \sum_{j>i} \frac{1}{r_{ij}} \right\} \quad \text{(in atomic units)}, \qquad (11.1)$$

where the indices i and j label the N electrons and α labels the M nuclei. Z_α is the charge on nucleus α, and $r_{i\alpha}$ is the distance from electron i to nucleus α.

In LCAO–MO–SCF theory, we assume that the wavefunction can be written as a Slater determinant of molecular orbitals. If the orbitals are all doubly occupied, then this determinant has the form

$$
\psi = \frac{1}{\sqrt{N!}}
\begin{vmatrix}
\phi_1(1)\alpha_1 & \phi_1(1)\beta_1 & \phi_2(1)\alpha_1 & \phi_2(1)\beta_1 \cdots \\
\phi_1(2)\alpha_2 & \phi_1(2)\beta_2 \cdots & & \\
\cdots & & & \\
\cdots & & & \\
\phi_1(N)\alpha_N & \phi_1(N)\beta_N \cdots & &
\end{vmatrix}
\tag{11.2}
$$

where the ϕ_ks $(k = 1, \ldots, N/2)$ are the molecular orbitals (MOs), written as a linear combination of atomic orbitals (LCAO) b_ν via the formula

$$
\phi_k = \sum_\nu c_{k\nu} b_\nu.
\tag{11.3}
$$

Note that we use Greek indices to label the atomic orbitals; the $c_{k\nu}$ are coefficients that are treated as variational parameters.

In the Hartree–Fock approach, the orbitals ϕ_k are obtained by solving the equation

$$
\hat{f}_k \phi_k = \varepsilon_k \phi_k,
\tag{11.4}
$$

where the Fock operator is [see Eq. (9.21)]

$$
\hat{f}_k = -\frac{1}{2}\nabla_k^2 - \sum_\alpha \frac{Z_\alpha}{r_{k\alpha}} + \sum_j \left(2J_j(k) - K_j(k)\right),
\tag{11.5}
$$

in which J_j and K_j are the coulomb and exchange operators, as in Chapter 9.

If the LCAO expansion given by Eq. (11.3) is substituted into Eq. (11.4), one can convert the latter equation into a secular equation by following a procedure similar to that used for Eqs. (10.14–10.20). In the first step, after dropping the index k in Eqs. (11.3) and (11.4), we get

$$
\sum_\nu \hat{f} b_\nu c_\nu = \varepsilon \sum_\nu b_\nu c_\nu.
\tag{11.6}
$$

Multiplying this equation by b_μ and integrating, we get

$$
\sum_\nu \left(f_{\mu\nu} - \varepsilon S_{\mu\nu}\right)c_\nu = 0,
\tag{11.7}
$$

where

$$
S_{\mu\nu} = \langle b_\mu | b_\nu \rangle
\tag{11.8}
$$

and

$$
f_{\mu\nu} = \langle b_\mu | \hat{f} | b_\nu \rangle.
\tag{11.9}
$$

Eq. (11.7) is a set of linear homogeneous equations whose solutions (other than $c_\nu = 0$) must satisfy the secular equation (see Appendix A.3)

$$|\mathbf{f} - \varepsilon \mathbf{S}| = 0, \tag{11.10}$$

where $\mathbf{f}$ and $\mathbf{S}$ are matrices formed from Eqs. (11.8) and (11.9). By substituting Eq. (11.5) into Eq. (11.9), we may express the Fock matrix

$$f_{\mu\nu} = h_{\mu\nu} + 2J_{\mu\nu} - K_{\mu\nu}, \tag{11.11}$$

where

$$h_{\mu\nu} = \left\langle b_\mu \left| -\frac{1}{2}\nabla^2 - \sum_\alpha \frac{Z_\alpha}{r_{i\alpha}} \right| b_\nu \right\rangle \tag{11.12}$$

is the one-electron part of the Fock operator and

$$J_{\mu\nu} = \left\langle b_\mu \left| \sum_j J_j \right| b_\nu \right\rangle = \sum_j \sum_{\sigma\lambda} \langle \mu\nu \,|\, \sigma\lambda \rangle c_{j\lambda} c_{j\sigma} \tag{11.13}$$

and

$$K_{\mu\nu} = \left\langle b_\mu \left| \sum_j K_j \right| b_\nu \right\rangle = \sum_j \sum_{\sigma\lambda} \langle \mu\lambda \,|\, \sigma\nu \rangle c_{j\lambda} c_{j\sigma} \tag{11.14}$$

are the two electron terms, in which $\langle \mu\nu \,|\, \sigma\lambda \rangle$ is shorthand notation for the repulsion integral

$$\langle \mu\nu \,|\, \sigma\lambda \rangle = \left\langle b_\mu(1) b_\sigma(2) \left| \frac{1}{r_{12}} \right| b_\nu(1) b_\lambda(2) \right\rangle. \tag{11.15}$$

Note that the coulomb and exchange terms depend on the orbital coefficients, and therefore, it is necessary to solve the Hartree–Fock equations self-consistently by first assuming values for the coefficients in Eqs. (11.13) and (11.14), then solving Eq. (11.6) to get new coefficients, and so on.

The Hartree–Fock theory that we have developed so far refers to orbitals that are assumed to be doubly occupied. This theory is sometimes called *restricted Hartree–Fock (RHF)* theory, for it restricts the spatial orbitals so that α and β spins are identical. For open-shell atoms and molecules, it is often better to allow the spatial orbitals to be different for different spins. This assumption leads to *unrestricted Hartree–Fock (UHF)* theory.

EXERCISE 11.1   Show that if we set all $\langle \mu\nu \,|\, \sigma\lambda \rangle$ terms to zero (or equivalently, if we ignore all effects of interelectronic repulsion), then the Hartree–Fock equations become separable, so that no iterative solution is needed, and the total energy is the sum of the one-electron energies.

Answer   The Fock operator is

$$f_{\mu\nu} = h_{\mu\nu} + 2J_{\mu\nu} - K_{\mu\nu}.$$

If all $\langle \mu\nu \,|\, \sigma\lambda \rangle$ terms vanish, then, from Eqs. (11.13) and (11.14), $J_{\mu\nu} = K_{\mu\nu} = 0$. It follows that

$$f_{\mu\nu} = h_{\mu\nu}(i) = \left\langle b_\mu \left| -\frac{1}{2}\nabla_i^2 - \sum_\alpha \frac{Z_\alpha}{r_{i\alpha}} \right| b_\nu \right\rangle,$$

for the ith electron. But this is just

$$h_{\mu\nu}(i) = \langle b_\mu \,|\, h_i \,|\, b_\nu \rangle.$$

The $h_{\mu\nu}$ can then be evaluated directly from the matrix elements of the nuclear attraction and the kinetic energy. These are one-electron terms, so $H = \sum_i h_i$ and $E = \langle H \rangle = \sum_i \langle h_i \rangle$. Hence, if the interelectron repulsion is ignored, finding E is very easy. This approach is adopted in the Hückel model of Chapter 12 and related models.

11.2 ATOMIC ORBITALS

In LCAO calculations, it is necessary to choose a set of functions b_μ to represent the atomic orbitals. This set is usually called the *basis set* for the calculation.

In principle, the atomic orbitals can be chosen to be hydrogenlike wavefunctions, or even Hartree–Fock orbitals, for the atoms. However, neither of these is often used, as their complicated functional form (with lots of nodes near the nucleus for high n and small ℓ functions) makes them cumbersome for evaluating integrals. A more commonly used set of functions is the set of *Slater orbitals*, which have the form

$$b = Ae^{-\zeta r}r^{n^*-1}Y_{\ell m}(\theta, \phi) \tag{11.16}$$

and thus are a lot like hydrogen orbitals, but without the complicated nodal structure. The parameters n^* and ζ are chosen to make the large-r (valence) part of the orbitals look like atomic Hartree–Fock orbitals.

Slater orbitals have often been used for Hartree–Fock calculations on linear molecules, and they are commonly used in semiempirical MO calculations, but the complexity of doing multicenter electron repulsion integrals [Eq. (11.15)] makes their more general use for molecular Hartree–Fock calculations difficult.

The most commonly used atomic orbitals are *Gaussian orbitals*, which have the form

$$g = x^a y^b z^c e^{-\alpha r^2}Y_{\ell m}(\theta, \phi), \tag{11.17}$$

where a, b, and c are integers and α is a parameter that is usually fixed. Equation (11.17) defines a *primitive* Gaussian function. Normally, several of these Gaussians are summed to define more realistic AO basis functions, as in the formula

$$b_\mu = \sum_p k_{\mu p} g_p. \tag{11.18}$$

The coefficients $k_{\mu p}$ in this expansion are chosen to make the basis functions look as much like Slater orbitals as possible. In the simplest version of this basis, n Gaussians are superimposed with fixed coefficients to form one Slater-type orbital (STO). Such a basis is denoted STO–nG, and $n = 3, 4, \ldots$ have commonly been used. The exponent parameters in STO–nG basis sets are chosen so that one Gaussian is sharply peaked near the nucleus, thereby approximating the cusp in the STO. Other exponents are smaller, so as to describe the large-r parts of the wavefunctions.

Split-valence basis sets use sums of Gaussians such that there is more than one set of basis functions for each subshell. Thus, a 3–21G basis uses three Gaussians, grouped into two Gaussians that are summed with fixed relative coefficients and one that is used directly. Sometimes these are called "double-zeta-quality" basis functions, as they are something like summing two Slater functions with different ζ's. Note that the split-valence character of basis functions is applied only to the valence subshell. In a 3–21G basis, the inner-shell orbitals are represented using three Gaussians with fixed coefficients. Similarly, a 6–31G basis uses six Gaussians to represent the inner-shell orbitals and a split-valence set of four Gaussians (grouped into subsets of three and one) for the valence orbitals.

Generally, basis functions are optimized to reproduce the results of Hartree–Fock calculations. However, in recent years it has been realized that, for calculations which go beyond Hartree–Fock in treating electron correlation, improved results may be obtained using "correlation-consistent" basis sets that are based on high-quality (i.e., beyond Hartree–Fock) atomic wavefunctions.

Another improvement to STO–nG basis sets involves using *polarization functions*. These are Gaussians similar to Eq. (11.17), except that the orbital angular momentum ℓ is one (or more) larger than is appropriate for the orbital being described. Thus, in describing a $2p$ orbital on an atom, a polarization function would have $3d$ character. The purpose of a polarization function is to describe distortion of the orbital by the other atomic centers away from what would be expected for a spherically symmetric atomic environment. We have already considered one example of a polarization function in Eq. (10.44), where we added a $2p_z$ function to the $1s$ orbital basis used to describe the H_2 molecule. In using Gaussian orbitals, it is typical to denote basis sets that allow for polarization functions using an asterisk (*), so a split-valence-plus-polarization basis might be 6–31G* (if a polarization function is to be added to atoms other than hydrogen) or 6–31G** (if polarization functions are to be added for hydrogen as well). A more explicit notation that differentiates between the orbital angular momentum used for H atoms with that for heavier atoms is 6–31G(kp, ℓd), where k and ℓ are integers that indicate the number of p (for H) and d (for other atoms) polarization functions included in the basis.

One other set of orbital basis functions that is commonly used is the set of *diffuse functions*—Gaussian orbitals that have very small exponent parameters α, so that they allow the wavefunction to extend far from the nucleus. Diffuse functions are important in describing weakly bound electronic states, such as occur for negative ions. A commonly used notation for diffuse functions is 4–31G+, where the "+" indicates that one diffuse function is included for each valence orbital. If one uses both polarization and diffuse orbitals, then the designation would be 4–31G*+.

The following are some other kinds of basis functions that are occasionally used:

1. *Bond-centered functions.* These are Gaussian orbitals that are located at the center (midpoint) of the bond between atoms. They are used to improve the description of the valence orbitals for geometries where bonding is most important.

2. *Floating spherical Gaussians.* In this basis, the position of the center of the orbital and the Gaussian width are variationally optimized.

3. *Plane waves.* These are completely delocalized functions of the form $e^{i\mathbf{k}\cdot\mathbf{r}}$.

4. *Numerical atomic Hartree–Fock orbitals.* These are products of spherical harmonics and numerically tabulated radial functions that are obtained from atomic Hartree–Fock calculations.

5. *Correlation-consistent orbitals.* These are Gaussian orbitals, expressed using Eq. (11.18), in which the coefficients are determined by solving the Schrödinger equation for atoms with a high level of correlation included. Examples of calculations based on these basis sets are given in Chapter 14.

6. *Point-grid basis.* With these basis functions, one transforms the Schrödinger equation into a set of algebraic equations on a grid. This can be done with finite-element methods, splines, discrete variable representations, and other methods.

EXERCISE 11.2   Because the coulomb attraction between nuclei and electrons becomes infinitely negative for $r_i = 0$, the wavefunction must obey the so-called *cusp condition*, wherein the function has a sharp point (with a discontinuous slope) at $r_i = 0$. (See Fig. 10.3 for an example.) Which of the following basis sets obey the cusp condition?

1. Slater-type orbitals
2. Hydrogen-atom eigenfunctions
3. Primitive Gaussians
4. STO–3G
5. 6–311G**

Answer   Slater and hydrogen-atom functions do, but the rest do not. Any orbitals based on Gaussians have a slope of zero at the origin.

11.3 HARTREE–FOCK CALCULATIONS

11.3.1 Application to Water

In this section, we present the results of a routine Hartree–Fock calculation for the molecule H_2O. These calculations can be carried out with many commonly available electronic structure programs, such as SPARTAN, GAMESS, Gaussian, Jaguar, Qchem, and the like. Many details of these calculations are considered in Chapter 14, but for now we simply assume that we have done them using a 6–311G** basis set for a geometry in which the molecule has C_{2v} symmetry, the OH bond distance is 0.941 Å, and the HOH internal angle is 105.4°. Table 11.1 gives the Cartesian coordinates of the atoms.

TABLE 11.1 CARTESIAN COORDINATES OF H_2O MOLECULE (IN Å)

Atom	X-coordinate	Y-coordinate	Z-coordinate
H_1	0.7487	0.0	0.4559
O	0.0	0.0	−0.1140
H_2	−0.7487	0.0	0.4559

The 6–311G** basis set in this case consists of 30 functions. For each H atom, there are $3s$ functions (designated by the "311" part of the basis set) and a single set of p's $\left(p_x, p_y, p_z\right)$ that correspond to the polarization functions. This gives a total of 6 basis functions for each H atom. For the O atom, there is a core s function (composed of a sum of 6 Gaussians) and then 3 valence s functions, 3 valence sets of p's, and a single set of d's (polarization functions). This means that there are 18 functions in all on the O atom.

The converged Hartree–Fock calculation gives a total electronic energy of −76.047012 hartrees, and the molecular orbital energies and orbital coefficients (the $c_{k\nu}$'s of Eq.11.3) are as indicated in Table 11.2. Figure 11.1 presents contours of the lowest seven molecular orbitals. To understand the results of the calculation, note first that there are 10 electrons in water, so the lowest five orbitals are occupied. The lowest energy orbital in Fig. 11.1 is strongly localized near the nucleus of the oxygen atom, indicating that that orbital is a $1s$ orbital on the oxygen atom. Note that this orbital has no nodal planes, so it is the lowest orbital of the system, and it has negligible H-atom coefficients, so it is not involved in bonding. Orbital 2 has only one nodal plane, namely, a nearly spherical surface around the oxygen atom. (This is hard to see in the figure.) Consequently, orbital 2 is similar to the $2s$ orbital of the isolated oxygen atom. Although orbital 2 extends out to the region where the H atoms are localized, Table 11.2 indicates that the orbital coefficients associated with the H atoms for this orbital are relatively small. Thus, this orbital participates only weakly in bonding. The most important bonding orbitals are orbitals 3 and 4, both of which have large coefficients on both oxygen and hydrogen atoms and neither of which has any

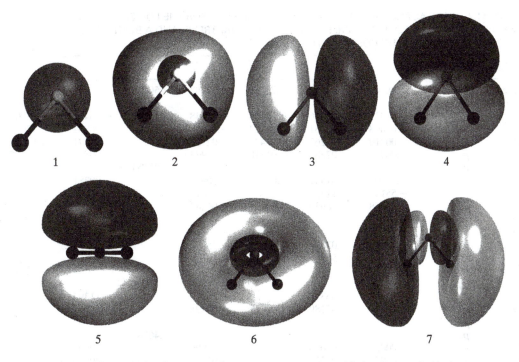

Figure 11.1 Contours of lowest seven molecular orbitals of water. Orbital
numbers are the same as in Table 11.2. Filled black circles connected by lines
are used to denote the positions of the atoms.

nodal planes between these atoms. The coefficients indicate that orbital 3 is similar to an oxygen p_x orbital, while orbital 4 is similar to a p_z orbital. Orbital 5,
which is the highest occupied molecular orbital (HOMO), is very close to an
oxygen p_y orbital and has very little H-atom content. Accordingly, it can be considered to be an oxygen *lone pair* orbital. Because it has a nodal plane that is co-
incident with the plane of the molecule, it will mix only with the polarization
functions on the H atoms, and even then, the mixing is weak. Orbital 6 is the
lowest unoccupied molecular orbital (LUMO), and it has predominately oxygen
$3s$ character, while orbital 7 has the same appearance as an oxygen $3p_x$ function, although it is mostly localized on the H atoms. These unoccupied orbitals
have nodal planes that cross the OH bonds, which means that they are of antibonding character. This suggests that the lowest two excited states of the water
molecule might be unbound (i.e., they will dissociate spontaneously into
H + OH), and indeed, that is what is found.

The energy of an orbital can often be a useful guide to the nature of the orbital. Orbital 1 has a much lower energy than the others, and although it is responsible for much of the total electronic energy, it does not contribute to bonding.
Orbital 2 is mostly oxygen $2s$, and the energy of this is still well removed from that
of isolated hydrogen atoms $(-0.5E_h)$, so there is little interaction. Orbitals 3 and

TABLE 11.2   ORBITAL ENERGIES AND ORBITAL COEFFICIENTS FOR THE LOWEST SEVEN HARTREE–FOCK ORBITALS IN WATER. ENERGIES ARE IN HARTREES, AND A BLANK CELL INDICATES THAT THE CORRESPONDING COEFFICIENT IS ZERO.

MO:		1	2	3	4	5	6	7
Energy:		−20.541	−1.3492	−.71721	−.57287	−.50066	.15259	.21857
Atom	Function							
H_1	s	−.00021	.09656	.15126	−.08748		−.03413	−.02364
H_1	s	−.00008	.08131	.21224	−.14604		.06832	.13008
H_1	s	−.00010	−.00267	.05449	−.02727		−.84441	−1.58014
H_1	p_x	.00007	−.02386	−.01870	.02973		−.00513	−.00885
H_1	p_y					.03154		
H_1	p_z	.00001	−.01441	−.02567	−.00956		.00381	−.00819
O	s	−.55143	−.11336		−.03815		−.03350	
O	s	−.47168	−.18936		−.06487		−.05463	
O	p_x			.22740				.12024
O	p_y					.29169		
O	p_z	−.00178	.03804		−.25567		.07159	
O	s	−.00557	.53789		.19649		.10269	
O	p_x			.34889				.12747
O	p_y					.43668		
O	p_z	.00062	.06315		−.37815		.11396	
O	s	.00046	.37189		.33491		.85269	
O	p_x			.21181				.49547
O	p_y					.46589		
O	p_z	−.00009	.02078		−.33996		.19994	
O	d_{z2}	.00002	.00276		−.01699		.00456	
O	d_{y2-z2}	.00012	.00804		−.00506		.00367	
O	d_{xy}							
O	d_{xz}			.02933				.00553
O	d_{yz}					.01714		
H_2	s	−.00021	.09656	−.15126	−.08748		−.03413	.02364
H_2	s	−.00008	.08131	−.21224	−.14604		.06832	−.13008
H_2	s	−.00010	−.00267	−.05449	−.02727		−.84441	1.58014
H_2	p_x	−.00007	.02386	−.01870	−.02973		.00513	−.00885
H_2	p_y					.03154		
H_2	p_z	.00001	−.01441	.02567	−.00956		.00381	.00819

4 have energies comparable to those of the isolated hydrogen atoms and thus can participate strongly in bonding. Orbital 5 also has an energy close to the hydrogen ground-state energy, but, as described in the previous paragraph, the nodal structure of this orbital doesn't allow it to participate effectively in bonding. Orbitals 6 and 7 have positive energies, which indicates that adding an electron would destabilize the water molecule. In fact, water does not have bound negative ion states.

11.3.2 Quality of Hartree–Fock Results

In this section, we describe more generally what one gets from Hartree–Fock calculations. For molecules, these calculations are typically done for a variety of basis functions of increasing complexity, in order to study the convergence behavior of the properties of interest. In the limit of a large enough basis, one recovers the exact solution of the Hartree–Fock equations, the so-called *Hartree–Fock limit*. However, this limit is still far from the exact description of the electronic structure for many properties. For molecules composed of atoms in the second row of the periodic table (Li–Ne), one finds the following results from Hartree–Fock calculations with modest basis sets:

1. Equilibrium bond distances are accurate to 0.05 Å, usually being too large.
2. Vibrational frequencies are high by 10–15%, on average. This result is so commonly found that it is customary to multiply all Hartree–Fock frequencies by a factor such as 0.89 before comparing the calculation with experiment.
3. Dissociation energies are usually terrible, typically low by a substantial fraction, like 50%. Barriers to chemical reactions are even worse, high by 100% to several hundred percent.

Examples that illustrate these results are given in Chapter 14.

11.4 BEYOND HARTREE–FOCK

The results given in the previous section indicate that Hartree–Fock does poorly for many properties, so it is common to use methods that are "beyond Hartree–Fock." This means that the correlation energy is actively included in some sense. In this section, we describe several of the most frequently used methods.

11.4.1 Perturbation Theory

Standard time-independent perturbation theory (see Section 5.4) is used to estimate the correlation energy perturbatively. The perturbation Hamiltonian in this case is the difference between the exact and Hartree–Fock Hamiltonians. Typically, such a treatment is called Møller-Plesset (MP) perturbation theory, and two commonly used levels are the second order (MP2) and fourth order (MP4). In second-order perturbation theory, the sum over intermediate states in Eq. (5.58) involves states in which two electrons are excited relative to the ground configuration (so-called double excitations), while fourth-order perturbation theory includes excitations of up to four electrons relative to the ground configuration.

MP2 calculations involve relatively modest additional effort over Hartree–Fock, and they produce reasonable improvements in results, including (for typical stable organic molecules) harmonic frequencies that are good to 5–10% and dissociation energies that are good to 20–100%. MP4 calculations take a good deal more effort, but produce a significant improvement in results.

11.4.2 Configuration Interaction

Configuration interaction (CI) is a traditional approach to describing electron correlation in molecules in which one writes the wavefunction as a sum of many Slater determinants, with coefficients that are to be variationally optimized. Thus, one writes

$$\psi = \sum_k c_k D_k, \tag{11.19}$$

where the determinants D_k include the Hartree–Fock ground state and excited determinants obtained from the ground state by exciting one or more electrons from occupied to unoccupied orbitals. Determinants are usually labelled by the number of electrons that are excited; thus, *single excitations* refer to determinants in which one electron is moved from an occupied orbital to an unoccupied orbital, *double excitations* involve moving two electrons, and so on. A CI expansion in which all possible excitations are included is called a *full CI* (FCI), and one in which only single and double excitations are included is called a *singles–doubles CI* (SDCI). FCI gives an exact answer, within the chosen basis of orbitals.

Variational optimization of the coefficients in Eq. (11.19) leads to the secular equation

$$|\mathbf{H} - E\mathbf{S}| = 0, \tag{11.20}$$

where the Hamiltonian matrix element $H_{k\ell}$ involves the complete many-electron Hamiltonian between determinants D_k and D_ℓ, and the overlap matrix $\mathbf{S}$ is usually the identity matrix. Although CI energies are always better than HF energies, to obtain results that are reasonably converged with respect to the number of terms in Eq. (11.19) usually requires a lot of configurations. As a result, the dimensions of Eq. (11.20) are often quite large (in the millions to billions), so special procedures are required to determine eigenvalues and eigenfunctions.

11.4.3 Multiconfiguration SCF and MCSCF–CI

The basic idea behind a multiconfiguration SCF calculation is to use a wavefunction that consists of a sum of determinants (typically, tens to thousands of determinants) and to optimize variationally both the coefficients in front of each determinant and the orbitals in each determinant. To do this, one needs to solve equations for the orbitals that are a lot like the Hartree–Fock equations (and hence they must be solved self-consistently), but with two-electron repulsions and exchanges that reflect the multi-determinantal nature of the wavefunction. Once self-consistency is achieved, it is possible to use the resulting molecular orbitals as the basis for configuration interaction calculations (i.e., MCSCF–CI) in which one adds in the effects of determinants that are not included in the MCSCF wavefunction.

MCSCF calculations are especially good for describing bond breaking and forming. In fact, one type of MCSCF calculation is based on summing determinants that would normally be used in a valence-bond description of the molecular electronic structure. Variants on this method are often called generalized valence-bond (GVB) calculations. Another type of MCSCF calculation is called a complete active-space SCF (CASSCF) calculation. In this calculation, one includes all possible determinants that can be constructed from the occupied and empty valence orbitals that are most important to the bond breaking and forming steps.

11.4.4 Coupled-Cluster Methods

Coupled-cluster (CC) methods use a wavefunction that is derived from the Hartree–Fock state Φ_0 using the expression

$$\Psi_{CC} = e^T \Phi_0, \tag{11.21}$$

where

$$T = T_1 + T_2 + T_3 + \cdots \tag{11.22}$$

describes the excitations. Here, T_1 represents single excitations, T_2 double excitations, T_3 triple excitations, and so forth. The advantage of this form of the wavefunction over a CI expansion is that Eq. (11.21) includes some excitations to all orders and thus is capable of describing more of the correlation energy. The exponential form of Eq. (11.21) also simplifies the evaluation of matrix elements of the Hamiltonian operator, compared to the evaluation of other wavefunctions that involve summing many determinants.

The most commonly used CC method is one in which single and double excitations are included in Eq. (11.22) (leading to the so-called CCSD method). This level of calculation is significantly more time consuming than the calculation that is performed in Hartree–Fock, but is typically much less expensive than MCSCF-CI calculations. The accuracy of CCSD may be improved using perturbation theory to estimate the contribution to the correlation energy arising from missing terms in Eq. (11.22).

11.5 DENSITY FUNCTIONAL THEORY METHODS

Density functional theory (DFT) methods are based on the Hohenberg–Kohn (HK) theorem, which states that the ground-state electronic energy E of an atom or molecule can be expressed exactly as a functional of the electron density ρ of the molecule. The term "functional" refers to a function of a function, which in the present case means that the energy has a functional dependence on ρ, and then ρ is a function of the coordinates of the electrons. To define what the HK theorem means, we first write the usual expression for the total electronic energy, viz.,

$$E = T + V_{\text{nucl}} + V_{\text{rep}} + E_{\text{xc}}, \tag{11.23}$$

where T is the electronic kinetic energy, V_{nucl} is the attraction of the electrons to the nuclei, V_{rep} is the interelectronic coulomb repulsion, and E_{xc} is the exchange-correlation energy. The second and third terms in this expression are familiar to us from earlier in the chapter:

$$V_{\text{nucl}} = -\sum_\alpha \int \frac{Z_\alpha \rho(1)}{r_{1\alpha}} \, d\tau_1; \tag{11.24}$$

$$V_{\text{rep}} = \frac{1}{2} \iint \frac{\rho(1)\rho(2)}{r_{12}} \, d\tau_1 \, d\tau_2. \tag{11.25}$$

Note that both V_{nucl} and V_{rep} are functionals of ρ. The kinetic energy T may also be expressed in terms of the electron density, although the general expression is quite complicated and not completely known. An expression that is appropriate for electrons in boxlike potentials (a free-electron gas) is

$$T = \frac{3}{10} \left(3\pi^2\right)^{2/3} \int \rho^{5/3} d\tau. \tag{11.26}$$

This expression, and improvements thereto that take into account the gradient of ρ (so-called gradient-corrected DFT theories), are rarely used in applications to the electronic structure of molecules. Instead, the more traditional expression,

$$T = -\frac{1}{2} \sum_i \int \psi_i \nabla^2 \psi_i d\tau, \tag{11.27}$$

involving the wavefunction is used. A method for determining the orbitals ψ_i is given shortly.

The last term in Eq. (11.23) represents the effects of electron exchange and correlation on the energy. There is no known exact expression for E_{xc}, but a variety of approximate expressions have been developed that lead to a variety of methods. The simplest DFT approximation is known as the *local density approximation* (LDA) and is given by

$$E_{\text{xc}} = -\frac{9}{8} \left(\frac{3}{\pi}\right)^{1/5} \alpha \int \rho(1)^{4/3} d\tau_1, \tag{11.28}$$

where α is equal to unity for a free-electron gas and values of about 0.7 are commonly used for molecules. The LDA approximation is not generally of high enough accuracy to be useful for determining structural properties and dissociation energies of molecules. However, it is possible to improve the quality of the results substantially by adding correction terms to Eq. (11.28) that depend on the gradient of the electron density. Several gradient-corrected density functional methods [generalized gradient approximation (GGA) methods] have appeared in recent years that (after empirical calibration) yield structures and energetic properties of molecules which are significantly better than Hartree–Fock properties. Unfortunately, there is no systematic procedure for approaching the exact solution to the Schrödinger equation through the use of successively more accurate gradient corrections. This is an important difference compared to wavefunction-based methods (such as SCF–CI), where increasing the size of the basis set and the number of configurations included in the wavefunction ultimately approaches the exact result.

EXERCISE 11.3   The variational principle guarantees that $\langle \psi | H | \psi \rangle$ for any approximate, normalized wavefunction will yield an energy E_{trial} that is an upper limit to the true ground-state energy. Therefore, any electronic structure method that is guaranteed to give an upper bound to the energy is called variational. Which of the following methods are variational, and why?

1. HF
2. CI
3. MCSCF
4. DFT–LDA
5. MP2
6. MP4

Answer   HF, CI, and MCSCF all use a wavefunction and optimize parameters in it. So all of them are variational. MP2 and MP4 correct the energy by using perturbation theory. Therefore, they are not variational. DFT–LDA uses an approximate energy functional, so it, too, is not variational.

If Eqs. (11.24), (11.25), (11.26), and (11.28) are substituted into Eq. (11.23), the energy is expressed only in terms of the electron density, which means that one does not need to determine wavefunctions. However, it is difficult to obtain high accuracy on the basis of this approach, so in practice it is common to determine the density from wavefunctions that are obtained from self-consistent field calculations. The relevant theory is from Kohn and Sham and involves solving the equation

$$F\psi = \varepsilon\psi, \tag{11.29}$$

where

$$F(1) = -\frac{1}{2}\nabla_1^2 - \sum_\alpha \frac{Z_a}{r_{1a}} + \sum_j J_j(1) + V_{\text{xc}} \tag{11.30}$$

and

$$V_{\text{xc}} = \frac{\partial E_{\text{xc}}}{\partial \rho}. \tag{11.31}$$

Once the Kohn–Sham orbitals ψ_i are determined, the electron density is obtained from the sum over occupied orbitals:

$$\rho = \sum_i |\psi_i|^2. \tag{11.32}$$

A key advantage of density functional methods based on the Kohn–Sham methodology is that the effort required to determine the electron density for the evaluation of Eqs. (11.24), (11.25), (11.27), and (11.28) is similar to (actually, even somewhat less than) that needed for a Hartree–Fock calculation, but the accuracy of the results is comparable to those obtained from MP2 (for an appropriate choice of gradient correction in the evaluation of E_{xc}). One way to understand the higher accuracy of Kohn-Sham DFT compared to Hartree–Fock is in terms of consistency: In Hartree–Fock, the exchange term is treated properly, but all correlation effects are ignored; in DFT, both exchange and correlation are included in the functional E_{xc}. Furthermore, since the precise form of E_{xc} is unknown, it has become popular to use expressions for E_{xc} which depend on parameters that are empirically adjusted to optimize agreement with experiment for a selection of molecules. Such hybrid functionals will be discussed further in Section 14.1.

SUGGESTED READINGS

D. Feller and E. Davidson, in *Reviews of Computational Chemistry*, ed. K. B. Lipkowitz and D. Boyd (VCH Publishers, New York, 1999), pp. 1–43. (Reviews basis sets.)

W. J. Hehre, L. Radom, P. v. R. Schleyer, and J. A. Pople, *Ab initio Molecular Orbital Theory* (Wiley, New York, 1986).

R. G. Parr and W. Yang, *Density-functional Theory of Atoms and Molecules* (Oxford, New York, 1989).

W. G. Richards and D. L. Cooper, *Ab Initio Molecular Orbital Calculations in Molecular Electronic Structure Theory* (Oxford University Press, Oxford, U.K., 1984).

J. Simons and J. Nichols, *Quantum Mechanics in Chemistry* (Oxford, New York, 1997).

A. Szabo and N. S. Ostlund, *Modern Quantum Chemistry* (McGraw-Hill, New York, 1989).

S. Wilson, *Electron Correlation in Molecules* (Clarendon, Oxford, 1984).

PROBLEMS

11.1 True or False?

    (a) MCSCF is an improvement on Hartree–Fock that describes correlation energy using perturbation theory.

    (b) 3–21G is a Gaussian basis set in which each valence orbital is represented as the fixed sum of three Gaussian functions.

    (c) For NaCl, the valence-bond wavefunction is more accurate than the molecular orbital wavefunction.

    (d) Slater orbitals are solutions of the Schrödinger equation for the hydrogen atom.

11.2. Which of the following basis sets have a nonzero slope at $r = 0$?

    (a) Slater.

    (b) 6–31G.

    (c) A numerical Hartree–Fock atomic wavefunction.

    (d) Diffuse Gaussian functions.

11.3. Which of the following are *ab initio* methods?
 (a) FEMO
 (b) MCSCF
 (c) CI
 (d) MP4

11.4. Which of the following properties can realistically be calculated using Hartree–Fock theory for a molecule like acetylene?
 (a) CH bond distance, to within 0.02 Å.
 (b) CH bond energy, to within 0.10 eV.
 (c) Lowest excitation energy, to within 0.05 eV.
 (d) Energy of carbon $1s$ orbital, to within 1 eV.
 (e) CCH bend frequency, to within 20%.

11.5. Consider the following orbital configuration in a molecule:

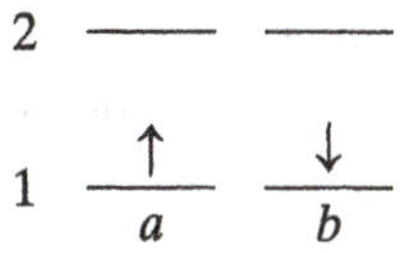

$$2 \quad \underline{\hspace{1.5em}} \quad \underline{\hspace{1.5em}}$$

$$1 \quad \underline{\uparrow}_{a} \quad \underline{\downarrow}_{b}$$

Note that the orbitals are labeled $1a$, $1b$, $2a$, and $2b$, as indicated in the diagram, and the electron spins are indicated by up and down arrows.
 (a) Write down a Slater determinant corresponding to this configuration.
 (b) Does this wavefunction describe a singlet or triplet spin state? Explain.

11.6. The quality of an electronic structure calculation depends both on the chosen basis set (the one-electron problem) and on the level of treatment of electronic correlation (the many-electron problem). One way to categorize calculations, then, is in terms of a diagram such as that sketched in Figure 11.2, sometimes called a Pople diagram. Insert the following labels $(a, b, c, ...)$ into the boxes shown:
 (a) STO–3G
 (b) MP2/6–31G
 (c) 6–31G
 (d) 6–311G*

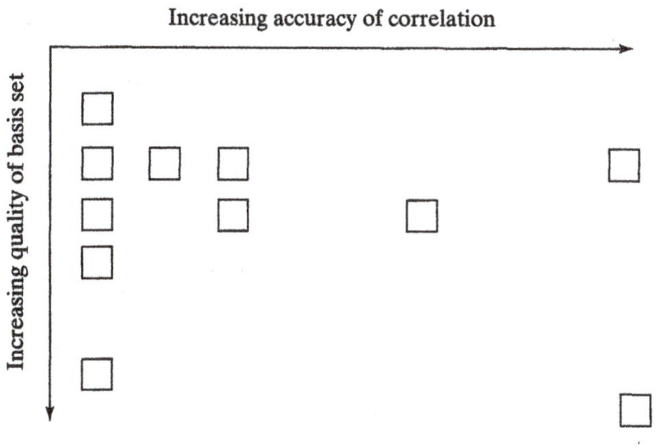

Figure 11.2 Pople diagram.

(e) MP4/6–31G
(f) Hartree–Fock limit
(g) Exact solution of Schödinger equation
(h) FCI/6–31G
(i) 6–311G$(2p,2d,f)+$
(j) MP4/6–311G*
(k) CCSD(T)/6–311G*

11.7. Show that the expectation value of a one-electron operator R_i summed over all electrons in a molecule (i.e., $R = \Sigma R_i$) for a doubly occupied Slater determinant ψ is

$$\langle R \rangle = \left\langle \psi \mid \sum_i R_i \mid \psi \right\rangle = 2 \sum_{j=1}^{N/2} \int \phi_j^*(a) R_a \phi_j(a) d\tau_a,$$

where $\phi_j(a)$ is the jth MO containing the ath electron and the sum over j is over the $N/2$ occupied MOs. (*Hint:* First show that this equation works for the ground state of H_2. Then do the general case, expressing the determinant in terms of a sum over cofactors.)

12

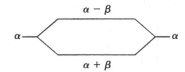

Semiempirical Methods

The *ab initio* methods that we discussed in the previous chapter all use the exact electronic Hamiltonian, but an approximate wavefunction, to solve the Schrödinger equation. In *semiempirical methods*, by contrast, one approximates the Hamiltonian to make it easier to deal with, and then often it is possible to solve the Schrödinger equation exactly, or at least much more accurately than with *ab initio* methods. There are many different kinds of semiempirical methods, depending on how the Hamiltonian is approximated. In the sections that follow, we start with the simplest possible semiempirical methods and then gradually move to increasing sophistication.

An alternative to semiempirical methods is the collection of *molecular mechanics* (MM) methods. In the approach these methods take, one abandons the Schrödinger equation and instead expresses the potential-energy surface [Eq. (10.12)] in a simple mathematical form in terms of empirically determined parameters. For example, the potential governing a CH single bond in a molecule might be represented as a harmonic oscillator, $1/2\, k(r - r_e)^2$, with the bond distance r_e and force constant k adjusted to match selected experimental data. This procedure is significantly simpler to use than even semiempirical methods, and for the structural properties of molecules that have simple bonding, it is often more accurate. However, it is difficult to use MM methods to describe bond breaking, fractional bonding, and the change in the potential-energy surface that occurs when a single bond is converted into a double bond, so semiempirical methods have a broad range of usefulness. Also, semiempirical methods can be used to determine many properties that MM methods cannot, such as dipole moments, electron spin densities, and orbital energies.

12.1 HÜCKEL MODEL

The Hückel model treats only the π-electrons in a planar conjugated, aromatic or heterocyclic, molecule. The basis set consists of one $p\pi$-orbital on each aromatic C, N, O, S, etc. The σ-framework is assumed frozen, with no σ–π interaction. In addition, electron–electron repulsion is neglected, so that the π-Hamiltonian is simply a sum:

$$\hat{H} = \sum_i h_i. \tag{12.1}$$

This equation implies that the wavefunction may be written as a product of spatial orbitals:

$$\psi = \phi_1(1)\phi_2(2)\cdots\phi_N(N). \tag{12.2}$$

We could also have used a Slater determinant of orbitals, but in the absence of electron–electron repulsion, there is no difference between the energies of the spatial product and the Slater determinant.

EXERCISE 12.1   For the specific case of H_2 in a minimum basis of $\{1s_a, 1s_b\}$, prove directly that if the approximation

$$H = \sum_i h_i$$

is made, the ground-state energy is the same no matter whether one uses a simple product or a Slater determinant.

Answer   The simple product wavefunction is $\psi_{\text{prod}} = \sigma_g(1)\overline{\sigma}_g(2)$, and if we assume that σ_g is normalized (i.e., $\langle\sigma_g(1)|\sigma_g(1)\rangle = \langle\overline{\sigma}_g(2)|\overline{\sigma}_g(2)\rangle = 1$), then ψ_{prod} also is normalized, and we have

$$E = \langle\psi|H|\psi\rangle = \langle\sigma_g(1)\overline{\sigma}_g(2)|h_1 + h_2|\sigma_g(1)\overline{\sigma}_g(2)\rangle$$
$$= \langle\sigma_g(1)|h_1|\sigma_g(1)\rangle + \langle\overline{\sigma}_g(2)|h_2|\sigma_g(2)\rangle.$$

Now consider the Slater determinant:

$$\psi_1 = \frac{1}{\sqrt{2}}\begin{vmatrix}\sigma_g(1) & \overline{\sigma}_g(1)\\ \sigma_g(2) & \overline{\sigma}_g(2)\end{vmatrix} = \frac{1}{\sqrt{2}}(\sigma_g(1)\overline{\sigma}_g(2) - \overline{\sigma}_g(1)\sigma_g(2)).$$

Direct substitution yields

$$E = \frac{1}{2}\langle[\sigma_g(1)\overline{\sigma}_g(2) - \overline{\sigma}_g(1)\sigma_g(2)]|h_1 + h_2|[\sigma_g(1)\overline{\sigma}_g(2) - \overline{\sigma}_g(1)\sigma_g(2)\rangle$$

$$= \frac{1}{2}\langle\sigma_g(1)\overline{\sigma}_g(2)|h_1 + h_2|\sigma_g(1)\overline{\sigma}_g(2)\rangle + \frac{1}{2}\langle\overline{\sigma}_g(1)\sigma_g(2)|h_1 + h_2|\overline{\sigma}_g(1)\sigma_g(2)\rangle$$

$$= \langle\sigma_g(1)\overline{\sigma}_g(2)|h_1 + h_2|\sigma_g(1)\overline{\sigma}_g(2)\rangle.$$

In going from the first line to the second in this equation, we have used the facts that spin orbitals with different spins are orthogonal and the Hamiltonian is not dependent on spin. This independence of spin also makes it possible to go from the second to the third line, and the latter is identical to what we get using ψ_{prod}.

We next assume that each MO may be represented in terms of a LCAO expansion

$$\phi_i = \sum_{\mu} C_{i\mu} p_{\mu}, \qquad (12.3)$$

where p_{μ} is a p-orbital on the μth atom and $C_{i\mu}$ are coefficients that will be variationally optimized. If this expansion is substituted into the Schrödinger equation for each electron i—that is, if

$$h_i \phi_i = E \phi_i, \qquad (12.4)$$

and then the result is multiplied by p_{μ} and integrated, we generate the usual secular equation [see Eqs. (10.20) and (11.10)], viz.,

$$|\mathbf{H} - E\mathbf{S}| = 0, \qquad (12.5)$$

where H is the Hamiltonian matrix and S is the overlap matrix. In the case where the only π-orbitals are associated with carbon atoms, the Hamiltonian matrix is assumed to have the form

$$
\begin{aligned}
H_{\mu\nu} &= \langle p_{\mu} | h | p_{\nu} \rangle \\
&= \begin{cases} \alpha \text{ for } \mu = \nu \\ \beta \text{ for atom } \mu \text{ bonded to atom } \nu \;, \\ 0 \text{ otherwise} \end{cases}
\end{aligned} \qquad (12.6)
$$

and the overlap matrix is assumed to be the identity matrix; that is,

$$S_{\mu\nu} = \langle p_{\mu} | p_{\nu} \rangle = \delta_{\mu\nu}. \qquad (12.7)$$

The parameters α and β are constants that may be adjusted to match experimental values. If heteroatoms are present in the molecule, then different α's and β's may be associated with different atoms.

To illustrate this method, we first consider the very simple example of the ethylene molecule C_2H_4, pictured in Figure 12.1. The secular equation in this case is

$$\begin{vmatrix} \alpha - E & \beta \\ \beta & \alpha - E \end{vmatrix} = 0. \qquad (12.8)$$

Before solving this equation, let us divide each row by β and then define $x = (\alpha - E)/\beta$ to get

$$\begin{vmatrix} \dfrac{\alpha - E}{\beta} & 1 \\ 1 & \dfrac{\alpha - E}{\beta} \end{vmatrix} = \begin{vmatrix} x & 1 \\ 1 & x \end{vmatrix} = x^2 - 1 = 0. \qquad (12.9)$$

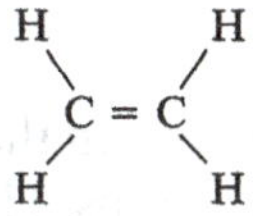

Figure 12.1   Structure of ethylene molecule.

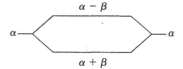

Figure 12.2 Hückel energy levels for ethylene.

The solutions are $x = \pm 1$, or

$$E_{\pm} = \alpha \pm \beta. \tag{12.10}$$

Since the energy levels would both have been $E_{\pm} = \alpha$ in the absence of bonding, we can construct the energy-level diagram shown in Figure 12.2 to reveal how π-bonding changes the energies.

Note that we have assumed that the $\alpha + \beta$ energy level lies below $\alpha - \beta$. This means that β must be negative, which turns out to be the choice for which electron delocalization (i.e., bonding) is favored over localization. To show that a negative choice of β favors electron delocalization, let us calculate the wavefunctions associated with $E_{\pm}$. This is done by writing the linear equations for the coefficients, which leads to the secular equation, which, in the general case, is

$$\sum_{\nu} (H_{\mu\nu} - ES_{\mu\nu})C_{\nu} = 0. \tag{12.11}$$

In the particular case of ethylene, this equation becomes

$$\begin{pmatrix} \alpha - E & \beta \\ \beta & \alpha - E \end{pmatrix}\begin{pmatrix} C_1 \\ C_2 \end{pmatrix} = 0, \tag{12.12}$$

where C_1 is the coefficient of carbon atom 1 and C_2 the coefficient of carbon atom 2. Substituting $E_+ = \alpha + \beta$ into this formula gives

$$\begin{pmatrix} -\beta & \beta \\ \beta & -\beta \end{pmatrix}\begin{pmatrix} C_1 \\ C_2 \end{pmatrix} = 0, \tag{12.13}$$

which implies that $C_1 = C_2$. After normalization (remember that $\langle p_1 | p_2 \rangle = 0$), the resulting wavefunction is

$$\phi_+ = \frac{1}{\sqrt{2}}(p_1 + p_2). \tag{12.14}$$

The corresponding wavefunction for $E_- = \alpha - \beta$ is

$$\phi_- = \frac{1}{\sqrt{2}}(p_1 - p_2). \tag{12.15}$$

We see that ϕ_- has a node between the two carbons, while ϕ_+ maximizes the electron density between the carbons. This shows that the choice $\beta < 0$ causes the bonding orbital to be stabilized. For ethylene, which has two π-electrons, the state ϕ_+ would be doubly occupied, which means that the *total* π-electron energy is

$$E = 2(\alpha + \beta). \tag{12.16}$$

One way to determine a value for the parameter β is to equate the lowest excitation energy in ethylene, namely, $2|\beta|$, to the measured value. This excitation occurs at about 200 nm, so the value of $|\beta|$ is about 70 kcal/mol. Other estimates of $|\beta|$ based on spectroscopic transitions in other molecules or based on thermodynamic stabilities are in the range 30–70 kcal/mol. If Hückel theory were exact, then β should be a constant, so the range of values that is obtained in fact provides an estimate of the validity of Hückel theory. Note that the parameter α is not important in determining spectra; in the absence of heteroatoms, it may be set equal to zero.

EXERCISE 12.2   The parameter α is physically best understood as the energy of an electron that is present in a $2p\pi$ basis function on one atom. For the simple situation of ethylene in the Hückel model, show directly that for optical spectra which involve a difference between two n-electron states ($n = 2$ for C_2H_4), the value of α is irrelevant. Is this true for the calculations of the ionization energy?

Answer   For optical excitation, we have

$$\Delta E_{opt} = E(\text{excited}) - E(\text{ground}) = (E_- + E_+) - (E_+ + E_+)$$

$$= [\alpha - \beta + \alpha + \beta] - [\alpha + \beta + \alpha + \beta]$$

$$= -2\beta, \text{ independent of } \alpha.$$

For the ionization energy, a similar calculation gives

$$\Delta E_{ion} = E(\text{ionized}) - E(\text{ground}) = (E_+) - (E_+ + E_+)$$

$$= [\alpha + \beta] - [\alpha + \beta + \alpha + \beta]$$

$$= \alpha + \beta, \text{ dependent on } \alpha.$$

This result is reasonable: Ionization removes an electron from a basis function of local energy α, whereas optical excitation just moves electrons.

As a second example, consider the allyl radical C_3H_5. Here we have three p-π orbitals, so the secular equation (or *topological determinant*) is

$$\begin{vmatrix} \alpha - E & \beta & 0 \\ \beta & \alpha - E & \beta \\ 0 & \beta & \alpha - E \end{vmatrix} = \begin{vmatrix} x & 1 & 0 \\ 1 & x & 1 \\ 0 & 1 & x \end{vmatrix} = 0. \tag{12.17}$$

This gives

$$E = \begin{cases} \alpha - \sqrt{2}\beta \\ \alpha \\ \alpha + \sqrt{2}\beta \end{cases} \tag{12.18}$$

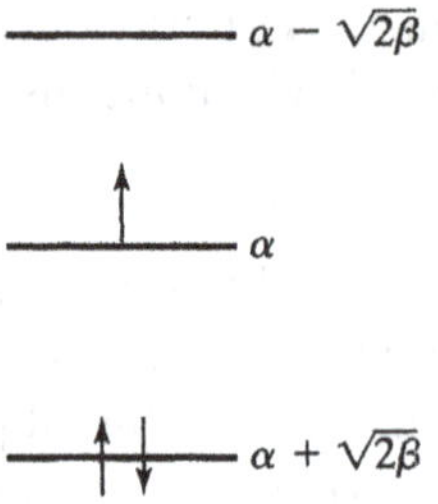

$$\underline{}\; \alpha - \sqrt{2}\beta$$

$$\underline{\uparrow}\; \alpha$$

$$\underline{\uparrow\downarrow}\; \alpha + \sqrt{2}\beta$$

Figure 12.3 Hückel energy levels for allyl.

with orbitals

$$\phi = \begin{cases} \dfrac{1}{2}(p_1 - \sqrt{2}p_2 + p_3) \\[2mm] \dfrac{1}{\sqrt{2}}(p_1 - p_3) \\[2mm] \dfrac{1}{2}(p_1 + \sqrt{2}p_2 + p_3) \end{cases} . \tag{12.19}$$

Figure 12.3 shows the orbital occupations.

The total electron energy is $3\alpha + 2\sqrt{2}\beta$. This energy is to be contrasted with what would be obtained for a hypothetical C_3H_5 molecule in which two adjacent carbons are π-bonded, as in ethylene (with energy $2\alpha + 2\beta$), and the third carbon is not bonded (energy $= \alpha$). The total energy of this hypothetical localized molecule is $3\alpha + 2\beta$. We can define the *delocalization energy* to be the difference between the calculated total energy and that of the localized molecule; that is,

$$E_{\text{deloc}} = 3\alpha + 2\sqrt{2}\beta - (3\alpha + 2\beta) = 2(\sqrt{2} - 1)\beta. \tag{12.20}$$

This provides a measure of the stability of the allyl radical that is due to π-conjugation. Alternatively, one can use measured values for the heat of formation of allyl to make estimates of β.

Another interesting example is the cyclopropenyl radical C_3H_3. Since this also involves $3 - p\pi$ orbitals, the secular equation is similar to that for allyl, except that now all three carbons are adjacent. Thus, we get

$$\begin{vmatrix} x & 1 & 1 \\ 1 & x & 1 \\ 1 & 1 & x \end{vmatrix} = 0. \tag{12.21}$$

The energy levels are

$$E = \begin{cases} \alpha - \beta \;\text{(twice)} \\ \alpha + 2\beta \end{cases}, \tag{12.22}$$

and the wavefunctions are

$$\phi_1 = \frac{1}{\sqrt{3}}(p_1 + p_2 + p_3),$$

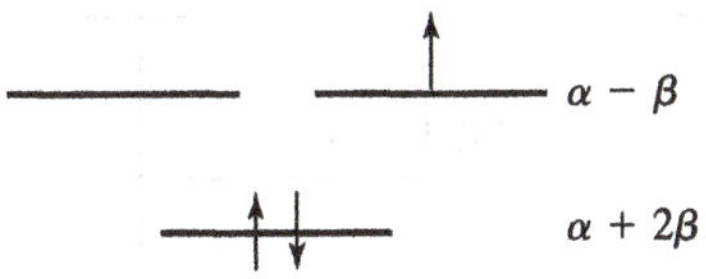

Figure 12.4 Hückel energy levels for cyclopropenyl.

$$\phi_2 = \frac{1}{\sqrt{2}}(p_1 - p_3), \tag{12.23}$$

and

$$\phi_3 = \frac{1}{\sqrt{6}}(p_1 - 2p_2 + p_3).$$

Note that the states ϕ_2 and ϕ_3 are the ones that correspond to the degenerate energy levels. Figure 12.4 shows the orbital occupations.

The total energy is $3\alpha + 3\beta$, which is lower than that for allyl ($3\beta + 2.83\beta$). This means that cyclopropenyl has greater π-electron stability. However, allyl is a much more stable molecule overall, because of ring strain in the σ-framework of cyclopropenyl. (This effect is absent in the π-electron Hückel model.)

EXERCISE 12.3 A general result [Eq. (2.28)] is that eigenfunctions with different eigenvalues are orthogonal. Show directly that the eigenfunctions for cyclopropenyl are orthogonal. An interesting feature of the two degenerate orbitals ϕ_2 and ϕ_3 is that they may be linearly combined in any way to generate functions that are still eigenfunctions with eigenvalue $\alpha - \beta$. Show that the combinations $\tilde{\phi}_2 = \phi_2 + \phi_3$ and $\tilde{\phi}_3 = \phi_3$ are still orthogonal to ϕ_1 (they have different eigenvalues), but are no longer orthogonal to each other.

Answer Remember that with the Hückel model we define $\langle p_\mu | p_\nu \rangle = \delta_{\mu\nu}$, so

$$\langle \phi_1 | \phi_2 \rangle = \frac{1}{\sqrt{6}} \langle p_1 + p_2 + p_3 | p_1 - p_3 \rangle = 0,$$

$$\langle \phi_1 | \phi_3 \rangle = \frac{1}{\sqrt{18}} \langle p_1 + p_2 + p_3 | p_1 - 2p_2 + p_3 \rangle = 0,$$

and

$$\langle \phi_2 | \phi_3 \rangle = \frac{1}{\sqrt{12}} \langle p_1 - p_3 | p_1 - 2p_2 + p_3 \rangle = 0.$$

But

$$\langle \tilde{\phi}_2 | \phi_1 \rangle = \langle \phi_2 + \phi_3 | \phi_1 \rangle = \langle \phi_2 | \phi_1 \rangle + \langle \phi_3 | \phi_1 \rangle = 0,$$

$$\langle \tilde{\phi}_3 | \phi_1 \rangle = \langle \phi_3 | \phi_1 \rangle = 0,$$

and

$$\langle \tilde{\phi}_2 | \tilde{\phi}_3 \rangle = \langle \phi_2 + \phi_3 | \phi_3 \rangle = \langle \phi_3 | \phi_3 \rangle \neq 0.$$

This confirms that only eigenfunctions with different eigenvalues are guaranteed to be orthogonal. Of course, it is always most convenient to choose the MOs so that they are orthogonal.

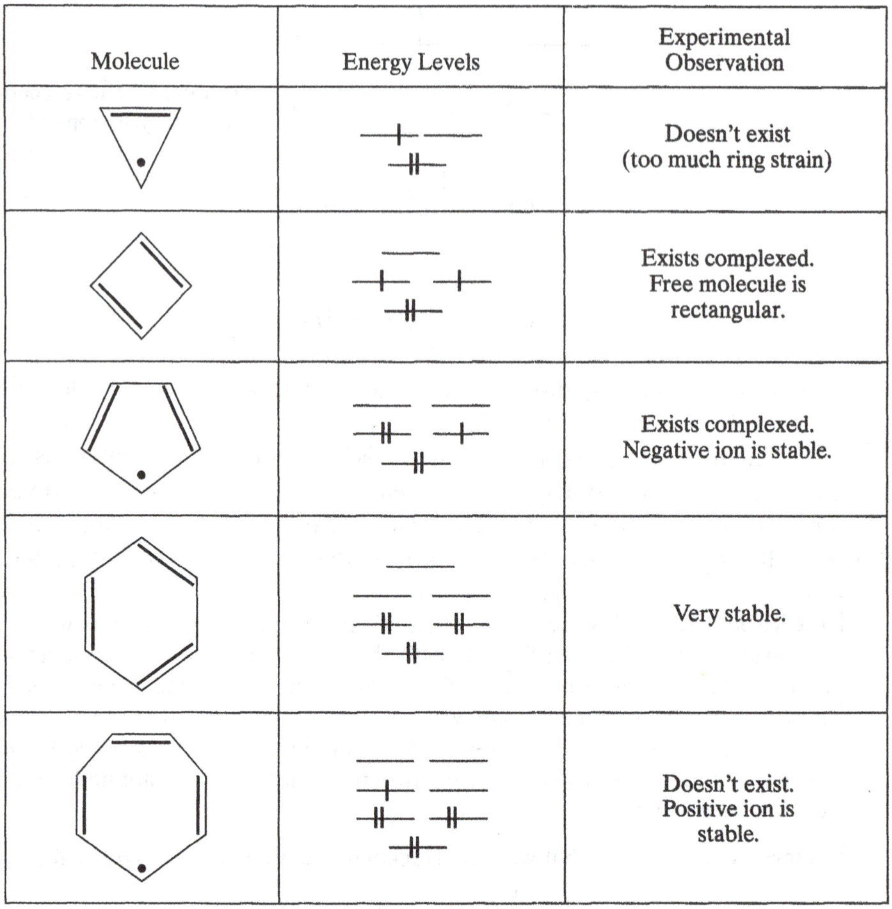

Molecule	Energy Levels	Experimental Observation
		Doesn't exist (too much ring strain)
		Exists complexed. Free molecule is rectangular.
		Exists complexed. Negative ion is stable.
		Very stable.
		Doesn't exist. Positive ion is stable.

Figure 12.5   Hückel energy levels for monocyclic aromatic hydrocarbons.

The cyclopropenyl radical is the simplest example of a conjugated monocyclic aromatic hydrocarbon whose general formula is C_nH_n. The application of Hückel theory to these molecules produces the energy levels and orbital occupations pictured in Figure 12.5. Note that the energy-level structure is identical to the positions of the vertices of each molecule. [Sometimes these diagrams are called *Frost–Musulin diagrams* (A.A. Frost and B. Musulin, *J. Chem. Phys.* 21, 572 (1953).] These molecules illustrate the *Hückel* $4n + 2$ *rule*: Maximum π-stability is associated with molecules having $4n + 2$ electrons, where $n = 0, 1, 2, \ldots$.

Other properties that may be obtained from the Hückel model are as follows:

a. The π-*charge density* at each atom is

$$q_\mu = \sum_i n_i |C_{i\mu}|^2, \tag{12.24}$$

where n_i is the occupation number of each MO (i.e., $n_i = 0, 1,$ or 2) and $C_{i\mu}$ are the orbital coefficients. The *net charge* f_μ on each atom is simply the difference between the net nuclear charge (after removing the σ-electrons) and the π-charge density. Thus,

$$f_\mu = 1 - q_\mu. \tag{12.25}$$

b. The π-*bond order* between atoms μ and ν is

$$p_{\mu\nu} = \sum_i n_i C_{i\mu}^* C_{i\nu}. \tag{12.26}$$

The total π-bond order at atom μ is $\sum_\nu p_{\mu\nu}$. The maximum possible value of this sum for a molecule with no heteroatoms is $\sqrt{3}$ [as occurs at the trigonal carbon in $C(CH_2)_3$]. The difference between $\sqrt{3}$ and $\sum_\nu p_{\mu\nu}$ is called the *free valence* F_μ; that is,

$$F_\mu = \sqrt{3} - \sum_\nu p_{\mu\nu}. \tag{12.27}$$

F_μ provides a measure of the ability of each carbon to form additional π-bonds.

EXERCISE 12.4   Show that, for the allyl radical,

$$q_1 = q_2 = q_3 = 1;$$

$$p_{12} = p_{23} = \frac{1}{\sqrt{2}}, \quad p_{13} = 0;$$

and

$$F_1 = F_3 = 1.025, \quad F_2 = 0.318.$$

Answer   From the orbitals of Eq. (12.19) and the definitions given by Eqs. (12.24)–(12.27), we have

$$q_1 = 2c_{11}^2 + c_{21}^2 = 2\left(\frac{1}{4}\right) + \frac{1}{2} = 1,$$

$$q_2 = 2c_{12}^2 + c_{22}^2 = 2\left(\frac{1}{2}\right) + 0 = 1,$$

$$q_3 = 2c_{13}^2 + c_{23}^2 = 2\left(\frac{1}{4}\right) + \frac{1}{2} = 1,$$

$$P_{12} = 2c_{11}c_{12} + c_{21}c_{22} = 2\left(\frac{1}{4}\right)(\sqrt{2}) + 0 = \frac{1}{\sqrt{2}},$$

$$P_{13} = 2c_{11}c_{13} + c_{21}c_{23} = 2\left(\frac{1}{2}\right)\left(\frac{1}{2}\right) + \left(\frac{1}{\sqrt{2}}\right)\left(-\frac{1}{\sqrt{2}}\right) = 0,$$

and

$$P_{23} = 2c_{12}c_{13} + c_{22}c_{23} = 2\left(\frac{\sqrt{2}}{2}\right)\left(\frac{1}{2}\right) + 0 = \frac{1}{\sqrt{2}}.$$

(This makes sense chemically! The p_1–p_2 and p_2–p_3 bonds are the same, and there is no p_1–p_3 bond.) The free valences are

$$F_1 = \sqrt{3} - P_{12} - P_{13} = \sqrt{3} - \frac{1}{\sqrt{2}} = 1.025,$$

$$F_2 = \sqrt{3} - P_{23} - P_{21} = \sqrt{3} - P_{23} - P_{12} = \sqrt{3} - \frac{2}{\sqrt{2}} = 0.318,$$

and

$$F_3 = \sqrt{3} - P_{32} - P_{31} = \sqrt{3} - P_{23} - P_{13} = \sqrt{3} - \frac{1}{\sqrt{2}} = 1.025.$$

The π-bond order can be used to make rough estimates of C–C bond lengths using the empirical formula

$$R_{\mu\nu} = 1.52 \text{ Å} - 0.186 \text{ Å} \cdot p_{\mu\nu}. \tag{12.28}$$

The formula gives $R_{\mu\nu} = 1.33$ Å for ethylene and 1.52 Å for ethane, corresponding to π-bond orders of 1 and 0, respectively. For allyl the formula predicts that $R_{\mu\nu} = 1.39$ Å.

The Hückel method may sometimes be improved by using the Wheland–Mann method of *charge iteration*. In this method, the α's are allowed to adjust for polarization of each atom through the flow of electrons to other atoms via the formula

$$\alpha_\mu = \alpha_\mu^0 + f_\mu |\beta| \omega, \tag{12.29}$$

where α_μ^0 is the initial value of α_μ (i.e., for $f_\mu = 0$), f_μ is the net charge, and ω is an empirically determined constant that is typically chosen to be 1.4. Note that since f_μ depends on the MOs, and these are obtained by solving the secular equation, the values of α_μ must be determined by iteration.

12.2   EXTENDED HÜCKEL METHOD

The extended Hückel method treats the σ and π electrons on an equal footing, using a minimal-basis-set expansion, typically with Slater orbitals. Electron repulsion is neglected, so the secular equation has the same appearance as in Hückel theory:

$$|\mathbf{H} - E\mathbf{S}| = 0. \tag{12.30}$$

A key difference between the extended Hückel and Hückel methods is in the parametrization of **H** and **S**. In the extended Hückel approach, one calculates the overlap **S** accurately using the assumed atomic orbital basis. Then the Hamiltonian is given by

$$H_{\mu\nu} = \begin{cases} \alpha_{\mu\nu} & \text{for } \mu = \nu \\ \beta_{\mu\nu} & \text{for } \mu \neq \nu \end{cases} \qquad (12.31)$$

where the α_μ's are equated to the (negative of the) gas-phase ionization potentials and $\beta_{\mu\nu}$ is often determined using the Wolfsberg–Helmholtz formula

$$\beta_{\mu\nu} = \frac{1}{2} K(\alpha_{\mu\mu} + \alpha_{\nu\nu})S_{\mu\nu}, \qquad (12.32)$$

with K being a "fudge factor" whose value is between 1.0 and 2.0 (often taken to be 1.75). The extended Hückel scheme is useful for both organic and inorganic molecules and even for solids.

12.3 PPP METHOD

The Pariser–Parr–Pople (PPP) method is the simplest example of a semiempirical method that includes electron–electron repulsion. The method is similar to Hückel theory in that only the π-orbitals of a molecule are considered. The Hamiltonian for each electron is similar to the Hartree Hamiltonian in that exchange is neglected. We therefore write

$$H_{\text{PPP}} = H_{\text{Hückel}} + \text{Repulsion}, \qquad (12.33)$$

where the two-electron coulomb integrals $\langle \mu\nu \,|\, \sigma\lambda \rangle$ [Eq. (11.15)] that appear in the repulsion term are taken to be parameters rather than being calculated. In particular, we take

$$\langle \mu\nu \,|\, \sigma\lambda \rangle = \gamma_{\mu\lambda}\delta_{\mu\nu}\delta_{\sigma\lambda}, \qquad (12.34)$$

where the $\gamma_{\mu\lambda}$'s are parameters.

12.4 NDO METHODS

The methods that neglect differential overlap (NDO) represent an extension of PPP to the treatment of both σ and π electrons. The differential overlap is just the integrand in the coulomb and exchange integrals of Eqs. (11.13) and (11.14); that is,

$$\text{differential overlap} = b_\mu(1)b_\sigma(2)\frac{1}{r_{12}}b_\nu(1)b_\lambda(2)d\tau_1 d\tau_2, \qquad (12.35)$$

for $\mu \neq \nu$ and $\sigma \neq \lambda$.

There are several different types of NDO methods. PPP is a π-orbital-based NDO method in which only the differential overlaps consistent with Eq. (12.34) are included. Two other NDO methods are the following:

   a. *CNDO* (*complete neglect of differential overlap*). This method generalizes PPP to include all orbitals, not just π-orbitals, in the calculations. Thus, all terms in Eq. (12.35) with $\mu \neq \nu$ and $\sigma \neq \lambda$ are neglected, and the remaining two-electron repulsion integrals are treated as parameters. There is no exchange in this approach.

   b. *INDO* (*intermediate neglect of differential overlap*). In this approach, one neglects all $\mu \neq \nu$ and $\sigma \neq \lambda$, except the exchange terms $\mu = \lambda$ and $\nu = \sigma$ in which the orbitals are on the *same atom*. All nonzero integrals are treated as parameters that are adjusted to reproduce Hartree–Fock calculations on test molecules. Variants on this idea include INDO/S, wherein the parameters are adjusted to reproduce measured molecular spectra, and MINDO, AM1, and PM3, in which the parameters are adjusted to reproduce experimental enthalpies of formation.

All of these semiempirical schemes involve model Hamiltonians in which the values of the α, β, and γ parameters are assumed rather than calculated. For the semiempirical models, such as PPP or CNDO, that retain some electron repulsion terms, Hartree–Fock or correlation-type methods are used to compute properties.

Methods like AM1 and PM3 have proven to be especially popular, as they are capable of reproducing the properties of a wide range of molecules to within a useful tolerance. To illustrate their applicability, in Chapter 14 we include AM1 in our applications of electronic structure methods.

SUGGESTED READINGS

T. A. Albright and J. K. Burdett, *Problems in Molecular Orbital Theory* (Oxford, New York, 1992).

J. N. Murrell and A. J. Harget, *Semi-empirical Self-consistent-Field Molecular Orbital Theory of Molecules* (Wiley-Interscience, New York, 1972).

J. N. Murrell, S. F. A. Kettle, and J. M. Tedder, *The Chemical Bond* (Wiley, New York, 1978).

J. A. Pople and D. L. Beveridge, *Approximate Molecular Orbital Theory* (McGraw-Hill, New York, 1970).

W. Thiel, *Advances in Chemical Physics*, 93, 703 (1996).

M. Zerner, *Reviews of Computational Chemistry*, 2, 313 (1991).

PROBLEMS

12.1 Which of the following give different energies for singlet and triplet states?
   (a) MINDO
   (b) PPP
   (c) MCSCF
   (d) Hückel

12.2 Which of the following neglect electron repulsion?
(a) extended Hückel
(b) DFT
(c) INDO
(d) coupled clusters

12.3 Which of the following defines an orthonormal basis?
(a) harmonic oscillator functions
(b) hydrogen orbitals (for an atom)
(c) Slater orbitals (for an atom)
(d) Gaussian orbitals (for an atom)

12.4 Which of the following describes the properly delocalized wavefunctions of a conjugated molecule like benzene?
(a) FEMO
(b) Hückel
(c) PPP
(d) valence bond

12.5 Which of the following will predict the bond energy of H_2 to within 1 eV?
(a) extended Hückel
(b) restricted Hartree–Fock
(c) PPP
(d) MC–SCF–CI

12.6 Which of the following methods take account of electron correlation?
(a) HF
(b) INDO
(c) DFT
(d) extended Hückel

12.7 For the Hückel model, the total energy is written as

$$E = \sum_k n_k e_k,$$

where n_k is the number of electrons in the kth MO, whose energy is e_k. Is this relationship true for
(a) extended Hückel?
(b) PPP?
(c) CNDO?
(d) MINDO?

12.8 Suppose we apply the Hückel model to butadiene:

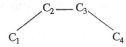

Assume that the standard Hückel assumptions are satisfied.
(a) Write down the secular equation.
(b) Suppose that the eigenvalues from this secular equation are

$$E = \alpha + 1/2\beta(1 \pm \sqrt{5})$$

and

$$E = \alpha + 1/2\beta(-1 \pm \sqrt{5}).$$

What is the lowest optical absorption energy, starting from the ground state of butadiene? Express this energy as a multiple of $|\beta|$.

(c) Suppose that the two doubly occupied molecular orbitals in butadiene are

$$\psi_1 = 0.371p_1 + 0.602p_2 + 0.602p_3 + 0.371p_4$$

and

$$\psi_2 = 0.602p_1 + 0.371p_2 - 0.371p_3 - 0.602p_4.$$

What is the π bond order between atoms 2 and 3?

(d) What is the π charge density on atom 2?

12.9 Set up and solve the Hückel secular equation for square cyclobutadiene. Assume the usual expressions for α and β. In addition to the orbital energies, calculate the following quantities:

(a) the molecular orbitals
(b) the π electronic energy
(c) the π electron bond order between each atom
(d) the effective charge on each atom
(e) the delocalization energy
(f) spin multiplicity of the ground state

12.10 The simplest fullerene is C_{60}. The Hückel model actually describes the properties of C_{60} quite well, but the Hückel $4n + 2$ rule does not work (60 is not $4n + 2$, but C_{60} is more stable than C_{60}^{++}). Can you give a reason the $4n + 2$ rule works for cyclic polyenes and aromatics, but not for C_{60}?

12.11 (a) One way to deal with heteroatoms or substitution within the Hückel model is to modify the α term at the substituted center. For allyl ($H_2C=CHCH_2$), assume that the α on the middle carbon is 0.85 times the α on the left carbon. Solve for the total energy and the optical transition energy, in terms of β and α.

(b) Now calculate the ground- and excited-state orbital energies, using first-order perturbation theory, with H_0 being the equal α Hamiltonian and $V = H(\text{allyl}) - H_0$. Compare your answer with the exact result you obtained in Part (a).

12.12 Extension of conjugation generally results in redshifted absorption spectra, because HOMO and LUMO levels become closer together. Demonstrate that this is true by using the Hückel energies for C_2H_4, C_3H_5 and C_4H_6.

If, on the other hand, twisting occurs around a π bond, the conjugation can be destroyed, and the spectra will change accordingly. To see a limit of this phenomenon, suppose that the left C–C bond in allyl is twisted so extremely that H_{12} changes its value from β to zero, because [from the Wolfsberg–Helmholtz result (Eq. 12.32)] the overlap vanishes. Compare the optical absorption frequency for twisted and untwisted allyl.

12.13 Consider an extended Hückel calculation on H_2O, using a basis set of $1s$ orbitals on the H's and $2s$ and $2p$ orbitals on O. Let the diagonal Hamiltonian matrix elements be denoted α_H, $\alpha_{O'}$, and α_O for the H $1s$, O $2s$, and O $2p$ orbitals, respectively. The only nonzero off-diagonal matrix elements are

$$\langle 2p_x|H|1s_A\rangle = \langle 2p_x|H|1s_B\rangle = \beta\cos(\theta/2),$$
$$\langle 2p_y|H|1s_A\rangle = -\langle 2p_y|H|1s_B\rangle = \beta\sin(\theta/2),$$

and
$$\langle 2s|H|1s_A\rangle = \langle 2s|H|1s_B\rangle = \beta',$$
where θ is the internal angle and β and β' are parameters.

(a) Using the AO basis in terms of the α's and β's, write down the complete secular equation. Assume that the overlap matrix is the unit matrix.

(b) By considering the symmetry of the orbitals with respect to reflection through the xz symmetry plane (i.e., $y \rightarrow -y$), show that the secular equation factors into three separate secular equations involving 1×1, 2×2, and 3×3 matrices. [*Hint*: Consider basis sets that are symmetric and antisymmetric combinations of $1s_A$ and $1s_B$, such as $(1s_A \pm 1s_B)/\sqrt{2}$.]

(c) Solve the secular equation in Part (b) under the assumption that the $2s$ orbital lies so far in energy below the others that it can be omitted from consideration in the bonding orbitals. Also, assume that $\alpha_H = \alpha_O$, and constrain θ to be greater than $90°$. What is the total electronic energy, and how does it vary with θ? What is the value of θ at the minimum?

(d) Now solve the secular equation as in (c), but including the $2s$ orbital with the assumptions $\alpha_H = \alpha_O = \alpha_{O'}$ and $\beta = \beta'$. What is the equilibrium value of θ?

12.14 Consider an extended Hückel calculation on H_3^+. Parametrize the Hamiltonian by assuming that the standard Hückel formulas apply to the overlaps and Hamiltonian matrix elements.

(a) Write down the secular equation for H_3^+ in an equilateral triangle geometry. Solve the equation to determine orbitals and energies. What is the total electronic energy?

(b) Consider H_3^+ in a linear geometry. Write down the secular equation and solve it for orbitals, energies, and the total energy. Assume that the same α and β parameters apply.

(c) Is linear H_3^+ more or less stable than equilateral H_3^+?

12.15 The Wolfsberg–Helmholtz relation (12.32) relates $\beta_{\mu\nu}$ to $S_{\mu\nu} = \langle\mu|\nu\rangle$. For the case where μ and ν refer to $1s$ orbitals on adjacent nuclei (as in the hydrogen molecule), how would you expect $S_{\mu\nu}$ to vary with $R_{\mu\nu}$? (What happens for $R_{ab} = 0$? For $R_{ab} \rightarrow \infty$?) Is this variation in accord with the actual calculated result [Eq. (10.23a)]?

The general dependence of π bond energy on bond order and bond distance (shorter bonds are stronger) is often referred to as the bond-energy–bond-order (BEBO) relationship. For a simple Hückel calculation on C_2H_4, does the π bond order change when β is changed?

12.16 Semiempirical models are often utilized for situations in which only selected orbitals are used in the basis and only some features of the electronic structure are of interest. For example, the so-called Creutz–Taube ion, $(NH_3)_5Ru\text{-}pyz\text{-}Ru(NH_3)_5^{+5}$, contains 11 d-electrons and is a classic example of mixed valence (pyz is pyrazine). Consider the two d_{xy} orbitals on the two Ru atoms. These orbitals can be represented by a Hückel-type Hamiltonian, with an effective Hückel β that is (in fact) modulated by the pyz. In the +5 ion, the two d-orbitals contain three electrons.

(a) In this Hückel model, calculate the energies of the two MOs. In terms of α and β, what is the expected optical transition energy for the +5 ion?

(b) When the +5 ion is reduced to the +4 state, will the transition from Part (a) be observed? This β is on the order of 0.2 eV. In what spectral region will the +5 absorption occur?

13

A_1 (Symmetric Stretch)

A_1 (Bend)

B_1 (Antisymmetric Stretch)

Applications of Group Theory

The symmetry of molecules provides both labels for the behavior of the wavefunction and constraints on how wavefunctions are constructed. Group theory is important for molecules exhibiting spatial symmetries, and its labels are quite commonly used in chemistry. In this chapter we present an abbreviated introduction to group theory, in the context of molecular electronic structure. Although primary consideration is given to the symmetry properties of molecular orbitals, we also discuss wavefunctions associated with many-electron molecules. The application of group theory to molecular vibrations is mentioned briefly.

13.1 GROUP THEORY FOR POINT GROUPS

The complete Hamiltonian of a molecule is unchanged by many kinds of symmetry operations. The most generally applicable of these operations are the *permutation* of the coordinates of identical particles (either electrons or nuclei) and the *inversion* of the coordinates of all particles through a coordinate origin (i.e., $x \to -x$, $y \to -y$, $z \to -z$ for all coordinates). In the Born–Oppenheimer approximation, one is interested in the symmetry properties of the electronic Hamiltonian $H = T_e + V_{eN} + V_{ee}$ [from Eq. (10.10)], for *fixed* nuclear positions. In this case, H is invariant with respect to the symmetry operations that interrelate equivalent nuclei for the rigid molecule. Among these symmetry operations are reflection about symmetry planes (e.g., the xy symmetry plane corresponds to

$x \rightarrow x, y \rightarrow y$, and $z \rightarrow -z$), inversion through a center of symmetry, and rotation about an axis by an angle that is a multiple of $2\pi/n$, where n is an integer. Except for linear molecules, the possible symmetry operations for a rigid molecule are always finite in number, and (mathematically) the collection of all such operations for a given molecule forms a *group* that is known as a *point* group. The treatment of linear molecules is described in some of the texts listed at the end of the chapter.

In what follows, we consider the consequences of point group symmetry for the electronic wavefunctions of molecules. Symmetry plays an important role in providing labels for electronic wavefunctions, so we shall focus on determining these symmetry labels for molecular orbitals and for many-electron wavefunctions. Since symmetry applies equally well to highly accurate and only approximate electronic wavefunctions, we will use the latter in order to keep the description as simple as possible.

13.1.1 Symmetry Operations and Point Groups

We begin by briefly reviewing some of the basic elements of group theory as applied to point groups. Possible symmetry elements include the following:

E: the identity operation;

C_n: n-fold rotation (by an angle $2\pi/n$);

σ: a symmetry plane;

S_n: an n-fold improper rotation (C_n followed by reflection perpendicular to the axis of rotation);

i: inversion.

Often, the operations σ are given labels that tell us something about the direction of the symmetry plane relative to an axis of rotation that also defines a symmetry operation for the molecule of interest. Thus, σ_v indicates a symmetry plane that contains a symmetry axis, while σ_h is a symmetry plane that is perpendicular to the axis.

If all of the symmetry elements of a given molecule are considered collectively, it is not difficult to show that they form a *group*. This means that

a. any *products* of two symmetry elements are also elements of the group,

b. there is always an *identity* symmetry element,

c. every element has an inverse, and

d. multiplication of symmetry elements is *associative* (i.e., $A(BC) = (AB)C$).

Molecular point symmetry groups are labeled by symbols (called *Schoenflies symbols*) that indicate the most important symmetry elements in the group. Thus, the symmetry group associated with H_2O is labeled C_{2v}, as this group contains the C_2 rotation as well as two σ_v symmetry planes that contain the associated rotation axis. Schoenflies symbols for other commonly used groups will be introduced later.

EXERCISE 13.1 Prove that $S_n^2 = C_n^2$, $S_n^n = \sigma_h$ (n odd), $S_n^n = E$ (n even), and $S_2^1 = i$.

Answer Reflecting twice is an identity operation $(\sigma^2 = E)$. Also, performing the S_n^2 operation is the same as performing $\sigma C_n \sigma C_n$. However, the order of the reflections and rotations doesn't matter, so we can write $S_n^2 = \sigma C_n \sigma C_n = C_n \sigma^2 C_n = C_n^2$. If we apply this same line of reasoning n times, we can write $S_n^n = (\sigma C_n)^n = C_n^n(\sigma)^n$. Since $C_n^n = E$, we conclude that $S_n^n = \sigma^n$, and since $\sigma^2 = E$, σ^n is just σ for n odd and E for n even. Finally, the operation S_2^1 involves rotation by $180°$ (which means that $x \to -x$ and $y \to -y$, but z is not changed), followed by σ (which makes $z \to -z$, but leaves x and y alone), so the net result is the same as inversion ($x \to -x$, $y \to -y$, $z \to -z$).

13.1.2 Representations

For any symmetry group, it is always possible to find square matrices that multiply the same way that the group's elements do. That is, if $AB = C$ for any three elements of the group, then $M_A M_B = M_C$ for the matrices M corresponding to A, B, and C. We then say that these matrices form a *representation* of the group. For the symmetry groups of rigid molecules (point groups), an easy way to generate a representation involving 3×3 matrices is to consider the changes in the three Cartesian coordinates x, y, z of a point that arise from applying each symmetry operation. In general, after such an operation, the point (x, y, z) is transformed to a new location (x', y', z'), and the relation between the two points can be represented with the use of a matrix. For the identity operation, for example, we have

$$\begin{pmatrix} x' \\ y' \\ z' \end{pmatrix} = \begin{pmatrix} 1 & 0 & 0 \\ 0 & 1 & 0 \\ 0 & 0 & 1 \end{pmatrix} \begin{pmatrix} x \\ y \\ z \end{pmatrix}, \tag{13.1}$$

so that the identity matrix forms a representation of E. Likewise, a C_n operation about the z-axis would lead to the transformed coordinates

$$x' = x \cos \frac{2\pi}{n} + y \sin \frac{2\pi}{n},$$

$$y' = -x \sin \frac{2\pi}{n} + y \cos \frac{2\pi}{n},$$

$$z' = z,$$

so the matrix representing C_n is

$$\begin{pmatrix} \cos 2\pi/n & \sin 2\pi/n & 0 \\ -\sin 2\pi/n & \cos 2\pi/n & 0 \\ 0 & 0 & 1 \end{pmatrix}. \tag{13.2}$$

EXERCISE 13.2   What is the 3 × 3 matrix representation associated with S_n? Show that $S_n = \sigma_h C_n = C_n \sigma_h$.

Answer   By analogy with the argument leading to Eq. (13.2), one would expect that the matrix representing S_n is

$$\begin{pmatrix} \cos 2\pi/n & \sin 2\pi/n & 0 \\ -\sin 2\pi/n & \cos 2\pi/n & 0 \\ 0 & 0 & -1 \end{pmatrix}.$$

By a similar argument, one would expect that the matrix representing σ_h (i.e., $x' = x$, $y' = y$, $z' = -z$) is

$$\begin{pmatrix} 1 & 0 & 0 \\ 0 & 1 & 0 \\ 0 & 0 & -1 \end{pmatrix}.$$

Now use matrix multiplication to show that

$$\begin{pmatrix} \cos 2\pi/n & \sin 2\pi/n & 0 \\ -\sin 2\pi/n & \cos 2\pi/n & 0 \\ 0 & 0 & -1 \end{pmatrix} = \begin{pmatrix} \cos 2\pi/n & \sin 2\pi/n & 0 \\ -\sin 2\pi/n & \cos 2\pi/n & 0 \\ 0 & 0 & 1 \end{pmatrix}\begin{pmatrix} 1 & 0 & 0 \\ 0 & 1 & 0 \\ 0 & 0 & -1 \end{pmatrix}$$

$$= \begin{pmatrix} 1 & 0 & 0 \\ 0 & 1 & 0 \\ 0 & 0 & -1 \end{pmatrix}\begin{pmatrix} \cos 2\pi/n & \sin 2\pi/n & 0 \\ -\sin 2\pi/n & \cos 2\pi/n & 0 \\ 0 & 0 & 1 \end{pmatrix}.$$

13.1.3 Similarity Transformations

Once we have a matrix representation of a group, it is not difficult to generate other representations via *similarity transformations*. If A_i is a matrix representing element i of a group, the similarity transformation generates a new set of matrices B_i via the equation

$$B_i = Q^{-1} A_i Q.$$

The matrices B_i form the new representation of the group, which means that they obey the same multiplication rules as the A_i's. The matrix Q is arbitrary as long as Q^{-1} exists. To show that this works, suppose that three matrices A_1, A_2, and A_3 satisfy $A_1 A_2 = A_3$. Then it is easy to see that $B_3 = Q^{-1} A_3 Q = Q^{-1} A_1 A_2 Q = Q^{-1} A_1 Q Q^{-1} A_2 Q = B_1 B_2$, which is the same multiplication rule.

Two important examples of similarity transformations are found when

1. Q is an element of the $\{A_i\}$ group representation and the B_i that is obtained is also an element of the $\{A_i\}$ representation. In this case, we say that B_i and its counterpart A_i both belong to the same *class*.
2. for some Q, we obtain B_i's that are all *block diagonal* in the same sense. A block diagonal matrix is a matrix in which the only nonzero elements

are grouped into square blocks along the diagonal. An example of such a matrix is

$$\begin{pmatrix} x_{11} & x_{12} & 0 \\ x_{21} & x_{22} & 0 \\ 0 & 0 & x_{33} \end{pmatrix},$$

where the x_{ij}'s are nonzero. When all the elements of a representation have the same block structure (the same dimensions and the same locations), each of the blocks forms a representation of smaller dimension than the full matrix, and we say that the first representation is *reducible*. If a representation cannot be further reduced, we say that it is *irreducible*. Irreducible representations play a significant role in describing the properties of symmetry groups.

EXERCISE 13.3   Use 3×3 matrix representations to show that the operations C_n and C_{-n} (i.e., rotation by $\pm 2\pi/n$) are related by a similarity transformation that employs a matrix Q which is the representation of a symmetry plane σ_v that contains the n-fold rotation axis. To simplify the derivation, assume that σ_v is the xz-plane.

Answer   With 3×3 matrix representations, the product $B_i = Q^{-1} A_i Q = \sigma_v^{-1} C_n \sigma_v$ becomes

$$\begin{pmatrix} 1 & 0 & 0 \\ 0 & -1 & 0 \\ 0 & 0 & 1 \end{pmatrix} \begin{pmatrix} \cos 2\pi/n & \sin 2\pi/n & 0 \\ -\sin 2\pi/n & \cos 2\pi/n & 0 \\ 0 & 0 & 1 \end{pmatrix} \begin{pmatrix} 1 & 0 & 0 \\ 0 & -1 & 0 \\ 0 & 0 & 1 \end{pmatrix}$$

$$= \begin{pmatrix} \cos 2\pi/n & -\sin 2\pi/n & 0 \\ \sin 2\pi/n & \cos 2\pi/n & 0 \\ 0 & 0 & 1 \end{pmatrix} = C_{-n}.$$

This indicates that C_n and C_{-n} belong to the same class for groups that contain a σ_v symmetry plane. Of course, it also requires that C_n and C_{-n} be distinct operations, so the result does not apply to $n = 2$, for which C_2 and C_{-2} are equivalent operations.

13.1.4 Irreducible Representations

Let us now consider the matrices associated with the irreducible representations of a group. We will use the symbol Γ to denote these matrices, and since there will be a different Γ for each symmetry element R and for each irreducible representation i, we use the notation $\Gamma_i(R)$ to label these different symmetry elements and representations. The individual matrix elements are denoted $\Gamma_i(R)_{mn}$. We also let h be the order of the group (the number of operations therein) and ℓ_i be the dimension of the ith irreducible representation.

We define the *character* (or trace) $\chi_i(R)$ of a given irreducible representation i for a given operation R via the formula

$$\chi_i(R) = \sum_n \Gamma_i(R)_{nn}. \tag{13.3}$$

Characters have a number of useful properties, one of which is that they are *independent of a similarity transformation*. To show this, we consider the similarity transformation $B = Q^{-1}AQ$. Then the proof is as follows:

$$\chi(B) = \sum_n B_{nn} = \sum_n (Q^{-1}AQ)_{nn} = \sum_n \sum_k \sum_\ell Q^{-1}_{nk} A_{k\ell} Q_{\ell n}$$

$$= \sum_k \sum_\ell A_{k\ell} \sum_n Q_{\ell n} Q^{-1}_{nk}$$

$$= \sum_k \sum_\ell A_{k\ell} \delta_{k\ell} = \sum_k A_{kk} = \chi(A).$$

Since characters are independent of similarity transformations, but the matrices are not, the characters $\chi_i(R)$ of the irreducible representations are unique, while the matrix elements $\Gamma_i(R)_{mn}$ are not. This means that it is more useful to consider the properties of the characters, and in what follows we will do this almost exclusively. Of course, for one-dimensional irreducible representations ($h = 1$), the characters and the matrices are the same.

An additional property of the characters of representations is that *characters of elements in the same class are identical*. This property reduces the number of independent characters required for each representation to be equal to the number of classes.

One of the most useful properties of irreducible representations is that the characters form a set of mutually orthogonal vectors. This is a statement of the *Great Orthogonality Theorem*. Mathematically, this theorem states that the characters are related to each other by the formula

$$\sum_R \chi_i(R)\chi_j^*(R) = h\delta_{ij}. \tag{13.4}$$

[This is actually a simplified form of a more complicated theorem that applies to the $\Gamma_i(R)_{mn}$ matrix elements, but it is all we will need here.] Note that the sum over elements in Eq. (13.4) can be condensed to a sum over classes, as the characters are independent of class. Also, since the number of mutually orthogonal vectors cannot exceed the dimension of the vectors, the number of irreducible representations must equal the number of classes.

13.1.5 Character Tables

Armed with the Great Orthogonality Theorem, it is possible to determine most of the properties of the irreducible representations of point groups, including the characters $\chi_i(R)$. Let us illustrate the basic idea involved in constructing a character table by a simple example: the H_2O molecule. Referring to Figure 13.1, we see that the symmetry elements associated with H_2O at equilibrium are E, C_2 (about z), $\sigma_v(xz)$ and $\sigma_v(yz)$. These four elements make up the C_{2v} point group,

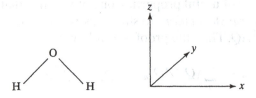

Figure 13.1 Coordinate system used for H_2O. Note that the plane of the molecule is chosen to be the xz-plane.

as mentioned in Section 13.1.1. It is not difficult to show that none of these elements belong to the same class, which implies that there are four classes and hence four irreducible representations. Since the order of the group is four, each of the representations must be one dimensional. From this information, Table 13.1 can be constructed. Note that all the representations satisfy $\sigma_v(xz)\sigma_v(yz) = \sigma_v(yz)\sigma_v(xz) = C_2(z)$, which explains why a representation with the characters $1, 1, 1, -1$ cannot occur.

TABLE 13.1 CHARACTER TABLE FOR C_{2v} POINT GROUP

C_{2v}	E	C_2	$\sigma_v(xz)$	$\sigma_v(yz)$
A_1	1	1	1	1
A_2	1	1	−1	−1
B_1	1	−1	1	−1
B_2	1	−1	−1	1

The symbols in the left-hand column of the table are the so-called *Mulliken symbols*. They describe the following characteristics of each irreducible representation:

1. *A* and *B* are used for one-dimensional representations, while *E* and *T* (or *F*) are used, respectively, for two- and three-dimensional representations.
2. *A* is used for representations that are symmetric with respect to the C_n operations, while *B* is used for antisymmetric representations.
3. The subscripts 1 and 2 refer to representations that are either symmetric or antisymmetric with respect to σ_v or a C_2 perpendicular to the main *n*-fold axis.
4. The superscripts ' and " refer to representations that are symmetric or antisymmetric with respect to σ_h (a mirror plane perpendicular to the *n*-fold axis).
5. The subscripts *g* and *u* refer to representations that are symmetric or antisymmetric with respect to *i* (inversion).

Note that a representation having all characters equal to unity is found for all groups and is called the *totally symmetric representation*.

Appendix C presents character tables for the most important point groups relevant to chemistry. The two right columns in these tables tell us which irreducible representations are associated with the coordinates (x, y, z), products of coordinates $(x^2, y^2, \text{etc.})$ and rotations (R_x, R_y, R_z). In Chapter 15 where we will use them to determine selection rules for optical spectra.

13.1.6 Direct Products

Earlier, we noted that a representation of any symmetry group could be generated by considering the effect of the symmetry operations on the three Cartesian coordinates (x, y, z). We would now like to consider what representation is generated when direct products of (x, y, z) are used as basis functions (such as $x^2, y^2, z^2, xy, yz, xz$), and how the representation thus generated is related to the one involving (x, y, z).

Let us introduce the notation $x_1 = x$, $x_2 = y$, $x_3 = z$, so that the transformation associated with the operator R can be written as

$$x_i' = Rx_i = \sum_j A_{ij}(R)x_j. \qquad (13.5)$$

Then the transformation associated with direct products such as $x_i x_j$ would be

$$Rx_i x_j = x_i' x_j' = \sum_k A_{ik}(R)x_k \sum_\ell A_{j\ell}(R)x_\ell = \sum_{k\ell} A_{ik}A_{j\ell}x_k x_\ell. \qquad (13.6)$$

Evidently, the forms of Eqs. (13.5) and (13.6) are identical, provided that we consider the indices ij and $k\ell$ as collective labels in the direct-product vector space. Thus, $C_{ij,k\ell}(R) = A_{ik}(R)A_{j\ell}(R)$ is a kind of super-matrix that forms a new representation of the symmetry group.

An important relation between the direct-product representation and the representations that make it up is provided by considering the character of C:

$$\chi_C(R) = \sum_{ij} C_{ij,ij} = \sum_{ij} A_{ii} A_{jj} = \chi_A(R)\chi_A(R). \qquad (13.7)$$

Eq. (13.7) shows that the character of the C representation is simply the product of the characters of the representations that make it up (in this case, A^2). This result applies generally to direct-product representations made up from any number of smaller dimensionality representations.

13.1.7 Clebsch–Gordan Series

Often, it is of interest to decompose a representation into irreducible representations. Doing this is equivalent to block diagonalizing the matrices that define the representation, but it can take considerably less effort than that. Note that the characters associated with the reducible representation should equal the sums of the characters of the blocks into which that representation can be decomposed. Thus,

$$\chi(R) = \sum_i a_i \chi_i(R), \tag{13.8}$$

where a_i is an integer coefficient that indicates how many times the irreducible representation i is contained in the reducible representation. Equation (13.8) is sometimes called the *Clebsch–Gordan series*. Multiplying it by $\chi_j^*(R)$, summing over R, and invoking the Great Orthogonality Theorem, we find that

$$a_j = \frac{1}{h} \sum_R \chi_j^*(R)\chi(R), \tag{13.9}$$

which provides a direct method for determining the coefficient a_j (and thereby decomposing the representation).

EXERCISE 13.4 Using Eqs. (13.7) and (13.9), show that the representation of the *direct product* C of two representations (say, A and B) will contain the totally symmetric representation only if A and B have at least one irreducible representation in common.

Answer To show the property in question, let's calculate the coefficient a_{TS} corresponding to the totally symmetric representation. Since $\chi_{TS}(R) = 1$ (all the totally symmetric characters are unity), Eq. (13.9) may be written as

$$a_{TS} = \frac{1}{h} \sum_R \chi_C(R) = \frac{1}{h} \sum_R \chi_A(R)\chi_B(R), \tag{13.10}$$

where, in the last expression, we have used the analogue of Eq. (13.7) to express the character of C in terms of the characters of A and B. If A and B are themselves irreducible, then the Great Orthogonality Theorem indicates that $a_{TS} = \delta_{AB}$, which completes the proof. If A and B are reducible, one must insert Eq. (13.8) for χ_A and χ_B. It then follows that a_{TS} will vanish, unless A and B have an irreducible representation in common. (Note that we have assumed that the χ's are real. This is usually true, but if it is not, an analogous statement can still be made.)

13.2 APPLICATIONS OF GROUP THEORY TO MOLECULAR QUANTUM MECHANICS

13.2.1 Symmetry–Adapted Linear Combinations

The utility of group theory in electronic structure calculations arises from the fact that molecular electronic wavefunctions can always be constructed to belong to irreducible representations. This is because (1) if R is a symmetry element of a molecule, R must commute with the electronic Hamiltonian H of that molecule, and (2) simultaneous eigenfunctions of two operators may be constructed if those operators commute [as was proved in the discussion of Eq. (2.31) on page 34.]

We learned in Chapter 12 that solutions of the electronic Schrödinger equation are often constructed by diagonalizing the matrix representation of the secular equation

$$|\mathbf{H} - E\mathbf{S}| = 0. \tag{13.11}$$

In this equation, the Hamiltonian and overlap matrices, $\mathbf{H}$ and $\mathbf{S}$, have elements $\langle \phi_i | H | \phi_j \rangle$ and $\langle \phi_i | \phi_j \rangle$, respectively, where ϕ_i is a basis function. If the ϕ_i's are constructed to belong to irreducible representations, $[H, R] = 0$ implies that the secular equation will automatically be block diagonalized. This is because the irreducible representation of the integrand in H_{ij} and S_{ij} must be totally symmetric if the resulting integral is to be nonzero. (Otherwise the integrand will be an odd function of at least one variable and would therefore integrate to zero.) From our previous discussion in Exercise 13.4, we know that the representation associated with S_{ij}, namely, the direct product $\Gamma(\phi_i) \times \Gamma(\phi_j)$, is totally symmetric only if ϕ_i and ϕ_j share irreducible representations. Since that will not be the case if ϕ_i and ϕ_j belong to different irreducible representations, S_{ij} must be block diagonal. H_{ij} is likewise block diagonal, because the Hamiltonian is totally symmetric and the representation generated by $\phi_i \times H \times \phi_j$ is the same as that of $\phi_i \times \phi_j$.

Thus, the computational effort in solving the Schrödinger equation can be reduced considerably if we use symmetry-adapted basis functions. Since molecular orbitals are conveniently represented as linear combinations of atomic orbitals, the symmetry-adapted basis functions are generated by taking symmetry-adapted linear combinations (SALCs) of atomic orbitals.

13.2.2 Construction of SALCs

SALCs are functions that are symmetric or antisymmetric with respect to the symmetry operations, as dictated by the character table. For multidimensional irreducible representations, they have more complicated symmetry properties, as will be discussed later. Often, it is a rather trivial matter to determine SALCs by linearly combining basis functions with the appropriate coordinates permuted. Thus, if we have a group with a σ_h operation, and we want a function that is antisymmetric with respect to that operation, then the function $\phi = f(x, y, z) - \sigma_h f(x, y, z)$ will have this property (i.e., $\sigma_h \phi = -\phi$) for any function $f(x, y, z)$.

For both one-dimensional and multidimensional irreducible representations, there is a simple method for constructing SALCs that uses the characters of the irreducible representations (from the character tables) to determine the appropriate linear combination of orbitals that gives a SALC. The appropriate formula (one that makes $R\phi_j = \chi_j \phi_j$) is

$$\phi_j = N \sum_R \chi_j(R) R f, \tag{13.12}$$

where N is a normalization factor, χ_j is a character from the character table, R is the symmetry operator, and f is an arbitrary function (typically, one of the atomic orbital basis functions). To show that this formula produces a SALC corresponding to the irreducible representation j, we note that the function f can, in general, be expressed in terms of a formula like Eq. (13.8) (i.e., f can be decomposed into a sum over SALCs belonging to all the irreducible representations of a group, as these form a "complete" set of all the possible symmetries that an arbitrary function may have). Thus, we would write

$$f = \sum_i a_i \phi_i. \tag{13.13}$$

If Eq. (13.13) is then substituted into Eq. (13.12), and we make use of the definition $R\phi_i = \chi_i\phi_i$, plus Eq. (13.4) (and assuming that the characters are real—see discussion after Eq. 13.10), we get

$$\phi_j = N \sum_R \chi_j(R) \sum_i a_i \chi_i(R)\phi_i$$

$$= N \sum_i a_i \sum_R \chi_j(R)\chi_i(R)\phi_i = Na_jh\phi_j = \phi_j, \tag{13.14}$$

provided that we take $N = 1/ha_j$. Actually, the choice of normalization factor isn't even important at this point, for in solving the Schrödinger equation, we will need to determine normalized SALCs that take into account the overlap properties of the atomic orbitals in the basis set (which is something that is not connected to group theory).

Now let us give an example of the use of Eq. (13.12). Employing the coordinate system in Figure 13.1, we consider the H_2O molecule. If we choose f to be a $1s$ orbital on one of the hydrogen atoms, then it is straightforward to show that

$$\phi_{A_1} = N(1s_1 + 1s_2),$$

$$\phi_{A_2} = 0,$$

$$\phi_{B_1} = N(1s_1 - 1s_2),$$

and

$$\phi_{B_2} = 0.$$

Note that two of the SALCs end up being zero. This means that the given irreducible representation isn't contained in the arbitrary function. That this will happen for water can be determined in advance by decomposing the representation of the two $1s$ orbitals. (See Section 13.2.3.)

One final point concerning the use of Eq. (13.12) is that, for multidimensional irreducible representations, that equation determines only one SALC. To determine others, it is necessary to apply additional procedures. This is best illustrated with an example that we consider in the next section.

Figure 13.2 Cyclopropenyl cation.

13.2.3 Applications to Hückel Theory

Let us now demonstrate the use of all of the group theory we have learned so far by considering an application to the cyclopropenyl cation (Figure 13.2), which has D_{3h} symmetry. Table 13.2 is the character table. Note that there are 12 symmetry operations, 6 classes and 6 irreducible representations in the D_{3h} point group.

TABLE 13.2 CHARACTER TABLE FOR D_{3h} POINT GROUP

D_{3h}	E	$2C_3$	$3C_2$	σ_h	$2S_3$	$3\sigma_v$		
A_1'	1	1	1	1	1	1		$x^2 + y^2, z^2$
A_2'	1	1	−1	1	1	−1	R_z	
E'	2	−1	0	2	−1	0	(x, y)	$x^2 - y^2, xy$
A_1''	1	1	1	−1	−1	−1		
A_2''	1	1	−1	−1	−1	1	z	
E''	2	−1	0	−2	1	0	(R_x, R_y)	xz, yz

Let us number the symmetry operations according to Figure 13.3 and assume that the molecule lies in the x, y-plane. In Chapter 12, we learned that in Hückel theory we associate a p_z orbital with each carbon atom and then take linear combinations of these orbitals to form MOs. The secular equation is given by Eq. (13.11), where

$$H_{\mu\nu} = \begin{cases} \alpha \text{ if } \mu = \nu \\ \beta \text{ if } \mu \text{ and } \nu \text{ are adjacent.} \\ 0 \text{ otherwise} \end{cases} \quad (13.15)$$

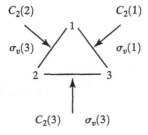

Figure 13.3 Symmetry operations for cyclopropenyl cation.

For $C_3H_3^+$, this becomes

$$\begin{vmatrix} \alpha - E & \beta & \beta \\ \beta & \alpha - E & \beta \\ \beta & \beta & \alpha - E \end{vmatrix} = 0. \tag{13.16}$$

This is only a 3×3 determinant, which is not difficult to evaluate explicitly. However, we can use group theory to simplify matters, and although that is of minor computational consequence, it provides significant insight into the structure of the molecular orbitals.

To use group theory to simplify, we must first determine irreducible representations spanned by AOs. To do this, we first construct the reducible representation, then reduce it. A simple set of rules for determining characters of the reducible representation is the following:

1. If an orbital is unchanged after applying a symmetry operation, count +1 per orbital to the character.
2. If an orbital changes sign, count -1.
3. If an orbital changes location, count 0.

Applying these rules to the cyclopropenyl cation, we get, for the reducible representation,

D_{3h}	E	$2C_3$	$3C_2$	σ_h	$2S_3$	$3\sigma_v$
Γ	3	0	-1	-3	0	1

Now, from Eq. (13.9), we calculate $a_i = \frac{1}{h} \sum_R \chi_i(R)\chi(R)$ to determine which irreducible representations are included in the basis of three p orbitals:

$$\left. \begin{aligned} a_{A_1'} &= \frac{1}{12}[3 - 1(3) - 3(1) + 3(1)] = 0 \\ a_{A_2'} &= \frac{1}{12}[3 - 1(3)(-1) - 3(1) + 1(3)(-1)] = 0 \end{aligned} \right\} \quad \begin{array}{l} \text{These indicate that} \\ \text{wavefunctions can't} \\ \text{have } A' \text{ symmetry.} \end{array}$$

$$a_{A_1''} = \frac{1}{12}(3 - 3 + 3 - 3) = 0,$$

$$a_{E'} = 0,$$

$$\left. \begin{aligned} a_{A_2''} &= \frac{1}{12}(3 + 3 + 3 + 3) = 1 \\ a_{E''} &= \frac{1}{12}(6 + 6) = 1 \end{aligned} \right\} \quad \text{These indicate that } \Gamma = A_2'' + E''.$$

TABLE 13.3   CHARACTER
TABLE FOR C_3 POINT
GROUP

C_3	E	C_3	C_3^2
A	1	1	1
$E\{$	1	ε	ε^*
	1	ε^*	ε

To construct SALCs, we first illustrate the method based on Eq. (13.12). The application of this method using the D_{3h} point group is difficult, due to the multidimensional irreducible representations in that group. A simpler procedure is therefore to use a subgroup of D_{3h} that still contains the important symmetry elements. The simplest subgroup that does this is C_3, which is easier to work with because all the irreducible representations are one dimensional. The character table for C_3 is shown in Table 13.3. Here

$$\varepsilon = e^{2\pi i/3},$$

$$\frac{1}{2}(\varepsilon + \varepsilon^*) = \cos 120° = -\frac{1}{2},$$

and

$$\frac{\varepsilon - \varepsilon^*}{2i} = \sin 120° = \frac{\sqrt{3}}{2}.$$

Note that the second and third representations are collectively labelled E even though they are one dimensional.

Applying the rules for characters, we find the table

	E	C_3	C_3^2
Γ	3	0	0

from which, by inspection,

$$\Gamma = A + E. \tag{13.17}$$

Note that the correlation between representations in C_3 and D_{3h} is as follows:

C_3	D_{3h}
A	A_2''
E	E''

Whenever such a correlation cannot be established, another subgroup must be used. Since all the representations are one dimensional, the generation of SALCs is easy. Almost by inspection (after normalization), we find that

$$\psi_A = \frac{1}{\sqrt{3}}(p_1 + p_2 + p_3), \tag{13.18}$$

$$\psi_E^{(1)} = \frac{1}{\sqrt{3}}(p_1 + \varepsilon^* p_2 + \varepsilon p_3), \tag{13.19}$$

and

$$\psi_E^{(2)} = \frac{1}{\sqrt{3}}(p_1 + \varepsilon p_2 + \varepsilon^* p_3). \tag{13.20}$$

Note that if we wish to avoid working with complex orbitals, we can take linear combinations of these functions to define real orbitals, as in

$$\psi_{E''}^{(1)} = \frac{1}{\sqrt{2}}\left(\psi_E^{(1)} + \psi_E^{(2)}\right)$$
$$= \frac{1}{\sqrt{6}}(2p_1 - p_2 - p_3) \tag{13.21}$$

and

$$\psi_{E''}^{(2)} = \frac{i}{\sqrt{2}}\left(\psi_E^{(1)} - \psi_E^{(2)}\right)$$
$$= \frac{1}{\sqrt{2}}(p_2 - p_3). \tag{13.22}$$

13.2.4 Energy Eigenvalues

In this section, we use the SALCs

$$\psi_1 = \psi_{A_2''}, \tag{13.23}$$

$$\psi_2 = \psi_{E''}^{(1)}, \tag{13.24}$$

and

$$\psi_3 = \psi_{E''}^{(2)}. \tag{13.25}$$

To apply Hückel theory, we can automatically separate out the A_2'' subblock of the Hamiltonian from the E'' block. With regard to A_2'', the secular equation is just 1×1 and is

$$\left|\langle\psi_1|H|\psi_1\rangle - E\langle\psi_1|\psi_1\rangle\right| = 0, \tag{13.26}$$

where $\langle\psi_1|H|\psi_1\rangle = \alpha + 2\beta$, so this immediately gives

$$E = \alpha + 2\beta. \tag{13.27}$$

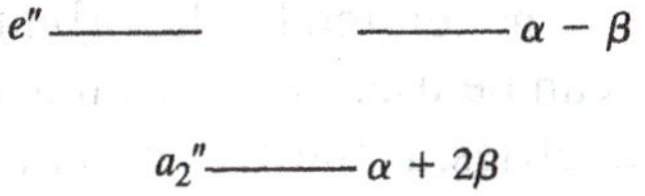

$$e'' \text{———}\qquad\text{———} \alpha - \beta$$

Figure 13.4 Energy-level
diagram for cyclopropenyl
cation.

$$a_2'' \text{———} \alpha + 2\beta$$

with regard to E'', the secular equation is 2×2 and is

$$\begin{vmatrix} \langle \psi_2 | H | \psi_2 \rangle - E & \langle \psi_2 | H | \psi_3 \rangle \\ \langle \psi_3 | H | \psi_2 \rangle & \langle \psi_3 | H | \psi_3 \rangle - E \end{vmatrix} = 0, \tag{13.28}$$

where

$$\langle \psi_2 | H | \psi_2 \rangle = \frac{1}{6} (6\alpha - 6\beta) = \alpha - \beta, \tag{13.29}$$

$$\langle \psi_3 | H | \psi_3 \rangle = \frac{1}{2} (2\alpha - 2\beta) = \alpha - \beta, \tag{13.30}$$

and

$$\langle \psi_2 | H | \psi_3 \rangle = 0 = \langle \psi_3 | H | \psi_2 \rangle. \tag{13.31}$$

Thus, the secular equation is

$$\begin{vmatrix} \alpha - \beta - E & 0 \\ 0 & \alpha - \beta - E \end{vmatrix} = 0, \tag{13.32}$$

which gives

$$E = \alpha - \beta (\text{twice}). \tag{13.33}$$

Note that the secular equation for the E'' block, Eq. (13.32), is diagonal, even though that is not required by symmetry. Generally, this does not happen, especially if the SALCs are not forced to be orthogonal; however, the eigenvalues of the E symmetry block will always be degenerate.

Note also that since these SALCs diagonalize the secular equation, they are by themselves the eigenfunctions of the Schrödinger equation. This is *not* generally true.

Assuming that $\beta < 0$, we can generate the energy-level diagram of Figure 13.4, with the two electrons occupying the a_2'' in the ground state. Notice that, in accordance with standard notation, we have used lowercase Mulliken symbols to label the one-electron orbitals in this diagram.

13.3 SYMMETRY PROPERTIES OF MANY-ELECTRON WAVEFUNCTIONS

13.3.1 Many-Electron Configurations; Terms

Up to this point, only one-electron orbital symmetries have been considered. In general, the many-electron wavefunction for a given electron configuration can be written in terms of Slater determinants of the one-electron orbitals. These determinants involve products of orbitals, and if the orbitals are SALCs, the representation generated by the Slater determinant will be derived from the direct product of the representations generated by the orbitals. In addition, simple models such as the Hückel one can be used to determine many-electron symmetries from the one-electron symmetries, as the symmetry properties of wavefunctions can be rigorously correct even for highly approximate functions.

13.3.2 Application to $C_3H_3^+$

For $C_3H_3^+$, we have a ground configuration of $(a_2'')^2$. The representation generated by this configuration is determined by applying the symmetry operations to the many-electron wavefunction, which in this case is just a Slater determinant. Since the Slater determinant involves a product of orbitals, the representation involves a direct product of the representations of the individual orbitals. For the ground state, this product is $a_2'' \times a_2'' = a_1'$. Thus, the ground term is $^1A'$. [Uppercase Mulliken symbols are used to label many-electron states, with the spin multiplicity (a singlet in this case) given as a left superscript.] Quite generally, the representation associated with a totally filled shell is always the *totally symmetric* one.

For the first excited configuration, $(a_2'')^1(e'')^1$, we find that $a_2'' \times e'' = e'$. Here, there is the possibility of either singlet or triplets. Since the electrons are in orbitals with different energies, there is no problem with the Pauli principle, and we get $^1E'$ and $^3E'$ as the allowed terms.

Now consider the configuration $(e'')^2$. Because there are two MOs, we can imagine several ways to arrange the electrons, but not all are consistent with the Pauli principle. Symmetry alone tells us that

$$e'' \times e'' = A_1' + A_2' + E'. \tag{13.34}$$

13.3.3 Incorporation of the Pauli Principle

An important question concerning Eq. (13.34) is which terms are singlets and which triplets? For the simple case of two electrons, the Slater determinant wavefunction can be written as a product of space and spin parts. This is not true for more than two electrons, but the two-electron result is important, so we will work it through in detail.

For this simple special case, then, we need to construct wavefunctions associated with A_1', A_2' and E' terms and then see which are symmetric and which are antisymmetric with respect to the interchange of spatial coordinates. The symmetric ones must be singlet and the antisymmetric ones triplet in order to obey the Pauli principle.

To construct wavefunctions, we can use operators given by Eq. (13.12). As before, it is easiest to use the C_3 group, although A_1' and A_2' are not distinguished thereby. Let

$$\chi_1 = \frac{1}{\sqrt{3}}(p_1 + p_2 + p_3), \tag{13.35}$$

$$\chi_2 = \frac{1}{\sqrt{3}} (p_1 + \varepsilon^* p_2 + \varepsilon p_3), \tag{13.36}$$

and

$$\chi_3 = \frac{1}{\sqrt{3}} (p_1 + \varepsilon p_2 + \varepsilon^* p_3). \tag{13.37}$$

Then

$$C_3 \chi_1 = \chi_1, \tag{13.38}$$

$$C_3 \chi_2 = \frac{1}{\sqrt{3}} (p_2 + \varepsilon^* p_3 + \varepsilon p_1) = \varepsilon \chi_2, \tag{13.39}$$

and

$$C_3 \chi_3 = \varepsilon^* \chi_3. \tag{13.40}$$

Now consider the following two-electron spatial wavefunctions (these are all that can be imagined for $(e'')^2$):

$$|1\rangle = \chi_2(1)\chi_2(2); \tag{13.41}$$

$$|2\rangle = \chi_3(1)\chi_3(2); \tag{13.42}$$

$$|3\rangle = \chi_2(1)\chi_3(2); \tag{13.43}$$

$$|4\rangle = \chi_3(1)\chi_2(2). \tag{13.44}$$

The characters generated from these functions are given in Table 13.4. Evidently, $|1\rangle$ and $|2\rangle$ form an E symmetry set, while $|3\rangle$ and $|4\rangle$ belong to A symmetry.

Since $|1\rangle$ and $|2\rangle$ are both spatially symmetric with respect to interchange of the two electrons, we immediately conclude that the E' symmetry term must be *singlet* ($^1E'$). For the A symmetry terms, we must differentiate A_1' from A_2'. This is most easily done by determining symmetry with respect to $\sigma_v^{(1)}$. Since $\sigma_v^{(1)} \chi_1 = \chi_1$, $\sigma_v^{(1)} \chi_2 = \chi_3$, and $\sigma_v^{(1)} \chi_3 = \chi_2$, we have

$$\sigma_v^{(1)} |3\rangle = |4\rangle. \tag{13.45}$$

TABLE 13.4 CHARACTERS FOR $(e'')^2$ CONFIGURATIONS OF $C_3H_3^+$

	E	C_3	C_3^2
$|1\rangle$	1	ε^*	ε
$|2\rangle$	1	ε	ε^*
$|3\rangle$	1	1	1
$|4\rangle$	1	1	1

———$^1A_1'$

———$^1E'$ **Figure 13.5** Energy-level
               diagram for $(e_2'')^2$ terms of
———$^3A_2'$ $C_3H_3^+$.

Thus, neither $|3\rangle$ nor $|4\rangle$ belongs to an irreducible representation of D_{3h}, although both are of A symmetry for C_3. It is easy to see that the following linear combinations of $|3\rangle$ and $|4\rangle$ generate the desired D_{3h} irreducible representations (projection operators could be used if this were not obvious):

$$|5\rangle = \frac{1}{\sqrt{2}}(|3\rangle + |4\rangle); \tag{13.46}$$

$$|6\rangle = \frac{1}{\sqrt{2}}(|3\rangle - |4\rangle). \tag{13.47}$$

Clearly, $|5\rangle$ is symmetric with respect to σ_v and hence must be A_1', whereas $|6\rangle$ is A_2'. Furthermore, $|5\rangle$ is symmetric with respect to the interchange of 1 and 2 and hence is singlet, while $|6\rangle$ is spatially antisymmetric and hence is triplet.

Overall, then, the allowed terms are

$$^1A_1', {}^3A_2', \quad \text{and} \quad {}^3E'.$$

These terms are all degenerate when electron repulsion is neglected, but when it is included, the degeneracy is removed. A generalization of Hund's rules introduced in Section 9.9 to determine the ground term and level of a many-electron atom) that would be applicable to this case is that

1. the highest spin multiplicity term has the lowest energy and
2. the highest spatial degeneracy term for a given multiplicity has the lowest energy.

The expected ordering is given in Figure 13.5. Note that the application of Hund's rules to this problem is not expected to be accurate, because there should be strong configuration interaction between the $(e'')^2$ configuration and other excited configurations that would be nearby in energy (but not included in the Hückel description). Hund's rule is usually reliable only for the lowest energy term.

13.4 SYMMETRY PROPERTIES OF MOLECULAR VIBRATIONS

The application of group theory to describing the symmetry properties of molecular vibrations begins with the development in Chapter 5. There we found that the molecular *normal coordinates* Q_k associated with mode k could be related to the Cartesian displacements $x_{i\alpha}$ [i labels the atom number, α the Cartesian coordinate (i.e., $\alpha = x, y, z$)] via the equation

$$Q_k = \sum_{i\alpha} m_i^{1/2} L_{i\alpha, k}(x_{i\alpha} - x_{i\alpha}^e), \tag{13.48}$$

where m_i is the mass of the ith atom and the $L_{i\alpha, k}$ are coefficients that can be grouped into a matrix $\mathbf{L}$ of dimension $3N \times 3N$ (i.e., the number of atoms times the number of Cartesian coordinates for each atom). This matrix is determined by diagonalizing the mass-weighted Hessian matrix $\mathbf{H}$ defined by

$$H_{i\alpha, j\beta} = \frac{F_{i\alpha, j\beta}}{\sqrt{m_i m_j}}, \tag{13.49}$$

where $\mathbf{F}$ is the force-constant matrix (the matrix of second derivatives of the potential-energy surface). If the eigenvalues of this matrix are denoted λ_k, then the eigenvalues and eigenvectors satisfy the relation

$$\mathbf{HL} = \mathbf{L} \wedge, \tag{13.50}$$

where $\wedge$ is a diagonal matrix with the λ_k's as diagonal elements. The vibrational frequency ω_k associated with mode k is related to λ_k via the formula $\omega_k = (\lambda_k)^{1/2}$.

The application of group theory to this problem is similar to that of electronic structure theory in that the goal is to represent the Hessian matrix in terms of symmetry coordinates rather than Cartesian coordinates. In that way, $\mathbf{H}$ is block diagonalized, as determined by the irreducible representations of the point group associated with the molecule in its equilibrium geometry. The symmetry coordinates S_n may be defined in terms of linear combinations of the Cartesian coordinates; that is,

$$S_n = \sum_{i\alpha} m_i^{1/2} B_{i\alpha, n}(x_{i\alpha} - x_{i\alpha}^e). \tag{13.51}$$

Determining the coefficient matrix $\mathbf{B}$ is very much like determining SALCs. The irreducible representations associated with the Cartesian displacements from equilibrium can be found by using Eq. 13.9 and the rules for characters given in Section 13.2.3. The appropriate linear combinations of Cartesian coordinates to define symmetry coordinates can sometimes be determined by inspection, but if not, then a formula analogous to Eq. (13.12) can be used. However, this is rarely done by hand, so we will omit the details. Once the symmetry coordinates have been defined, reexpressing $\mathbf{H}$ in terms of those coordinates involves the product of three matrices:

$$\mathbf{H}_{\text{sym}} = \tilde{\mathbf{B}}\mathbf{H}\mathbf{B}.$$

Subsequent diagonalization of $\mathbf{H}_{\text{sym}}$ leads to the same eigenvalues that are obtained with the unsymmetrized problem.

As an example, let us consider the water molecule. There are three atoms, so there are nine Cartesian coordinates. Applying the rules for characters leads to the following characters for the C_{2v} group (see the explanation of notation in Section 13.1.4):

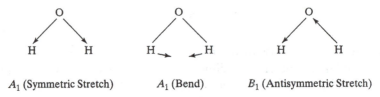

| A_1 (Symmetric Stretch) | A_1 (Bend) | B_1 (Antisymmetric Stretch) |

Figure 13.6   Vibrational normal modes of water.

	E	C_2	$\sigma_v(xz)$	$\sigma_v(yz)$
Γ	9	-1	3	1

Decomposing this table into irreducible representations (see Table 13.1) gives

$$\Gamma = 3A_1 + A_2 + 3B_1 + 2B_2.$$

One thing we have to remember in interpreting this result is that it includes the contributions from the three translational and three rotational modes. Using the C_{2v} character table in Appendix C, we note that these modes have the characters $A_1 + B_1 + B_2$ (for translations) and $A_2 + B_1 + B_2$ (for rotations). If these terms are subtracted from the overall result, we obtain $\Gamma = 2A_1 + B_1$. This equation displays the irreducible representations associated with the three vibrations in water, and what we see is that two are totally symmetric and one is antisymmetric with respect to $\sigma_v(yz)$. The antisymmetric mode is just the antisymmetric stretch of water, while the two symmetric modes are the symmetric stretch and bend modes. Figure 13.6 shows what all three modes look like. Since the symmetric stretch and bend modes belong to the same irreducible representation, they are not cleanly separated into purely stretch and purely bend motions.

SUGGESTED READINGS

The "bible" of group theory, as discussed in this chapter, is F. A. Cotton, *Chemical Applications of Group Theory* (Wiley-Interscience, New York, 1971).
Other useful texts are the following:

D. B. Chesnut, *Finite Groups and Quantum Theory* (Wiley-Interscience, New York, 1974).

H. Eyring, J. Walter, and G. Kimball, *Quantum Chemistry* (Wiley, New York, 1944).

F. W. Pilar, *Elementary Quantum Chemistry*, 2nd ed. (McGraw-Hill, New York, 1990).

G. C. Schatz and M. A. Ratner, *Quantum Mechanics in Chemistry* (Prentice-Hall, Englewood Cliffs, NJ, 1993).

M. Tinkham, *Group Theory and Quantum Mechanics* (McGraw-Hill, New York, 1964).

PROBLEMS

13.1 What terms from the $(e'')^2$ configuration of the cyclopropenyl cation discussed in Section 13.2.3 are optically connected to the ground state (i.e., are dipole allowed)?

13.2 Apply Hückel theory to the p_z orbitals in cyclopentadienyl (assumed to be planar and symmetrical).
(a) What is the symmetry group?
(b) Give the secular equation, using AOs.
(c) Find the irreducible representations spanned by the orbitals.
(d) Construct SALCs.
(e) Find the one-electron energies and wavefunctions.
(f) What are the many-electron symmetry and spin multiplicity of the ground state?
(g) What is the total electronic energy?

13.3 Consider the four-center exchange reaction

$$H_2 + D_2 \rightarrow 2HD.$$

Assume that the reaction coordinate involves the following three configurations of the atoms (note that we ignore the difference between H and D in determining the point group):

(I)   Reagents (D_{2h})

H   D
⋮   ⋮
H   D

(II) Transition state (D_{4h})

H   ⋯   D
⋮   ⋮
H   ⋯   D

(III) Products (D_{2h})

H   ⋯   D

H   ⋯   D

(a) Apply the extended Hückel method for I, II, and III, making the Hückel-like assumptions

$$H_{\mu\nu} = \begin{cases} \alpha & \text{if } \mu = \nu \\ \beta & \text{if } \mu \text{ is adjacent to } \nu \\ 0 & \text{otherwise} \end{cases}$$

and

$$S_{\mu\nu} = \delta_{\mu\nu}.$$

Obtain orbital energies and wavefunctions. Determine the ground-state electron configuration and the many-electron symmetry and spin multiplicity. (*Hint:* It is easiest to set up and solve the secular equation for D_{2d} symmetry and then evaluate for the three specific cases only at the end.)

(b) Now use symmetry to determine the correlation of the electronic orbitals as one progresses along the reaction coordinate. Draw a correlation diagram (energy level versus reaction coordinate). Show that if the electrons do not change orbitals as the reaction proceeds, then the reaction is *Woodward–Hoffmann forbidden* (meaning that the ground state of the reagents correlates to an excited state of the products). This is a specific example of a general set of orbital symmetry conservation rules called the *Woodward–Hoffmann rules*.

13.4 Consider the square planar complex $PtCl_4^{-2}$ (with Pt in a d^8 configuration). Let us use the extended Hückel method to describe σ bonding between metal and ligands. The AO basis consists of p orbitals on each Cl^- directed towards the Pt and the d orbitals on Pt. Assume that the only nonzero Hückel parameters are

$$\langle d_j | H | d_j \rangle = \alpha_{Pt} \quad \text{(independent of } j\text{),}$$

$$\langle p_i | H | p_i \rangle = \alpha_{Cl} \quad \text{(independent of } i\text{),}$$

and

$$\langle p_i | H | d_j \rangle = \beta_{ij}.$$

Choose $\beta_{ij} = 0$ for $j = d_{xy}, d_{yz}, d_{xz}$. For the remainder, use the values in the following table:

$i =$	1	2	3	4
$j = d_{z^2}$	β_A	β_A	β_A	β_A
$j = d_{x^2-y^2}$	β_B	$-\beta_B$	β_B	$-\beta_B$

Assume that $\alpha_{Pt}\rangle\alpha_{Cl}$ and $|\beta_B|\rangle|\beta_A|$.
(a) What is the symmetry group?
(b) Give the secular equation, using AOs.
(c) Find the irreducible representations spanned by the orbitals.
(d) Construct SALCs.
(e) Find the one-electron energies.
(f) Which orbitals are occupied in the ground state?
(g) What is the lowest energy optically allowed transition?

13.5 Consider an application of extended Hückel theory to the H_3 molecule, using the $1s$ orbitals on each nucleus as a basis set. Suppose that we don't know the geometry of this molecule, so we vary it in order to minimize the total electronic energy. To do this, assume that the bend angle θ defined in the diagram at the right is a variable to be determined and that the Hamiltonian matrix elements are as follows:

$$\langle i | H | i \rangle = \alpha (i = 1, 2, 3);$$

$$\langle 1 | H | 2 \rangle = \langle 2 | H | 3 \rangle = \beta;$$

$$\langle 1 | H | 3 \rangle = 2\beta(1 - \sin(\theta/2)).$$

(a) Construct SALCs and evaluate the one-electron energies for this basis.

(b) What energy levels do the electrons occupy? What is the total electronic energy? In the range $60° \langle \theta \langle 180°$, for which value of θ is the total energy minimized? What is the total energy for the ion H_3^+, and how does this energy vary with θ?

(c) Construct the molecular orbitals for arbitrary θ.

(d) Show how the symmetry labels of the orbitals correlate with θ between $\theta = 180°$ and $\theta = 60°$ (i.e., show how the linear molecule and the isosceles and equilateral triangle geometry point group labels interrelate).

(e) What are the many-electron-term symbols for $\theta = 60°$ for H_3, H_3^+, and H_3^-?

13.6 Cyclooctatetraene (C_8H_8) is not planar, but tub shaped (as suggested by the $4n + 2$ Hückel rule). The dication $C_8H_8^{++}$ is planar and is a regular octagon. Use the Hückel model and point group arguments to construct the eight MOs and the energy-level diagram for $C_8H_8^{++}$.

13.7 Determine the irreducible representations associated with vibrations in the methyl radical CH_3, which is planar. Describe qualitatively which atoms' motions should be involved in each mode. Now consider the perfluoro methyl radical CF_3, which has C_{3v} symmetry. How do the modes correlate with those in methyl?

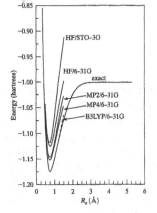

14

Applications of Electronic Structure Theory

In this chapter, we demonstrate the use of electronic structure theory methods (i.e., *ab initio*, semiempirical, and density functional theory methods) in studying the energetic, structural, and spectroscopic properties of molecules and chemical reactions. In general, we emphasize methods that can be implemented using widely available quantum chemistry programs, such as GAMESS, Gaussian, Hyper-Chem, Jaguar, QChem, and SPARTAN, but since we are not discussing any particular program, we do not go into any details associated with setting up input data or reading the output. Students should consult the user manual associated with their specific software for these details.

We do not present timings for these applications, but the choice of methods and basis sets is such that most of the applications require less than 5 minutes on a typical workstation for each run. As a result, the emphasis of the chapter is on methods that are commonly used throughout the field of chemistry as opposed to computations that push the forefront of current computer hardware.

The chapter is subdivided according to the type of physical properties being considered, classified into the following major categories:

1. Potential-energy functions.
2. Optimized geometries and frequencies.
3. Raman and infrared intensities.
4. Barriers and reaction paths.

5. Excited states.

6. Molecular clusters.

We use relatively simple molecules $[H_2, H_3, H_2O, (HF)_2,$ pyridine$]$ to illustrate the properties, as our primary goal is to provide some simple examples that can be used as the starting point for more ambitious projects such as are presented in the set of problems at the end of the chapter. Much of the theory needed for this chapter has already been presented, particularly in Chapters 11 and 12, but a few new concepts are explained where needed.

14.1 POTENTIAL-ENERGY FUNCTIONS

Suppose that we wanted to know the potential-energy function $V(R)$ for the ground state of the H_2 molecule. This sort of information would be of use in understanding the infrared spectrum of H_2 (which, although only weakly allowed through a quadrupole mechanism, has been observed in emissions from the clouds in the interstellar medium). To find $V(R)$, we need merely select a method (like Hartree–Fock), select a basis set, and do a series of electronic structure calculations on a grid of internuclear distances.

Figure 14.1 presents results of this type for a number of methods and basis sets, plotting energy in hartrees versus internuclear distance in angstroms

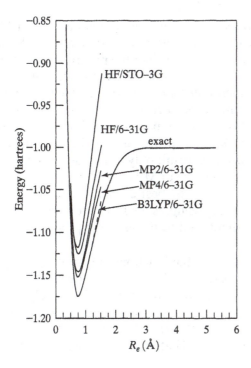

Figure 14.1 Potential-energy curves for H_2.

$(1\,a_0 = 0.529177\,\text{Å})$. The commonly used notation "(method)/(basis set)" is used to identify the curves. Thus, HF/STO-3G means that we are using a Hartree–Fock wavefunction (a Slater determinant of molecular orbitals), as described in Section 11.1, and an STO–3G basis set (a sum of three Gaussians with coefficients chosen to mimic a Slater-type orbital, as described in Section 11.2). The other results presented include Moller–Plesset perturbation theory in the second and fourth orders (MP2 and MP4) and a gradient-corrected density functional method known as B3LYP, in which the exchange-correlation energy in Eq. (11.23) is expressed [A. D. Becke, *J. Chem. Phys.* 98, 5648 (1993)] as a weighted sum of local exchange and correlation terms plus gradient corrections to these terms, with weighting coefficients adjusted to fit the properties of several well-known molecules. The gradient corrections to the exchange terms are based on a mixture of a Hartree–Fock exchange functional due to Becke and a correlation functional from Lee et al. [C. Lee, W. Yang and R. G. Parr, Phys Rev. B37, 785 (1988)]. B3LYP is one of a large number of gradient-corrected DFT methods that have been developed in recent years, and it is probably the most successful.

The atomic basis sets in Figure 14.1 include STO–3G and 6–31G (described in Section 11.2). Note that for hydrogen, the 6–31G basis only has valence orbitals, so the "6" is irrelevant. We will explore other, more sophisticated, basis sets later. Also included in the figure are the results of Kolos and Wolniewicz [W. Kolos and L. Wolniewicz, *J. Chem. Phys.* 49, 404 (1968)], which are of such high accuracy that they can be considered to be exact for the present purposes.

Note in the figure that the exact curve has its minimum at $-1.17444 E_h$. Since the energy of two separated hydrogen atoms is -1.0 exactly (see Section 10.5), the dissociation energy is 0.17444. If we consider the progression of results from HF, MP2, and MP4 for a given basis set, we notice that the energy at the minimum gradually decreases with increasing sophistication in the treatment of electron correlation. This demonstrates that correlation is important in determining the bond energy, and it also shows that the perturbation theory result systematically approaches the exact result as one proceeds to higher and higher order. (Rigorously, it will approach the exact result only if the basis set is complete.) Even lower total-energy results than those of MP4 are obtained using B3LYP, so one might conclude that B3LYP gives the best estimate of the dissociation energy. However, this should not be surprising, as the B3LYP exchange correlation functional was empirically parametrized to fit the exact properties of a number of molecules, including H_2.

Another feature of the potential functions in Figure 14.1 is that the HF results rise up above the exact asymptote. This behavior demonstrates an important flaw in the molecular orbital picture of bonding in H_2, namely, that dissociation is to a mixture of covalent and ionic configurations, rather than to the purely covalent configuration that is correct for H + H. In fact, the energy associated with the HF/6–31G results with the two H atoms separated is -0.71093, which is well above the correct result. This problem can be fixed by using an unrestricted Hartree–Fock method (see problem 14.11) in which the spatial orbitals of electrons with different spins are allowed to be different.

Table 14.1 summarizes the equilibrium geometry and harmonic vibrational frequency that are obtained from the curves in Figure 14.1. We could have obtained these data by fitting a polynomial function to the points and then using that function to determine the energy minimum and force constant. However with many electronic structure methods, it is possible to obtain these data from the electronic structure calculations directly, often based on analytical gradients and hessians (second derivatives) that are determined simultaneously with the energy; this is what we have done in generating the table. Note that even if analytical gradients or hessians are not available (as is often the case for the highest level calculations), many codes have the ability to generate numerical derivatives that can be used for minimization or normal coordinate analysis.

Table 14.1 also contains a few results based on methods that are not shown in Figure 14.1, including Austin Model 1 (AM1) INDO semiempirical-method calculations and coupled-cluster calculations that include double excitations (CCD). In addition, the table presents some more sophisticated basis functions: 6–311G(p) [three independently optimized Gaussians, augmented by polarization functions (p-functions)] and cc-pVTZ [correlation-consistent polarized valence triple-zeta basis functions].

The table shows that all the methods except AM1 yield good estimates of the equilibrium geometry (accurate to 0.012 Å or better). However, the frequencies are more sensitive, with HF/6–31G being off by 6%, MP2/6–31G by 3%, and MP4/6–31G by 0.4%. Notice the systematic improvement in the results as one goes from HF to MP2 to MP4 for a fixed basis set, indicating convergence of the perturbation approximation. Also, improving the basis set from 6–31G to 6–311G(p) gives a significantly better total energy and a slightly better R_e, but no change in ω_e. The B3LYP/6–31G geometry and frequency are somewhere between those of MP2 and MP4 in accuracy. CCD involves a higher level treatment of correlation effects than MP4 does, and the results are consistent, but note that

TABLE 14.1 EQUILIBRIUM ENERGIES, GEOMETRIES, AND VIBRATIONAL FREQUENCIES FOR H_2

Method	Energy	R_e (Å)	ω_e (cm^{-1})
AM1		0.6766	4,342
HF/6–31G	−1.12683	0.7301	4,644
MP2/6–31G	−1.14414	0.7325	4,534
MP2/6–311G(p)	−1.16027	0.7384	4,533
MP4/6–31G	−1.15096	0.7443	4,412
B3LYP/6–31G	−1.17548	0.7428	4,453
CCD/6–31G	−1.15163	0.7454	4,384
CCD/cc–pVTZ	−1.17220	0.7420	4,423
Experiment	−1.174474	0.7416	4,395

even the CCD total energy is noticeably off with the 6–31G basis set. However, the correlation-consistent cc-pVTZ basis set gives a significant improvement in the total energy (but not frequency), indicating that the limitations in the basis set are especially important to the determination of total energies.

14.2 OPTIMIZED GEOMETRIES AND FREQUENCIES

Now let us consider the water molecule, with special emphasis on the geometry of the molecule and its vibrational frequencies. (The molecular orbitals were presented in Chapter 11.) Table 14.2 presents the results of electronic structure studies of these properties, obtained by using a number of commonly available methods described in the previous section, as well as by experiment. The methods include the semiempirical AM1 method, the gradient-corrected density functional B3LYP method, standard HF theory, and second- and fourth-order perturbation theory. Basis sets include 6–31G, 6–31G(d, p) and 6–311G(d, p), with the (d, p) signifying two types of polarization functions: d-functions for the oxygen atom and p-functions for the hydrogen atoms. The parameters considered are the OH bond distance R_e, the HOH angle θ, and the harmonic vibrational frequencies (discussed in Chapter 12) ω_1 (the symmetric stretch), ω_2 (the bend), and ω_3 (the asymmetric stretch). These frequencies are similar to the fundamental frequencies that one would observe in the infrared spectrum of water; however, they are not actually the same, due to the influence of anharmonicity. A detailed study of the vibrational eigenvalues is needed to determine accurate values of the harmonic frequencies from spectral information. Such studies have been done for water, so the experimental values in Table 14.2 are the true harmonic frequencies, but it is unusual to find this information for molecules much larger than triatomics.

TABLE 14.2   ENERGY, STRUCTURE AND VIBRATIONAL FREQUENCIES OF H_2O

Method	Energy (E_h)	$R_e(\text{Å})$	θ (deg)	$\omega_1\ (\text{cm}^{-1})$	$\omega_2\ (\text{cm}^{-1})$	$\omega_3\ (\text{cm}^{-1})$
AM1		0.9612	103.6	3,505	1,884	3,584
HF/STO–3G	−74.96590	0.9895	100.0	4,139	2,170	4,390
HF/6–31G	−75.98536	0.9496	111.6	3,988	1,737	4,146
MP2/6–31G	−76.11318	0.9748	109.3	3,654	1,663	3,830
MP2/6–31G(d,p)	−76.19685	0.9687	104.0	3,775	1,735	3,916
MP4/6–31G	−76.20725	0.9704	103.9	3,739	1,742	3,870
MP4/6–311G(d,p)	−76.27634	0.9590	102.4	3,879	1,680	3,977
B3LYP/6–31G	−76.40895	0.9685	103.6	3,730	1,713	3,851
Experiment	−76.48038[a]	0.9752[a]	104.5[a]	3,833[b]	1,649[b]	3,943[b]

[a] Energy is based on an analysis presented by A. Lüchow, J. B. Anderson, and D. Feller, *J. Chem. Phys.* 106, 7706 (1997).
[b] Experimental harmonic frequencies are based on the force field of A. R. Hoy, I. M. Mills, and G. Strey, *Mol. Phys.* 24, 1265 (1972).

Several points can be made concerning the results in the table. First, note that AM1 and HF give the least accurate bond distances and frequencies, with MP2 and MP4 providing systematic improvements on HF. The HF/6–31G frequencies are larger than those obtained from experiment by about 4–5%. This error in the frequencies is very commonly found, and in fact (as we pointed out in Chapter 11), it is customary to scale HF/6–31G results by a factor of about 0.89 before comparing them with experiment [J. A. Pople, H. B. Schlegel, R. Krishnan, D. J. Defrees, J. S. Binkley, M. J. Frisch, and R. A. Whiteside, *Int. J. Quan. Chem. Sym.* 15, 269 (1981)]. A somewhat smaller scaling factor (0.95) has been suggested for MP2 results. MP4, in a fairly large basis $(6{-}311G(d,p))$, gives the most accurate frequencies of all the methods considered, being off by 46, 31, and 34 cm^{-1}, respectively, for ω_1, ω_2, and ω_3, corresponding to errors of 1.1%, 1.8%, and 0.9%, in turn. B3LYP is somewhat less accurate than MP4, but gives results that are only a little less accurate than MP2 for the same basis set. The approximate equivalence of MP2 and B3LYP is another commonly found result that has steered many people in favor of the latter, computationally simpler, method in recent years. Note that because of the empirical calibration used for B3LYP, larger basis sets will not necessarily give better results. Note also the anomalous result that the MP4/6–311G(d,p) method gives a rather poor bond length. Unexpectedly large errors of this type are unusual, but do occur.

Now consider a more complicated molecule, namely, pyridine (C_5H_5N). In dealing with this molecule, we will bypass some of the steps done with water and jump directly to the determination of frequencies, using the optimized geometry for each electronic structure approach considered. Table 14.3 shows the results of several calculations, ranging from HF/6–31G to B3LYP in a much larger basis set that includes diffuse functions (the ++ notation). The use of diffuse functions is not especially important for frequency determinations, but is sometimes important in the calculation of excited-state energies that will be discussed later. Note that the calculations presented here are not difficult by today's standards, although the B3LYP/6–311++G calculation takes more than an hour on many workstations.

The frequencies in the table have been ordered by the value of the B3LYP frequencies. This means that the frequencies are not always shown in numerical order for the other methods. Pyridine has C_{2v} symmetry, so the vibrational modes belong to one of the four irreducible representations of the C_{2v} group, and we have indicated the labels in the table. The experimental results are based on a detailed analysis of the infrared and Raman spectra of pyridine and its deuterated counterparts [H. D. Stidham and D. P. Dilella, *J. Raman Spectrosc.* 8, 180 (1979); 9, 90, 239, 247 (1980)], but the effects of anharmonicity have not been included, so the frequencies are not the harmonic frequencies.

The table shows that many of the frequencies get closer to experiment as one progresses from left to right across the columns (leaving out the column labeled "scaled"). A comparison of the 6–31G and 6–311++G basis sets reveals relatively small changes in frequencies, with the results generally closer to

TABLE 14.3 VIBRATIONAL FREQUENCIES OF PYRIDINE[a]

Symmetry	HF/ 6–31G	HF/ 6–311++G	HF/ 6–311++G (scaled)	B3LYP/ 6–31G	B3LYP/ 6–311++G	Experiment
A2	450	444	400	391	386	380
B1	481	478	430	433	430	406
A1	680	674	607	629	629	603
B2	737	731	658	682	681	654
B1	807	840	756	735	732	703
B1	856	796	716	774	765	747
A2	1,022	1,008	907	918	915	884
B1	1,096	1,078	970	974	969	941
A1	1,094	1,077	969	1,005	996	991
A2	1,160	1,135	1,022	1,018	1,010	980
B1	1,183	1,153	1,038	1,042	1,029	1,007
A1	1,138	1,125	1,013	1,059	1,049	1,030
B2	1,175	1,156	1,040	1,093	1,079	1,069
A1	1,192	1,178	1,060	1,109	1,099	1,069
B2	1,253	1,223	1,101	1,212	1,199	1,146
A1	1,354	1,339	1,205	1,262	1,249	1,217
B2	1,336	1,317	1,185	1,300	1,271	1,227
B2	1,522	1,507	1,356	1,416	1,405	1,355
B2	1,614	1,594	1,435	1,492	1,478	1,437
A1	1,659	1,639	1,475	1,526	1,512	1,483
B2	1,776	1,752	1,577	1,625	1,604	1,574
A1	1,785	1,759	1,583	1,630	1,610	1,581
A1	3,369	3,332	2,999	3,198	3,163	3,025
B2	3,382	3,345	3,011	3,200	3,168	3,034
A1	3,391	3,353	3,018	3,210	3,178	3,057
B2	3,402	3,366	3,029	3,222	3,188	3,079
A1	3,412	3,375	3,038	3,233	3,199	3,070

[a] HF/6–311++G and experimental results adapted from W.-H. Yang and G. C. Schatz, *J. Chem. Phys.* 3831 (1992).

experimental values for the bigger basis sets. The HF results are systematically higher than those obtained from experiment, which is consistent with what we learned for H_2O, so in the fourth column in the table we have scaled the third column by 90% (as suggested by W.-H. Yang and G. C. Schatz, *J. Chem. Phys.* 97, 3831 (1992)). This yields results that are in very good agreement with experiment, with an rms deviation of 27 cm^{-1} (omitting the CH stretch modes). The B3LYP/6–311++G results are in very good agreement with experiment, even without scaling (30 cm^{-1}, omitting the CH stretches), although the CH stretches are systematically high.

14.3   IR SPECTRA

Figure 14.2 shows the infrared spectrum of pyridine, calculated using the HF and B3LYP approaches with a variety of basis sets. The STO–3G result is considered obsolete by modern standards, and indeed, we see that the spectrum is seriously

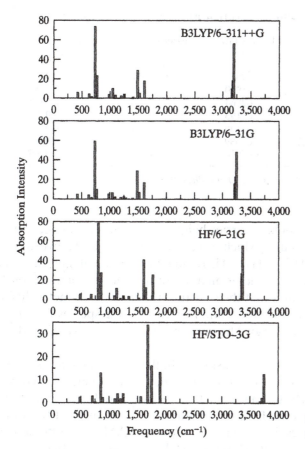

Figure 14.2 Infrared spectra of pyridine.

off compared to the other approaches in both the frequencies of the transitions and the intensities of the lines. The other three results are much closer, with the primary difference being in the frequencies. The B3LYP/6–311++G results are in good agreement with experiment.

14.4 BARRIERS TO REACTION

An important use of electronic structure calculations is the determination of barrier heights for chemical reactions, because such barriers play a crucial role in determining the rates of the reactions. As noted in Chapter 11, it is significantly more difficult to calculate accurate barriers than equilibrium properties, so higher quality methods must in general be used in the calculations. It is also important to note that, for a reaction which shows an Arrhenius dependence of the rate constant on temperature, a 1-kcal/mol error in a barrier height leads to a change of a factor of 5 in the rate constant at 300 K, so it is highly desirable to be able to determine barriers to at least this level and, preferably, to 0.1 kcal/mol.

TABLE 14.4   ENERGY, BARRIER HEIGHT, AND GEOMETRY OF H_3 SADDLE POINT

Method	Energy (E_h)	Barrier Height (E_h)	H–H distance (Å)
AM1		−0.00489	0.8141
HF/6–311G	−1.59650	0.02857	0.9341
MP2/6–311G	−1.61990	0.02579	0.9259
MP4/6–311G	−1.62864	0.02384	0.9324
B3LYP/6–311G	−1.67193	0.00686	0.9293
CCSD(T)/6–311G	−1.63166	0.02165	0.9431
CCSD(T)/Aug-cc-pVTZ	−1.65689	0.01557	0.9317
Exact[a]	−1.65916	0.01531	0.9298

[a] D. L. Diedrich and J. B. Anderson, *J. Chem. Phys.* 100, 8089 (1994).

In Table 14.4, we tabulate barrier height and geometry for the species H_3 (a linear symmetric molecule), which serves as the barrier to the hydrogen atom transfer reaction $H + H_2 \rightarrow H_2 + H$. To generate the results in this table, we first applied each method to a determination of the energy minimum of the H_3 structure with the constraint that the structure retain $D_{\infty h}$ symmetry during the minimization. Then we performed a minimization for the species $H + H_2$, so as to determine the reagent energy. The barrier height is then determined by subtraction.

A wide range of methods is considered in the table, but generally, we have used larger basis sets (at least 6–311G) and higher levels of electron correlation [CCSD(T), or coupled-cluster theory, including single and double excitations and with perturbative inclusion of triple excitations] than in most of the work discussed previously. The results in the table show that AM1 generates a negative barrier height, which means that H_3 is more stable as a molecule than as $H + H_2$. This, of course, is not correct, but it shows one of the artifacts that can occur when low-quality methods are used. All the other barrier estimates in the table are positive, including a very accurate calculation (converged to as many digits as are quoted) that serves as the "exact" reference, $0.01531 E_h = 9.61$ kcal/mol. Note that the HF barrier is well above (almost double) the correct result and thus is not useful for estimating rate constants. As one includes correlation effects via perturbation theory, the barrier comes down. However, even the high-level MP4 result is well above the exact result. This overestimate arises because the electronic state of the saddle point is poorly described by a single Slater determinant (which serves as the starting point for many of the higher level methods). So it is only with the much higher level CCSD(T) result that a high-quality barrier is obtained. To show how well one can do, we present results using a very large basis set (augmented correlation-consistent polarized valence triple zeta). The resulting barrier height is high by only 0.2 kcal/mol compared to the exact result, a small enough error that the estimate is useful for rate constant calculations. This combination of CCSD(T) with an augmented correlation-consistent triple-zeta basis set has been demonstrated to give extremely high-quality results in a large variety of calculations, and accordingly, it is one of the most powerful methods of *ab initio* quantum chemistry that is in common use.

Table 14.4 shows that the B3LYP barrier is much lower than the correct value. This point was considered by Johnson et al. [B. G. Johnson, C. A. Gonzales, P. M. W. Gill, and J. A. Pople, *Chem. Phys. Lett.* 221, 100 (1994)], who found that density functional methods give barriers that are too low unless the so-called self-energy correction is included. This correction arises because the density functional expression for the exchange energy does not precisely cancel the part of the coulomb energy in which the interaction of the electron with its opposite spin-counterpart in the same orbital is double counted. The problem does not occur in Hartree–Fock-based theories, in which the coulomb and exchange terms are consistently calculated, but it is important in density functional theory calculations whenever energy differences are determined between structures that have significantly different bonding characteristics (such as between a transition state and the reactants). When this double counting is corrected for, the barrier rises by roughly $0.013E_h$.

14.5 EXCITED STATES

The lowest excited state of water involves the excitation of an electron from a lone-pair $(b_1$ symmetry) orbital [the highest occupied molecular orbital (HOMO)] to an antibonding $(a_1$ symmetry) orbital [the lowest unoccupied molecular orbital (LUMO)], yielding a state with overall $^1B_1$ symmetry. (See Chapter 11 for a description of the orbitals of water.) The resulting state is repulsive, so the potential surface is dissociative, and the molecule falls apart to OH + H directly. The spectrum shows a broad peak that can be understood using theory discussed in Chapter 15, from which the vertical excitation energy can be inferred through a theoretical analysis. Table 14.5 lists the energies obtained from CIS/6–311G(d) calculations, done at the equilibrium H_2O geometry, as well as those derived from other theoretical methods and experiment. Results are included for the $^1B_1$ transition as well as the next two singlet states, $^1A_2$ and $^1A_1$.

The Configuration Interaction—Singles (CIS) calculations involve a configuration interaction calculation in which the only excited determinants that are included involve single excitations relative to the ground state. This restriction makes the calculation feasible for a broad range of molecules, and in fact, CIS is among the few routinely available methods for determining excited-state energies. What the table shows is that CIS yields energies that are above the experimental results by roughly 1 eV. The other electronic structure calculations are essentially in agreement with experiment, but they involve more specialized methods. Generally, one needs to use a multireference configuration interaction method, such as CASSCF-SDCI, or a high-level coupled-cluster method to determine excited-state energies to a few tenths of an electron volt.

TABLE 14.5 VERTICAL EXCITATION ENERGIES (IN eV) ASSOCIATED WITH THE LOWEST EXCITED STATES OF THE WATER MOLECULE

State	CIS/6–311G(d)	Other theory	Experimental
$^1B_1$	8.74	7.52[a], 7.55[b], 7.60[c]	7.42[d], 7.49[e], 7.70[f]
$^1A_2$	10.55		9.1–10.2[g]
$^1A_1$	11.30		

[a] (CEPA), V. Staemmler and A. Palma, *Chem. Phys.* 98, 63 (1985).
[b] (CCSD), A. Balkava and R. J. Bartlett, *J. Chem. Phys.* 99, 7907 (1993).
[c] (CASSCF+SDCI), G. C. Schatz, A. Papaioannou, L. A. Pederson, L. B. Harding, T. Hollebeek, T.-S. Ho and H. Rabitz, *J. Chem. Phys.* 2340 (1997).
[d] H. T. Wang, W. S. Felps, and S. P. McGlynn, *J. Chem. Phys.* 67, 2614 (1977).
[e] K. Watanabe and A. S. Jursa, *J. Chem. Phys.* 41, 1650 (1964).
[f] M. L. Doublet, G. J. Kroes, E. J. Baerends, and A. Rosa, *J. Chem. Phys.* 103, 2538 (1995).
[g] This is the spectral range (including zero-point energy corrections) associated with the transition. See E. Segev and M. Shapiro, *J. Chem. Phys.* 77, 5604 (1982).

Table 14.6 presents excitation energies for pyridine from a number of electronic structure methods and from experiment. The one *ab initio* method in the list is again CIS, and we see that the energies derived from this calculation are mostly higher than those obtained from experiment by over 1 eV. Much better results are from two semiempirical methods, ZINDO and HAM, but of course, the two-electron integrals in these methods have been parametrized to give accurate spectra.

Note that the $^1A_2$ state is not observed in the measurements, as it is a forbidden transition. We will learn in the next chapter that in order for a transition to be allowed, the direct product

$$\Gamma(\text{initial state}) \times \Gamma(x \text{ or } y \text{ or } z) \times \Gamma(\text{final state})$$

TABLE 14.6 VERTICAL EXCITATION ENERGIES (IN eV) ASSOCIATED WITH THE LOWEST EXCITED STATES OF PYRIDINE

State	CIS/6–311G(d)	ZINDO[a]	HAM[b]	Experiment[c]
$^1B_1$	5.83	4.37	4.9	4.59
$^1B_2$	6.45	4.78	4.9	4.99
$^1A_2$	7.15	5.36		
$^1A_1$	6.74	5.72	6.2	6.38

[a] W.-H. Yang and G. C. Schatz, *J. Chem. Phys.* 97, 3831 (1992).
[b] E. Lindholm and A. Asbrink, *Molecular Orbitals and Their Energies Studies by the Semiempirical HAM Method* (Springer, Berlin, 1985).
[c] A. Bolovinos, P. Tsekeris, J. Philis, E. Pantos, and G. Audritsopoulos, *J. Mol. Spectrosc.* 103, 240 (1984).

must contain the totally symmetric representation. In the present application, the ground state is of $^1A_1$ symmetry and the coordinates x, y, and z belong to B_1, B_2, and A_1, respectively, so the direct product with the $^1A_2$ final state is not totally symmetric.

14.6 MOLECULAR CLUSTERS

Another important area of research in electronic structure theory is the structure and stability of molecular clusters. Clusters that are formed from closed-shell molecules, such as the HF dimer or the water dimer, are bound by the relatively weak electrostatic, van der Waals, and hydrogen bonding forces, so they have "floppy" structures that are difficult to characterize experimentally. Theory can therefore be useful for interpreting the experiments and for making predictions of structural properties in the absence of measurements. Such information is of interest not only in the characterization of gas-phase experiments, but also in the development of potential-energy information for modelling the properties of liquids and for characterizing molecular solvation.

To illustrate the use of electronic structure theory in studies of molecular clusters, consider the hydrogen fluoride dimer $(HF)_2$. Table 14.7 illustrates the properties of this dimer, as obtained from a variety of *ab initio* methods and from experiment. The geometrical parameters shown in the table are defined with reference to Figure 14.3, which shows that the two HF's are not equivalent; one of

TABLE 14.7   STRUCTURES AND ENERGIES ASSOCIATED WITH $(HF)_2$

Method	Energy	r_b	r_f	R_{FF}	θ_b	θ_f	D_e
MP2/6–31G	−200.23664	0.9501	0.9516	2.7112	14.3	73.8	0.0127
MP2/large[a]		0.9264	0.9233	2.747	6.0	67.6	0.0072
CCSD(T)/large[b]		0.923	0.921	2.74	7	70	0.0075
Experiment[c]		0.923	0.920	2.736	7	68	0.00730
							±0.0001

[a] The basis set is a variant on aug-cc-pVTZ, as reported by W. Klopper, M. Quack, and M. A. Suhm, *Chem. Phys. Lett.* 261, 35 (1996).
[b] C. L. Collins, K. Morihashi, Y. Yamaguchi, and H. F. Schaefer, III, *J. Chem. Phys.* 103, 6051 (1995).
[c] B. J. Howard, T. R. Dyke, and W. Klemperer, *J. Chem. Phys.* 81, 5417 (1984).

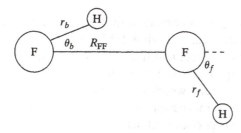

Figure 14.3   Structure of the HF dimer. Note that this structure is planar.

the H's interacts strongly with the F that it is not bonded to, while the other does not. Thus, the parameters in the table refer to "bound" and "free" HF's.

The methods used in Table 14.7 are all single-reference based, as would seem to be appropriate for the weak interactions between two closed-shell molecules. At the same time, it is important to use large basis sets in order to describe subtle distortions in the electron density that are induced by these weak interactions. Dissociation energies have been calculated by the simple difference

$$D_e = -[E(\text{dimer}) - 2E(\text{monomer})],$$

where $E(\text{monomer})$ is obtained from an independent calculation for HF at its minimum energy. This formula has occasionally been criticized as potentially overestimating the dissociation energy due to basis-set superposition errors [S. F. Boys and F. Bernardi, *Mol. Phys.* 19, 553 (1970)]. What that means is that at the dimer minimum, each HF can use basis functions on the other HF to lower the overall energy. A way to fix the situation, at least approximately, involves making estimates of the monomer energies using basis functions in which the other HF orbitals (but not the nuclei) are still present. The energy for this counterpoise (CP) corrected calculation is thus given by

$$D_e = -[E(\text{dimer}) - E(\text{monomer1}) - E(\text{monomer2})],$$

where each of the monomer energies is calculated at the geometry of the dimer and with the nuclei, but not the basis functions, of the other monomer left out.

To show the importance of the CP correction, consider the MP2/6–31G results in Table 14.7. In this case, the monomer energy is -100.11197, so with a dimer energy of -200.23664, one gets a dissocation energy of 0.0127. However, if calculations are done for the two monomers in the presence of the basis functions of the other monomer, one obtains energies of -100.11277 and -100.11527, with a dissociation energy of 0.0086. This smaller energy is closer to the experimental value shown in the table, suggesting that the CP correction gives improved results. The point is controversial, though, and more detailed studies have failed to find the systematic improvement that one might expect as the size of the basis set is varied (D. W. Schwenke and D. G. Truhlar, *J. Chem. Phys.* 82, 2418 (1985); M. J. Frisch, J. E. Del Bene, J. S. Binkley, and H. F. Schaefer, III, *J. Chem. Phys.* 84, 2279 (1986)).

Corrections for superposition errors in the basis set generally become less important as its size is increased, and what Table 14.7 shows is that the results systematically approach the correct result as the basis set and the level of treatment of correlation are improved. In fact, even the MP2/large result is in very good agreement with experiment, which shows that the ground-state properties of molecular clusters converge more rapidly with the level of inclusion of electron correlation than do other properties, such as excited-state energies.

14.7 REMARKS ON OTHER METHODS

Methods for calculating molecular electronic structure (that is, solving the electronic Schrödinger equation) have reached "chemical accuracy" of roughly 1 kcal/mol for small and medium-sized organics and for compounds containing elements in the first or second row of the periodic table. These methods are contained in available commercial and freeware codes and have proven to be of huge value to that part of the scientific community which is interested in molecules. Table 14.8 provides a schematic overview of what is available. The Hartree–Fock-based methods in the table have been examined both in Chapter 11 and in the current chapter, and we have arranged them in the first column of the table into groups according to their treatment of electron correlation. Note that all of these approaches provide a systematic procedure for improving the quality of the calculation by improving the orbital basis set and the treatment of electron correlation. To obtain "chemical accuracy", it is generally necessary to use highly accurate geometries, large basis sets, and correlation at the MP4 level or beyond. The orbital-based DFT methods—especially the parametrized hybrid methods such as B3LYP—provide an alternative (generally much simpler) approach that generates results with close to chemical accuracy; however, there is no systematic procedure for improving the quality of the calculation.

The methods at the bottom and right side of the table extend beyond the scope of this book. Density-based DFT methods, which use the kinetic-energy functional rather than the Kohn–Sham orbital-based approach, have only briefly been mentioned, in Chapter 11. These methods offer the promise of extending electronic structure calculations to molecules that are much larger than can be treated with other approaches, but accuracy has been a problem. The floating spherical Gaussian orbital method (also briefly discussed in Chapter 11) is an example of a method that works outside the realm of conventional electronic structure theory, but it is not very accurate. Quantum Monte Carlo methods, by contrast, which recast the Schrödinger equation as a diffusion equation in imaginary time (see Exercise 1.6), give extremely accurate total energies, better even than CCSD(T) with very large basis sets. However, they are difficult to apply to molecules with more than 10 electrons and thus are not used extensively.

TABLE 14.8   A SCHEMATIC CLASSIFICATION OF ELECTRONIC STRUCTURE METHODS BASED ON FIRST PRINCIPLES

Hartree–Fock-based	Density Functional	Other Approaches
Single determinant (HF)	Kohn–Sham (orbital-based) methods:	Floating spherical Gaussian orbital
MP2, MP3, MP4	LDA, GGA, hybrid schemes [B3LYP] (mix Hartree–Fock with DFT)	
CCSD, CCSD(T) CASSCF, MCSCF, MCSCF-CI CISD, FCI	Density-based methods (kinetic-energy functional)	Quantum Monte Carlo

While stable small molecules containing only atoms smaller than K can be treated with chemical accuracy, many aspects of the electronic structure problem remain only poorly solved. Issues such as relativistic calculations, excited states, and highly degenerate open-shell species remain to challenge researchers.

SUGGESTED READINGS

J. B. Foresman and A. Frisch, *Exploring Chemistry with Electronic Structure Methods* (2nd ed.) (Gaussian Inc., Pittsburgh, 1996).

W. J. Hehre, L. Radom, P. v. R. Schleyer, and J. A. Pople, *Ab initio Molecular Orbital Theory* (Wiley, New York, 1986).

W. J. Hehre, *Practical Strategies for Electronic Structure Calculations* (Wavefunction, Inc., Irvine, CA, 1995).

W. G. Richards and D. L. Cooper, *Ab Initio Molecular Orbital Calculations for Chemists* (Clarendon Press, Oxford, 1983).

J. Simons and J. Nichols, *Quantum Mechanics in Chemistry* (Oxford, New York, 1997).

A. Szabo and N. S. Ostlund, *Modern Quantum Chemistry* (McGraw-Hill, New York, 1989).

PROBLEMS

14.1 Use the CIS method to calculate the lowest four excited-state potential curves for H_2 with the coordinate range restricted to match that in Figure 14.1. Do your curves match those for the excited states expected from the simple arguments presented in Chapter 10? What about the comparison with experiment? [See K. P. Huber and G. Herzberg, *Constants of Diatomic Molecules* (Van Nostrand Reinhold, New York, 1979].

14.2 The molecule N_2H is of great interest in combustion chemistry due to its possible participation as an intermediate species in the removal of NO from exhaust gases by reaction with ammonia. However, it is not clear whether N_2H is stable enough to participate in kinetic processes at temperatures relevant to combustion (1,500 K). Use a variety of electronic structure methods, including AM1, HF, MP2, MP4, and B3LYP, to search for a minimum corresponding to N_2H. The smallest basis set to consider is 6–31G. If possible, study 6–311G(d,p). The minimum will be located at approximately N–N = 2.25 a_0, N–H = 2.02a_0, and an internal angle of 115°. Compare your results with the energy of N_2 + H to determine whether the molecule is stable. Compare the properties of this molecule with those reported by Walch [S. P. Walch, *J. Chem. Phys.* 93, 2384 (1990)]. Now try to find the saddle point for the dissociation of N_2H. This point should be located near N–N = 2.1a_0, N–H = 2.64a_0, and an internal angle of 125°.

14.3 Using HF, MP2, and B3LYP, calculate the infrared and Raman spectra of benzene, and compare your results with experiment. [See J. Koput, *J. Molec. Spectros.* 115, 438 (1986); J. M. Fernandez-Sanchez and S. Montero, *J. Chem. Phys.* 90, 2909 (1989); and J. E. Bertie and C. D. Keefe, *J. Chem. Phys.* 101, 4610 (1994)]. Why are the infrared allowed transitions all Raman forbidden and vice versa? (*Hint:* Use group-theoretical ideas.) Are there vibrational modes for which are forbidden in both IR and Raman?

14.4 Using HF, MP2, B3LYP, and a 6–31G basis set, find the minimum associated with the ground state of NO_2. Be sure to look for structures that are of lower symmetry than C_{2v}. What is the point group if the molecule is not C_{2v}? Which methods yield non-C_{2v} minima? What is the right answer? [See E. R. Davidson and W. T. Borden, *J. Phys. Chem.* 87, 4783 (1983).]

14.5 Using HF, MP2, B3LYP, and a 6–31G basis, determine the ground state of ozone (O_3). Then use CIS to determine the minima associated with the lowest two excited states. Do any of these excited states have a D_{3h} minimum? Rationalize the symmetry of the states in terms of the symmetries of the ground- and excited-state molecular orbitals. What orbitals are involved in these transitions? [See N. Naval and S. Pal, *J. Chem. Phys.* 111, 4051 (1999).]

14.6 The polarizabilities of molecules are commonly provided in calculations of vibrational frequencies. This is because the polarizability may be obtained from the second derivative of the energy with respect to applied electric field, thus requiring derivatives of the same order as those used in determining vibrational frequencies. What are the latter derivatives? Use an electronic structure code (HF, MP2, MP4) to determine the polarizabilities of the series of molecules HF, HCl, and HBr. Consider two basis sets 6–31G and 6–311G(d,p). Which has the larger polarizability component along the molecular axis and why? Often, it is stated that polarizabilities are proportional to molecular volume. Test this idea by comparing the polarizabilities (i.e., the average of the xx, yy, and zz polarizabilities) of the series of hydrocarbons methane, ethane, propane, and butane. This can be done by assigning polarizabilities per atom to H and C atoms and then using the first two molecules to determine values of these parameters, (i.e., methane = 4H + C, ethane = 6H + 2C). The other molecules can then be used to test the results.

14.7 Using HF, MP2, MP4, and B3LYP, determine the minimum energies of *cis* and *trans* planar hydrogen peroxide (HOOH) and the nonplanar global minimum. Examine results for different basis sets, (6–31G to cc-pVDZ) to determine the degree of convergence of the results. What labels (e.g. $n \rightarrow \pi^*$, $\pi \rightarrow \pi^*$) would you use to describe the HOMO-to-LUMO transition? Now compare your results with those of higher level theory and experiment. [See L. B. Harding, *J. Phys. Chem.* 93, 8004 (1989)].

14.8 Oxywater (H_2OO) is an isomer of hydrogen peroxide that is similar in structure to formaldehyde. Use HF, MP2, MP4, B3LYP, and CCSD to determine the stability of this isomer relative to HOOH. (See Problem 7.) If a stationary point is found, do a frequency calculation to verify that the structure is a minimum and to determine vibrational frequencies. Compare your results with some of those found in the literature [H. H. Huang, Y. Xie, and H. F. Schaefer, III, *J. Phys. Chem.* 100, 6076 (1996)]. Compare the HOMOs and LUMOs for oxywater and hydrogen peroxide.

14.9 Vinylidene (H_2CC) is an isomer of acetylene (HCCH) whose stability is uncertain. Use HF, MP2, MP4, and B3LYP (6–31G* basis) to determine the energy minimum of vinylidene compared with that of acetylene. How does your result compare with experiment? [See S. M. Burnett, A. E. Stevens, C. S. Feigerle, and W. C. Lineberger, *Chem. Phys. Lett.* 100, 124 (1983); T. Carrington, Jr., L. M. Hubbard, H. F. Schaefer and W. H. Miller, *J. Chem. Phys.* 80, 4347 (1984); M. M. Gallo, T. P. Hamilton, and H. F. Schaefer, *J. Am. Chem. Soc.* 112, 8714 (1990).]

14.10 Use HF, MP2, and B3LYP to determine minimum energy structures for the water dimer. The starting configuration is analogous to Figure 14.3, but there are several possible isomers, depending on where the second H atom in each water molecule is located relative to what is in the figure. Find as many of these isomers as you can. Which has the lowest energy? How do your results compare with those of other theories? Make sure that you do a frequency calculation for each stationary point, to verify that the structure is a minimum. [See M. J. Frisch, J. E. Del Bene, J. S. Binkley, and H. F. Schaefer, III, *J. Chem. Phys.* 84, 2279 (1986).]

14.11 HF-like methods can be defined either by restricting spin pairing in doubly occupied orbitals or by allowing α-spins and β-spins to have different spatial orbitals. These assumptions respectively lead to the restricted Hartree–Fock (RFH) and unrestricted Hartree Fock (UHF) methods that were discussed in Chapter 11. Similarly, there are RMP2 and UMP2 methods. Using UHF and RHF methods with a 6–31G basis, compute potential curves for the H_2 molecule. Compare your results with those of Fig. 14.1. Can you explain physically why UHF and RHF differ so strikingly? Now calculate the potential curve for the OH radical, using UHF, UMP2, B3LYP, and a 6–31G* basis. Compare the predicted geometries and frequencies with experiment. (See Huber and Herzberg, cited in Problem 14.1.)

14.12 Use UHF, UMP2, and B3LYP to calculate the difference in energy between the ground states of singlet and triplet methylene (CH_2). Be sure to optimize the geometry of each state separately. Do these states correspond to the same or a different electronic configuration? (You can answer this question by comparing the occupied molecular orbitals to see if they are qualitatively the same or different!) What is the zero-point energy correction to the singlet–triplet energy difference? How does the singlet-triplet energy difference compare with previous calculations and with experiment? [This question was a hot topic for many years, but has now been settled. See D. G. Leopold, K. K. Murray, and W. C. Lineberger, *J. Chem. Phys.* 81, 1048 (1984); L. B. Harding and W.A. Goddard, III, *J. Chem. Phys.* 67, 1777 (1977); and H. F. Schaefer, *Science,* 231, 1100 (1986).]

14.13 The biphenyl molecule ($H_5C_6 - C_6H_5$) consists of two benzene rings bound together by a C-C link. Use a reasonable calculation (6–31G at the SCF level) to plot the ground-state energy as a function of twist angle. Repeat these calculations at the MP2 level.

Now do CIS calculations with the same basis set. Compare the lowest excited state energy level to the ground state, by plotting the two energy curves as a function of the twist angle. Which states have planar minima, and which have staggered minima? From which geometry should the fluorescence line arise? Compare your results with M. Rubio et al. *Chem. Phys. Lett.* 234, 373 (1995), with the experimental study of Y. Takei et al., *J. Phys Chem.* 92, 578 (1998), and with the figure on the cover of this book.

15

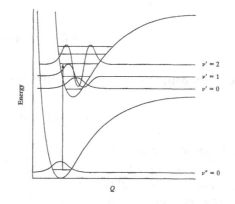

Time Dependence
and Spectroscopy

Much of chemistry is devoted to the study of the structure and the dynamical properties of molecules. Up to this point, we have concentrated on structural behavior, studying the properties of the wavefunctions and energies given by solutions to the time-independent Schrödinger equation. In this chapter, we examine the evolution of wavefunctions with time and the formulation of the spectroscopic response of molecules, via both electronic spectroscopy and vibrational spectroscopy.

Even more than in previous chapters, here we simplify a very elaborate and beautiful subject: time-dependent quantum mechanics. Far more extensive discussions are given in advanced texts.

15.1 TRANSITION PROBABILITIES AND THE GOLDEN RULE

When a molecule in a stationary state is subjected to a perturbation, the molecule can undergo transitions among its different stationary states. The perturbation may be due to an external electromagnetic field (as occurs in spectroscopy) or to collisions with surfaces, other atoms, molecules, neutrons, electrons, or other scattering partners. In this section, we derive the simplest and most useful formula approximating the rate of change of a molecular system subject to an external perturbation.

We begin with the time-dependent Schrodinger equation

$$H\Psi(\mathbf{x}, t) = i\hbar \frac{\partial \Psi}{\partial t}. \tag{15.1}$$

The total time-dependent wavefunction depends on all the electronic and nuclear coordinates, collectively described by $\mathbf{x}$. We shall describe the state of the molecule in terms of a set of stationary states of the unperturbed molecular structure. The molecule is assumed to be initially in one of these eigenstates, and then a perturbation causes transitions to other eigenstates. The stationary states, denoted ϕ_m, are eigenfunctions of a static, unperturbed Hamiltonian that we will call H_0; that is,

$$H_0\phi_m = E_m\phi_m. \tag{15.2}$$

The total Hamiltonian in the presence of the perturbation is written as

$$H = H_0 + V(t), \tag{15.3}$$

where $V(t)$ is the perturbation produced on the molecule by the external field. The state of the system as a function of time, then, is described as a superposition of the stationary states

$$\Psi(\mathbf{x}, t) = \sum_m C_m(t)\phi_m(\mathbf{x}), \tag{15.4}$$

where the time-dependent coefficients $C_m(t)$ can be physically interpreted as [see the discussion following Eq. (2.42)]

$$|C_m(t)|^2 = \text{probability of being in state } \phi_m \text{ at time } t. \tag{15.5}$$

Therefore, if we can solve for the coefficients in Eq. (15.4), we can describe the time-dependent probabilities of the molecule being in its different stationary states.

The formal analysis of this problem was originally given by Dirac. It actually uses an expansion very much like the one employed in Chapter 5 for describing time-independent perturbations, but now applied to the time-dependent perturbation $V(t)$. Details are given elsewhere. (See the suggested readings at the end of the chapter.) The most important result—one that is valid for the relatively weak perturbations that occur in spectroscopy between molecules and electromagnetic fields—is that the transition rate $w_{m \rightarrow n}$, the probability per unit time for the molecule to transfer from state ϕ_m to ϕ_n, is given by the Fermi "golden rule" as

$$w_{m \rightarrow n} = \frac{2\pi}{\hbar} |\langle \phi_m | V | \phi_n \rangle|^2 \delta(E_m - E_n). \tag{15.6}$$

The integral that appears squared as the second term in this expression is generally called the transition matrix element. The last term is the so-called Dirac δ function, a generalization of the Kronecker δ function that occurs in the discussion of wave-function orthogonality (Eq. 4.45). Physically, the Dirac δ function describes the fact that energy must be conserved; the function vanishes unless its argument, $E_m - E_n$, vanishes. Formally, and for interpretive purposes, it is useful to define the Dirac δ function as

$$\delta(E_m - E_n) \equiv \lim_{\Gamma \to 0} \frac{\Gamma/\pi}{(E_m - E_n)^2 + \Gamma^2}. \tag{15.7}$$

The right side of Eq. (15.7) is sketched in Figure 15.1. In the limit indicated in the equation, this line [the line-shape function on the right-hand side of Eq. (15.7) is referred to as a Lorentzian shape] becomes infinitely narrow and infinitely high, so that its amplitude vanishes for all values of $E_m - E_n$, except $E_m = E_n (E_m - E_n = 0)$. Note that the δ function has units of inverse energy.

Equation (15.6) is *extremely* important: It is the fundamental rule for describing how molecules evolve under a time-dependent perturbation. Physically, it states that a molecule can undergo a transition from state ϕ_m to state ϕ_n only if two things occur: First, the energy must be conserved as a result of the delta function; second, the transition will be permitted (that is, the transition rate $w_{m \to n}$ will not vanish) only if the external perturbation V has the correct form to mix state ϕ_m with state ϕ_n, so that the transition matrix element in Eq. (15.6) is nonzero.

The conditions required for Eq. (15.6) to obtain are actually quite complicated. The most important condition is simply that the final state ϕ_n be one of many states that lie close together in energy. When states do not lie close together in energy, the probability per unit time to go from state m to state n is no longer constant in time, and Eq. (15.6) will fail. (See Problem 2 at the end of the chapter.) For real molecules, the requirement of very close lying states will generally hold.

If there is a set of final states that do indeed conserve energy, then the rate $w_{m \to n}$ will depend on how many such states there are. A second form of the golden rule used in such a situation yields

$$w_{m \to n} = \frac{2\pi}{\hbar} |\langle m | V | n \rangle|^2 \rho(E_n), \tag{15.8}$$

for the rate, where $\rho(E_n)$ is the density of final states at energy $E_n = E_m$. This situation occurs in most spectroscopic processes. (See Problem 15.2 at the end of the chapter.)

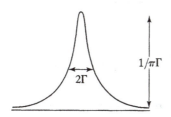

Figure 15.1 Lorentzian line-shape function.

The Fermi golden rule can be generalized to situations in which several perturbations occur at once, or in which the matrix elements in Eqs. (15.6) and (15.8) indeed vanish. We shall concern ourselves only with spectroscopic processes that are based on Eqs. (15.6) and (15.8).

15.2 ELECTRONIC SPECTROSCOPY OF MOLECULES

The most important experimental methods for studying molecules involve their interaction with electromagnetic radiation. Figure 15.2 shows the electromagnetic spectrum, characterized in terms of both wavelength and wave number $\bar{\nu}$. The energy of an electromagnetic wave is given as $E = h\nu = hc\bar{\nu} = hc/\lambda$, where h, c, and λ are, respectively, Planck's constant, the speed of light, and the wavelength of the radiation; $\bar{\nu}$ is the frequency in wave numbers (cm^{-1}). From the analogy between molecules undergoing transitions among states and playground swings being pushed, one might expect that the most efficient transitions between two molecular states will occur when the energy difference between those states is equal to the energy of the electromagnetic radiation (light) shined on the molecule.

When the molecule interacts with an electromagnetic field, the simplest semiclassical form for the perturbation Hamiltonian term is given by the time-dependent equation

$$V = V_{ext}(\mathbf{x}, t) = \mathbf{E} \cdot \boldsymbol{\mu} \cos \omega t, \tag{15.9}$$

where V_{ext} reminds us that the perturbation arises from an external field, acting at position $\mathbf{x}$. The terms $\mathbf{E}, \boldsymbol{\mu}$, and ω are, respectively, the electromagnetic field intensity vector, the dipole moment vector of the molecule, and the frequency of the electromagnetic radiation. The study of the evolution of molecules under perturbations of the type of Eq. (15.9) makes up the subject of molecular spectroscopy.

The dipole moment can be simply defined as

$$\boldsymbol{\mu} = \sum_{\lambda} Z_{\lambda} \mathbf{x}_{\lambda}, \tag{15.10a}$$

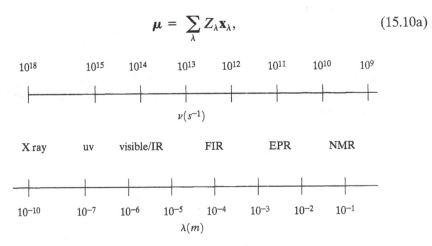

Figure 15.2   The electromagnetic spectrum.

where Z_λ and $\mathbf{x}_\lambda$ are, respectively, the charge of the particle indexed by λ and the position of that particle. Particles are either nuclei or electrons, and therefore, the total dipole moment can be written as the sum of the nuclear and electronic dipole moments:

$$\boldsymbol{\mu} = \boldsymbol{\mu}_{\text{ele}}(\mathbf{r}) + \boldsymbol{\mu}_{\text{nuc}}(\mathbf{R}), \tag{15.10b}$$

where, as in the discussion of the Born–Oppenheimer separation in Chapter 10, the notations $\mathbf{r}$ and $\mathbf{R}$ respectively denote the coordinates of all the electrons and all the nuclei.

To proceed, we make the usual Born–Oppenheimer approximation, so that the wavefunction of the molecule can be written as a product of electronic and nuclear terms:

$$\phi_m(\mathbf{x}) = \chi_m^{\text{nuc}}(\mathbf{R})\psi_m^{\text{el}}(\mathbf{r}; \mathbf{R}). \tag{15.11}$$

The notation reminds us that the state labeled by the quantum number m consists of the product of nuclear and electronic parts and that the nuclear part depends only on the nuclear coordinates, whereas the electronic part depends directly on the electronic coordinates and parametrically (only weakly) upon the nuclear positions. Ordinarily, the Born–Oppenheimer separation will be valid as long as the electronic-energy state separations are large compared to the nuclear-energy state separations; this will be the case, except when the electronic states are either degenerate or near degenerate.

We now wish to study the transition rate or transition probability, given by Eq. (15.6), of a molecule subject to electromagnetic radiation. If we apply the Born–Oppenheimer approximation of Eq. (15.11), then, assuming real wavefunctions, the matrix element in Eq. (15.6) becomes

$$\langle \phi_m | \boldsymbol{\mu} | \phi_n \rangle = \sum_\lambda Z_\lambda \int \chi_m \psi_m \mathbf{r}_\lambda \chi_n \psi_n \, d\mathbf{r} \, d\mathbf{R}. \tag{15.12}$$

The product of the electronic and nuclear parts of Eq. (15.12) is usually called the *dipole matrix element*. It is clear from Eq. (15.10b) that this consists of an electronic part and a nuclear part, so, apparently, light can interact with either the electrons or the nuclei. Because the electrons are much lighter, their energy levels are spaced by larger intervals than are the nuclear energy levels. We therefore first consider interaction with radiation of frequencies substantially above $3,000 \text{ cm}^{-1}$, corresponding to excitation of the electronic part of the wavefunctions. This subject is called *electronic* spectroscopy.

The electronic part of the dipole matrix element is then given by

$$\langle \phi_m | \boldsymbol{\mu}_{\text{el}} | \phi_n \rangle = \int d\mathbf{R} \chi_m(\mathbf{R}) \chi_n(\mathbf{R}) \int d\mathbf{r} \psi_m(\mathbf{r}; \mathbf{R}) \boldsymbol{\mu}_{\text{el}}(\mathbf{r}) \psi_n(\mathbf{r}; \mathbf{R}). \tag{15.13}$$

It is convenient to define an electronic transition moment

$$\mathbf{M}_{mn}(\mathbf{R}) \equiv \int d\mathbf{r} \psi_m(\mathbf{r}; \mathbf{R}) \boldsymbol{\mu}_{\text{el}} \psi_n(\mathbf{r}; \mathbf{R}). \tag{15.14}$$

Using this definition, we can simplify the overall electronic dipole matrix element to

$$\langle \phi_m | \boldsymbol{\mu}_{\mathrm{el}} | \phi_n \rangle = \int d\mathbf{R} \chi_m(\mathbf{R}) \chi_n(\mathbf{R}) \mathbf{M}_{\mathrm{mn}}(\mathbf{R}). \tag{15.15}$$

Equation (15.15) is still exact, assuming only the Born–Oppenheimer separation of Eq. (15.11).

At this point, a further approximation, called the Condon approximation, is generally made which assumes that the electronic transition moment of Eq. (15.14) does not vary with the nuclear position, so that it can be approximated as

$$\mathbf{M}_{\mathrm{mn}}(\mathbf{R}) \approx \mathbf{M}_{\mathrm{mn}}(\mathbf{R}_0). \tag{15.16}$$

This approximation is, in fact, not quantitatively accurate in general, as we might suspect from the discussion of the electronic wavefunctions of H_2^+ and H_2 in Chapter 10: Usually, electronic wavefunctions do vary with the nuclear coordinate, and therefore, transition moments among them also should. Corrections to Eq. (15.16) can be fairly easily made, but if the approximation is assumed, then the dipole matrix element can be rewritten, from Eqs. (15.15) and (15.16), as

$$\langle \phi_m | \boldsymbol{\mu}_{\mathrm{el}} | \phi_n \rangle = \mathbf{M}_{\mathrm{mn}}(\mathbf{R}_0) \int \chi_m(\mathbf{R}) \chi_n(\mathbf{R}) d\mathbf{R}. \tag{15.17}$$

This last term is simply an overlap matrix element between nuclear states, so that the overall golden-rule expression for the probability of transition can now be written directly as

$$w_{m \rightarrow n} = \frac{2\pi}{\hbar} |\mathbf{E} \cdot \mathbf{M}_{\mathrm{mn}}(\mathbf{R}_0)|^2 |\langle \chi_m | \chi_n \rangle|^2 \delta(E_m - E_n \pm \hbar\omega). \tag{15.18}$$

Note here that, in contrast to Eq. (15.6), the appropriate energy conservation constraint for this process includes the photon energy in the delta function. The plus-or-minus sign ($\pm$) refers to photon absorption or emission.

EXERCISE 15.1 The Condon approximation, Eq. (15.16), can break down. Consider the $\pi \rightarrow \pi^*$ optical excitation in ethylene, for which (Chapter 12) the transition energy is $2|\beta|$ in the Hückel approximation. Which of the following motions should lead to the greatest failure of the Condon approximation, and why?

1. C–H bond stretch.
2. H–C–H bond bend.
3. Twisting about the C=C double bond.

Answer The transition is of the $\pi \rightarrow \pi^*$ type. The $2p$–π atomic orbitals mix well (to a first approximation, the mixing H_{ij} is proportional to the overlap S_{ij}), as long as the molecule is planar. The C–H stretch and the HCH bend do not change planarity, but twisting about the C=C double bond can alter the π overlap, thus changing the transition moment. So Condon breakdown is most sensitive to the twist.

Equation (15.18) is of fundamental importance for understanding the electronic spectroscopy of molecules and therefore requires some discussion. The last term simply reflects the conservation of energy: It states that the transition probability will vanish unless the radiation absorbed or emitted by the molecule has precisely the right energy to balance the molecular energy differences; this is simply the Bohr frequency rule. The first term on the right side is the square of the electronic transition-moment matrix element; it has units of dipole moment times field squared and is evaluated at any convenient nuclear geometry R_0 (often taken to be the minimum of the ground-state electronic energy surface). The square of the overlap matrix element—that is, the second term on the right side—is often called a Franck–Condon factor. It is just the squared overlap of the vibrational wavefunctions in the m and the n electronic states, between which the optical transition occurs.

The energetics and time scales associated with electronic and vibrational motions are important. An electronic transition might occur near 20,000 cm^{-1}, whereas characteristic vibrations such as metal–ligand stretch vibrations in transition metal complexes would occur in the infrared, near 500 cm^{-1}. Using these frequencies, we can estimate the periods τ for the electronic and vibrational motions as involving femtoseconds and tenths of picoseconds—for example,

$$\bar{\nu}_{el} \cong 20{,}000 \text{ cm}^{-1} \Rightarrow \tau_{el} = \frac{1}{c \cdot \bar{\nu}_{el}} \cong 1.67 \cdot 10^{-15}/\text{s} \qquad (15.19)$$

and

$$\bar{\nu}_{vib} \cong 500 \text{ cm}^{-1} \Rightarrow \tau_{vib} = \frac{1}{c \cdot \bar{\nu}_{vib}} \cong 6.7 \cdot 10^{-14}/\text{s}. \qquad (15.20)$$

The ratio of these results yields a factor of 40 and is key to the validity of the Franck–Condon approach and to the Born–Oppenheimer separation of electronic and nuclear motions.

To interpret the transition probability given by Eq. (15.18), it is useful to construct the so-called configuration coordinate diagram, as in Figure 15.3, where we sketch, as a function of a generalized nuclear coordinate Q, the ground and excited electronic state energies of the molecule (the potential-energy curves). When the Born-Oppenheimer approximation is assumed, the total energy is the sum of the electronic and nuclear energies:

$$E^{tot} = E^{el} + E^{nuc}. \qquad (15.21)$$

The curved lines in the figure are the electronic energies, and the horizontal lines are the total energies. Notice that the separations are indeed much larger for electronic than for nuclear energies, as would be expected from Eqs. (15.19) and (15.20) and from the mass differences.

To discuss spectroscopy in slightly more detail, we must remember that the electronic and nuclear states of the molecule are actually described not by only one quantum number, but by a set of electronic quantum numbers and a set of nuclear quantum numbers. We therefore generalize the labeling according to

$$m \rightarrow m, \mu; \qquad (15.22a)$$

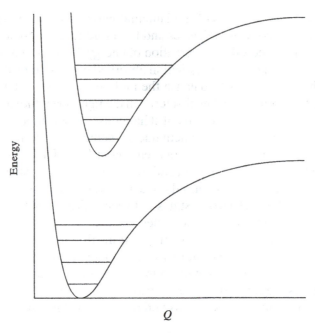

Figure 15.3   Potential curves for two different electronic states in a molecule.

$$n \rightarrow n, \nu, \tag{15.22b}$$

where the Latin letters describe electronic states and the Greek ones nuclear states. The total energy is then expressed as

$$E_{n\nu} = E_n^{el} + E_{n\nu}^{nuc}. \tag{15.23}$$

It is this total energy that actually is indicated by the horizontal lines in Figure 15.3.

At very low temperatures, the molecule at equilibrium is in the ground vibrational state of the ground electronic state; that is, its behavior is represented by the lowest horizontal line in the lower curve of Figure 15.3. The molecule's interaction with light occurs in a very brief time, roughly the period of Eq. (15.19), which corresponds to an electronic transition. In this brief time, the nuclei really cannot move, so that excitation occurs with no change in the nuclear coordinate Q. The transition is therefore characterized by the vertical line drawn between the two potential-energy curves (see Figure 15.4) and is accordingly called a *vertical* (or Franck–Condon) transition. The Franck–Condon principle, which derives directly from the Born–Oppenheimer principle, states that the electronic transition in a molecule occurs with no change in the nuclear geometry of the molecule. Formally, one can understand this behavior in terms of the sketch in Figure 15.4, where we have drawn in the forms of the vibrational wavefunctions on the ground and excited electronic states of the molecule (assuming that the electronic potential-energy curves can be approximated by harmonic curves at least for low energies, so that the vibrational eigenstates are simply those of the harmonic oscillator). Then the vibrational overlap term of Eq. (15.18) will be

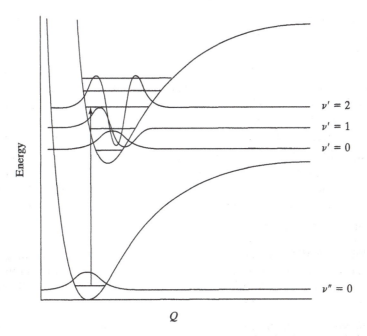

Figure 15.4   Vibrational wavefunctions used in Franck–Condon factor calculations.

maximal when the transition is indeed vertical in the figure. In the case chosen for the plot, where the molecule is initially in the $\nu = 0$ state at the bottom of the figure, the final state at the top will have $\nu = 2$.

From Eq. (15.18), we can differentiate two sorts of electronic transitions: allowed and forbidden. A transition is said to be *forbidden* when $w_{m \to n}$ vanishes. Since the electronic transition moment $\mathbf{M}_{nm}$ involves the dipole operator $\boldsymbol{\mu}_{el}$, the $m \to n$ transition is forbidden unless the states ψ_m and ψ_n are mixed by the dipole operator. This is an example of a selection rule: For electronic spectra, transitions are said to be *dipole allowed* if $\mathbf{M}_{mn} \neq 0$.

Group theory is useful for determining selection rules. For optical spectra, the transition is forbidden unless the transition moment $\mathbf{M}_{mn}$ contains a component that transforms like the totally symmetric representation of the point group. (See Problem 15.6 at the end of the chapter.)

From the Fermi golden rule for electronic spectroscopy, Eq. (15.18), and the fact that the total energy is the sum of the nuclear and electronic energy [Eq. (15.23)], it is clear that there should be transitions starting from the ground vibrational state in Figure 15.4 to all of the excited bound vibrations; the transition probabilities for each of these states (the line intensities) will be proportional to the square of the overall electronic matrix element (or transition moment) $\left(|\mathbf{M}_{nm}|^2 \right)$ and to the Franck–Condon vibrational overlap factors. Of course, the overlaps will differ for the different vibrational components of a given electronic transition. In Figure 15.4, for example, two transitions are of lower total energy difference (longer wavelength) than the Franck–Condon maximum transition; these can be denoted, in terms of increasing wavelength, as $0 \to 1$ and $0 \to 0$. This last is generally referred to as the $0 \rightarrow 0$ line, or the adiabatic transition; it is the vibrational line with the longest wavelength and the lowest energy in a given electronic absorption band for a molecule initially in its ground state.

The relative intensities of the different vibrational lines within an electronic transition depend on the ratios of the overlap (Franck–Condon) factors. Figure 15.5 shows a low-resolution electronic spectrum of the iodine molecule (I_2). We see clearly the different vibrational components of the overall electronic transition; in this case, the vertical transition is the $0 \rightarrow 25$ transition, because there is a large difference in the bond distance (and, more generally, the bonding) between the ground and excited states of this molecule. The structure near the maximum, around 19,000 cm^{-1}, is easily interpreted as a set of transitions $0 \rightarrow 22$, $0 \rightarrow 23, 0 \rightarrow 24$, etc., spaced by the I_2 stretch frequency. Hot bands (Problem 4) complicate the spectrum at lower frequencies.

EXERCISE 15.2 Suppose that, in the configuration coordinate diagram, the potential curve for the excited state is exactly the same as that for the ground state, except for an overall energy shift. What are the relative intensities of the $0 \rightarrow 0, 0 \rightarrow 1$, $0 \rightarrow 2$, and $0 \rightarrow 3$ transitions?

Answer The intensity, in the Franck–Condon treatment, depends on the vibrational overlap: $I_{\nu'' \rightarrow \nu'} \propto |\langle \chi_{\nu''} | \chi_{\nu'} \rangle|^2$, where χ is the vibrational wavefunction. If the curves are exactly the same, then the overlap is just $\langle \chi_{\nu''} | \chi_{\nu'} \rangle = \delta_{\nu'\nu''}$ (just as it is for states within one parabola). Therefore, $I_{0 \rightarrow 1}$, $I_{0 \rightarrow 2}$ and $I_{0 \rightarrow 3}$ are all zero.

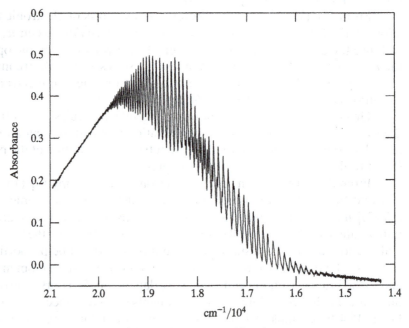

Figure 15.5 Electronic spectrum of I_2 at low temperature.

15.3   VIBRATIONAL (INFRARED) SPECTROSCOPY

Transitions among the different vibrational states (the parallel horizontal lines drawn within each state in Figure 15.3) within a given electronic state occur at a much lower frequency than the electronic transitions, as indicated in Eq. (15.20). Transitions of this type are caused by the interaction of the nuclear motions in the molecule with the electromagnetic radiation and formally arise from the second term in Eq. (15.10b). To understand vibrational spectroscopy, then, we must examine the golden rule for transitions arising from the interaction of the dipole moment of the molecular nuclei with the electromagnetic field. (We neglect the small contribution to the dipole matrix element that arises from the first term in Eq. (15.10b) due to the dependence of the electronic wavefunction on the nuclear coordinates.)

The Born–Oppenheimer wavefunction (15.11), with the generalized quantum number labeling of Eq. (15.22), is then rewritten as

$$\phi_{m\mu}(\mathbf{r}; \mathbf{R}) = \chi_{m\mu}(\mathbf{R})\psi_m(\mathbf{r}; \mathbf{R}). \tag{15.24}$$

The nuclear transition moments that must be substituted into the golden-rule expression of Eq. (15.6) to describe vibrational spectroscopy then become

$$\langle\phi_{m\mu}|\boldsymbol{\mu}_{\mathbf{nuc}}(\mathbf{R})|\phi_{n\nu}\rangle = \int d\mathbf{R}\chi_{m\mu}(\mathbf{R})\boldsymbol{\mu}_{\mathbf{nuc}}(\mathbf{R})\chi_{n\nu}(\mathbf{R})\int d\mathbf{r}\psi_m(\mathbf{r}; \mathbf{R})\psi_n(\mathbf{r}; \mathbf{R}). \tag{15.25}$$

An obvious simplification now occurs: The electronic states in the second integral of Eq. (15.25) are orthogonal. [They correspond to the different eigenfunctions of the electronic Hamiltonian of Eq. (10.10).] This means that

$$\int d\mathbf{r}\psi_m(\mathbf{r}; \mathbf{R})\psi_n(\mathbf{r}; \mathbf{R}) = \delta_{nm}, \tag{15.26}$$

where the Kronecker delta function vanishes unless the electronic quantum numbers m and n are identical. Then we can rewrite the nuclear transition moment as

$$\langle\phi_{m\mu}|\boldsymbol{\mu}_{\mathbf{nuc}}(\mathbf{R})|\phi_{n\nu}\rangle = \delta_{mn}\langle\chi_{m\mu}|\boldsymbol{\mu}_{\mathbf{nuc}}(\mathbf{R})|\chi_{n\nu}\rangle. \tag{15.27}$$

Physically, this simply means that, as long as the Born–Oppenheimer separation holds, the nuclear transition moment will induce transitions only among the vibrational levels of a given electronic state.

To proceed, we discuss the vibrational motions in terms of the normal coordinates of Section 5.3. The wavefunction is written as a simple product, as in Eq. (5.49), but using a simplified notation:

$$\chi_{m\mu} = \prod_{\sigma=1}^{N_m} u_{\mu_r}(Q_\sigma). \tag{15.28}$$

Here, σ labels the different vibrational normal coordinates, of which there are N_m. (Remember that, for general nonlinear molecules, N_m equals $3N - 6$, where N is the number of nuclei). In Eq. (15.28), the normal coordinates are denoted by

Q_σ and the vibrational eigenstates of Chapter 5 are called u_σ. We can now expand the nuclear dipole operator as

$$\boldsymbol{\mu}_{\mathbf{nuc}} = \sum_{\sigma=1}^{N_m} \mathbf{a}_\sigma Q_\sigma + \boldsymbol{\mu}_{\mathbf{nuc}}^0, \qquad (15.29)$$

where a_σ is an expansion coefficient that includes nuclear charge and mass factors and $\boldsymbol{\mu}_{\mathbf{nuc}}^0$ is the nuclear dipole moment at an arbitrary reference geometry.

Utilizing the normal coordinate form, we can write the nuclear transition dipole moment schematically as

$$\langle \chi_{m\mu} | \boldsymbol{\mu}_{\mathbf{nuc}} | \chi_{n\nu} \rangle = \sum_\sigma \mathbf{a}_\sigma \langle u_1(Q_1)u_2(Q_2) \cdots u_{N_m}(Q_{N_m}) | Q_\sigma | u_{1'}(Q_1)u_{2'}(Q_2) \cdots u_{N_m'}(Q_{N_m}) \rangle$$

$$+ \boldsymbol{\mu}_{\mathbf{nuc}}^{(0)} \langle u_1 u_2 \cdots u_{N_m} | u_{1'} u_{2'} \cdots u_{N_m'} \rangle. \quad (15.30)$$

If we are considering transitions among different vibrational states, then, by orthogonality, the second term vanishes. The multidimensional integral in the first term can be separated, because the operator in this matrix is a sum of the coordinates, one at a time. Using the orthogonality relations

$$\langle u_1(Q_1) | u_{1'}(Q_1) \rangle = \delta_{1,1'}, \qquad (15.31)$$

we find that the nuclear transition moment then becomes

$$\langle \chi_{m\mu} | \boldsymbol{\mu}_{\mathbf{nuc}} | \chi_{n\nu} \rangle = \sum_\sigma \mathbf{a}_\sigma \langle u_\mu(Q_\sigma) | Q_\sigma | u_\nu(Q_\sigma) \rangle. \qquad (15.32)$$

To simplify this expression, we use the normal coordinate matrix elements in terms of the raising and lower operators of Section 5.3. The matrix elements then vanish unless

$$\mu = \nu \pm 1. \qquad (15.33)$$

This effectively means that, for these vibrational transitions, selection rules occur in each mode and are given by

$$\begin{cases} \mu_\sigma = \nu_\sigma \pm 1 & \text{for any one mode } \sigma \qquad (15.34) \\ \mu_\sigma = \nu_\sigma & \text{for all other modes .} \qquad (15.35) \end{cases}$$

The interpretation of these quantum mechanical results, in terms of vibrational spectroscopy, is straightforward. The vibrational spectrum of a molecule in a given electronic state consists of a series of transitions, one along each normal coordinate. Energy is conserved, such that the energy difference between the initial and final vibrational states of the molecule exactly balances the energy absorbed from or emitted to the electromagnetic field. The former is measured by vibrational absorption spectroscopy, the latter by vibrational emission spectroscopy.

At room temperature, the Boltzmann factor for the vibrational population distribution shows that, for any vibrational frequency much greater than 500 cm^{-1}, the occupation populations for all vibrationally excited states will be very small. This means, effectively, that Eqs. (15.34) and (15.35) can be replaced by

$$\begin{cases} \mu_\sigma = 1 \text{ for any one mode } \sigma \\ \mu_\sigma = 0 \text{ for all other modes.} \end{cases} \quad (15.36)$$

Equation (15.36) implies that only vibrational absorption spectra will be observed, with one transition occurring from ground to first excited state for each normal mode. Formally, this means that the expression for the transition moment can be written

$$\langle \chi_{m\mu} | \mu | \chi_{m\nu} \rangle = \sum_\sigma \mathbf{a}_\sigma \langle 0 | Q_\sigma | 1 \rangle. \quad (15.37)$$

Most vibrational transitions occur in the infrared region of the electromagnetic spectrum, because most vibrational quanta are in that energy range. Infrared transitions are *allowed* if the transition probability in Eq. (15.37) does not vanish and *forbidden* if it does vanish. Generally, the amplitude factor of Eq. (15.37) determines whether a particular vibrational transition is allowed in the infrared. If the normal coordinate Q_σ results in a change in the dipole, as suggested by Eq. (15.29), then $\mathbf{a}_\sigma$ will not vanish and the transition will be allowed. This selection rule then states that transitions will be allowed only for normal coordinates that change the dipole moment. Figure 15.6 shows a low-resolution infrared spectrum of the carbon dioxide molecule. CO_2 has three normal coordinate vibrations, a symmetric stretch, an asymmetric stretch, and a bend coordinate. Of these, the asymmetric stretch and bend occur at frequencies of 2349 and 667 cm^{-1}, respectively. Note the two strong transitions in the spectrum that arise from these modes. For more general molecules, these vibrational selection rules in the infrared remain valid as long as the harmonic approximation is adequate for the vibrations and the Born–Oppenheimer and Franck–Condon principles hold.

Group theory can be used to predict which normal coordinates will give allowed transitions in the infrared spectrum. Here we use the symmetry properties of the normal coordinates that were discussed in Chapter 13, in addition to the symmetry properties of the electronic states. For example, the integral in Eq. (15.32) will vanish unless $\Gamma(u_\mu) \times \Gamma(u_\nu) \times \Gamma(Q_\sigma)$ contains $\Gamma(a_1)$, where a_1 is the totally symmetric representation. For CO_2 (point group $D_{\infty h}$), the symmetric stretch, asymmetric stretch, and bend belong to the σ_g, π_u, and σ_u representations, respectively. As pointed out in Section 13.2.3, the third column of the character tables lists the dipole components (x, y, z) and gives their representations. The dipole moment transforms (from Appendix C) like π_u or σ_u [for (x, y) and z, respectively]. For the triple direct product $\Gamma(u_\mu) \times \Gamma(u_\nu) \times \Gamma(Q_\sigma)$ to be totally symmetric, the double direct product $\Gamma(u_\mu) \times \Gamma(u_\nu)$ must contain either π_u or σ_u. Therefore, the integral of Eq. (15.32) will vanish for the σ_g symmetric stretch,

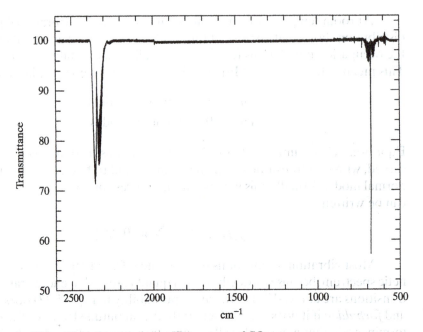

Figure 15.6 Low-resolution IR spectrum of CO_2.

but not for the bend or the asymmetric stretch. This is why only the latter modes, described as infrared active, are seen in Figure 15.6.

SUGGESTED READINGS

P. W. Atkins and R. S. Friedman, *Molecular Quantum Mechanics*, 3rd ed. (Oxford, New York, 1997).

H. Eyring, J. Walter, and G. Kimball, *Quantum Chemistry* (Wiley, New York, 1944).

W. H. Flygare, *Molecular Structure and Dynamics* (Prentice Hall, Englewood Cliffs, NJ, 1978).

J. Simons and J. Nichols, *Quantum Mechanics in Chemistry* (Oxford, New York, 1997).

G. C. Schatz and M. A. Ratner, *Quantum Mechanics in Chemistry* (Prentice Hall, Englewood Cliffs, NJ, 1993).

PROBLEMS

15.1 Use normalization of the wavefunction $\psi(\mathbf{x}, t)$ to prove that $\sum_m |C_m(t)|^2 = 1$, so that $|C_m(t)|^2$ can indeed be interpreted as a probability.

15.2 Suppose that there are only two states, ϕ_1 and ϕ_2, that a molecule can occupy and that these states are degenerate—that is, $E_1 = E_2$. Suppose also that at the initial time $t = 0$ the molecule is in state ϕ_1, so that $C_1(t = 0) = 1$ and $C_2(t = 0) = 0$. Prove that if the perturbation $V(t) = V_0$ is constant after $t = 0$, the time-dependent wavefunction $\psi(t) = \cos(at)\phi_1 + \sin(at)\phi_2$ satisfies the time-dependent Schrödinger equation (15.1). Then prove that $|C_2(t)|^2$ is proportional to t^2 for short times. This means that only for two states is there no transition probability proportional to time and therefore no time-independent transition probability per unit time, w. This is why the golden rule holds only when there are many final states that are close together (which is almost always true in chemical systems).

15.3 The $C_3H_3^+$ ion was examined in Chapter 13. Recall that its MO scheme included an a_1 orbital and a degenerate e orbital. Recall also that an electronic transition between states ψ_x and ψ_g is forbidden unless $\langle \psi_x | \mathbf{r} | \psi_g \rangle \neq 0$. Using full-state labels (e.g., $^1E'$) and not orbital labels, describe which optical transitions are allowed for $C_3H_3^+$.

15.4 So-called **hot bands** correspond to transitions that originate from vibrationally excited levels of the electronic ground state. They are called hot bands because they require population in the excited vibrational level. Assuming that both ground and excited electronic potential-energy surfaces are approximately harmonic, with the same frequency in the ground and excited states, show that there is a hot-band peak arising from the $\nu'' = 1$ (first excited) vibrational level of the ground electronic state and that this peak is of lower frequency than the adiabatic transition.

For the $^{127}I^{35}C\ell$ molecule, the $0 \rightarrow 0$ transition occurs at 12,658 cm^{-1}. The vibrational frequency is $\bar{\nu} = 384.2$ cm^{-1}. Compute the frequencies of the $0 \rightarrow 1$, $1 \rightarrow 0$, and $1 \rightarrow 1$ absorptions.

15.5 In pump–probe spectroscopy, one laser is used to pump a molecule to an electronic excited state, and another is used to probe that state, often by a second absorption. For the $^{127}I^{35}C\ell$ molecule of Problem 4, what laser wavelength can be used to probe the vibrational energy following a pump laser excitation at $\bar{\nu} = 14{,}194.8$ cm^{-1}? In a pump–dump sequence, the second laser probes the population of a vibrational level of the excited state by inducing a downwards transition. If the $IC\ell$ is pumped on the $0 \rightarrow 4$ transition, what wavelength of dump pulse can be used to probe the $\nu' = 3$ population using a $3 \rightarrow 0$ dump?

15.6 Transitions that are dipole forbidden can be quadrupole allowed, meaning that the electronic matrix element M_{mn} now involves the quadrupole operator, which has the general form

$$\hat{Q} = a_{xx}X^2 + a_{xy}XY + a_{xz}XZ + \ldots$$

(nine terms in all). For a molecule with inversion symmetry, use the point-group labels g and u for symmetric and antisymmetric states to prove the following statements:
(a) Dipole transitions can occur only between g and u symmetry states; $g \rightarrow g$ and $u \rightarrow u$ are forbidden (LaPorte rule).
(b) Transitions between two g states or between two u states are permitted by quadrupole radiation.
(c) For a molecule with an inversion center, all vibrational transitions are either Raman allowed or infrared allowed, but not both (exclusion rule). (Raman spectra depend on polarizability derivatives, which have the same symmetry as quadrupole radiation.)

15.7 For two harmonic potentials whose origins are displaced a distance $X_D = g\{\hbar/2\mu\omega\}^{1/2}$, the Franck–Condon factor $|\langle 0|n\rangle|^2 = \dfrac{(e^{-g^2}g^{2n})}{n!}$. Consider this displaced-oscillator model for the $^{12}C^{16}O$ molecule, whose frequency $\bar{\nu} = 2,170.2 \text{ cm}^{-1}$ and for which $X_D = .107\text{Å}$.

(a) Compute the relative intensities of the $0 \to 0, 0 \to 1, 0 \to 2$, and $0 \to 3$ transitions.

(b) Compute which excited-state vibrational level $\bar{n}$ should show the most intense absorption. (Maximize $(|\langle 0|\bar{n}\rangle|^2$.) Sketch the curves. Is this most intense level the one that the Franck–Condon principle suggests?

15.8 The simplest discussion of ESR or EPR spectroscopy involves a single electron spin that is subjected to two applied fields. A static magnetic field $\mathcal{H}_0$ is directed along the z-axis, and an oscillating radio-frequency field of strength $\mathcal{H}_1$ and frequency ω_0 lies in the (x, y) plane. Then the Hamiltonian can be written

$$H = \gamma \mathcal{H}_0 S_z + \gamma \mathcal{H}_{1x} S_x \cos \omega_0 t + \gamma \mathcal{H}_{1y} S_y \cos \omega_0 t,$$

where $\gamma = \dfrac{g_e \beta_e}{\hbar}$, with $g_e \cong 2.002$ and β_e is defined in Appendix D.

(a) For the single electron, what are the eigenstates of the first term $(\gamma \mathcal{H}_0 S_z)$? What are their energies, in terms of $\mathcal{H}_0$ and γ?

(b) Suppose the term $H - \gamma \mathcal{H}_0 S_z$ is a time-dependent perturbation. If the density of final states is denoted $T_2/\hbar$, find both the transition probability and the transition frequency for all allowed transitions.

APPENDICES

A | $\dfrac{d^2y}{dx^2} + k^2y = 0$

Mathematical Background

A goal of this book is to present quantum mechanics as applied to chemical problems with as little mathematical sophistication as is possible. Indeed, nearly all of the mathematics needed is covered in typical first-year calculus courses. The purpose of this appendix is to give a brief introduction to some of the most important concepts and to present a few formulas that go beyond first-year calculus.

A.1 COMPLEX NUMBERS

Complex numbers are important in quantum mechanics. A complex number may always be written as $z = x + iy$, where $i = (-1)^{1/2}$, and x and y are real numbers. The *complex conjugate* of a complex number is obtained by replacing i by $-i$ everywhere. Usually, the operation of complex conjugation is denoted using the superscript *. Thus, $z^* = x - iy$ is the complex conjugate of $z = x + iy$.

The absolute square of a complex number is just the product of the number and its complex conjugate. This means that if z is a complex number, then its absolute square is z^*z, which is often written as $|z|^2$. If $z = x + iy$, then we can write

$$|z|^2 = z^*z = x^2 + y^2 \tag{A.1}$$

and

$$|z| = (x^2 + y^2)^{1/2}. \tag{A.2}$$

Complex exponentials are simply exponentials of the form $\exp(z)$ where z is a complex number. Like real exponentials, their meaning is defined via the Taylor series expansion

$$\exp(z) = \sum_{n=0}^{\infty} \frac{z^n}{n!}. \tag{A.3}$$

If we consider the Taylor expansion of a purely imaginary exponential of the form $\exp(i\theta)$, where θ is real, and compare the result with the Taylor expansions of $\sin(\theta)$, and $\cos(\theta)$, then it is not difficult to prove the *Euler identity:*

$$\exp(i\theta) = \cos(\theta) + i\sin(\theta). \tag{A.4}$$

Special cases of this identity are

$$\exp(i\pi/2) = i, \tag{A.5a}$$

$$\exp(i\pi) = -1, \tag{A.5b}$$

and

$$\exp(2i\pi) = 1. \tag{A.5c}$$

A.2   A TINY BIT OF DIFFERENTIAL EQUATIONS

The only differential equations that we will need to deal with in quantum mechanics are *second-order linear* differential equations. A simple example of such an equation is:

$$\frac{d^2y}{dx^2} + k^2y = 0. \tag{A.6}$$

To solve this equation, we assume that

$$y = \exp(mx), \tag{A.7}$$

where m is a constant to be determined. Substitution into the differential equation then leads to

$$(m^2 + k^2)y = 0, \tag{A.8}$$

which means that $m = \pm ik$, and thus, the solutions are $y = \exp(\pm ikx)$. There are two independent solutions in this case, which is a general property of second-order differential equations. In addition, since the differential equation is linear, any linear combination of the two solutions is also a solution. Hence, we can write

$$y = A\exp(+ikx) + B\exp(-ikx), \tag{A.9}$$

where A and B are arbitrary constants. To determine these constants, we give the *boundary conditions* that y must satisfy. Clearly, there must be two boundary conditions to determine the two unknown constants, but there is significant freedom in how these may be specified. For example, one can specify y and dy/dx at some particular value of x, or one can specify just y at two different values of x. Both of these possibilities are found in the applications discussed elsewhere in the text.

A.3   MATRICES, DETERMINANTS, AND EIGENVALUES

Often in quantum mechanics we must solve linear equations of the form (here written for the specific case of three equations and three unknowns)

$$(a_{11} - e)x_1 + a_{12}x_2 + a_{13}x_3 = 0,$$
$$a_{21}x_1 + (a_{22} - e)x_2 + a_{23}x_3 = 0, \qquad (A.10)$$
$$a_{31}x_1 + a_{32}x_2 + (a_{33} - e)x_3 = 0,$$

where the a_{ij} are known constants and e and the x_i's are unknowns.

This equation is conveniently written in matrix-vector form as

$$(\mathbf{A} - e\mathbf{I})\mathbf{x} = \mathbf{0},$$

where $\mathbf{A}$ is a 3×3 matrix with elements a_{ij}, $\mathbf{x}$ is a vector of length 3 that contains the x_i's, and $\mathbf{I}$ is a 3×3 identity matrix, with elements given by

$$I_{ij} = \delta_{ij}, \qquad (A.11)$$

in which δ_{ij} is the Kronecker delta function $(\delta_{ij} = 0$ if $i \neq j$, and $\delta_{ij} = 1$ if $i = j)$.

For linear equations of this type, solutions other than the trivial solution $\mathbf{x} = \mathbf{0}$ can be obtained only if the determinant of the matrix of coefficients vanishes—that is, only if

$$|\mathbf{A} - e\mathbf{I}| = \mathbf{0}. \qquad (A.12)$$

Evaluation of this determinant gives a polynomial equation (the so-called *secular equation*) that can be used to solve for the allowed values of the parameter e. These values are called the *eigenvalues* of the matrix, and in the case of a 3×3 matrix $\mathbf{A}$, there would be three possible eigenvalues. If each eigenvalue is, in turn, substituted back into the original equation, it is possible to solve the resulting equations for the x_i's, which are called the *eigenvectors*. The absolute normalization of the eigenvectors is not determined in the process of obtaining them. However, normalization of the quantum wavefunction can always be used to define normalized eigenvectors.

EXAMPLE

Consider the matrix

$$\mathbf{A} = \begin{pmatrix} 1 & 1 & 0 \\ 1 & 1 & 1 \\ 0 & 1 & 1 \end{pmatrix}. \qquad (A.13)$$

The secular equation of this matrix is $(1 - e)^3 - 2(1 - e) = 0$. The eigenvalues are $e = 1$ and $e = 1 \pm \sqrt{2}$ and the (normalized) eigenvectors are as follows:

(a)   For $e = 1$, $x = \left(\dfrac{1}{\sqrt{2}}, 0, -\dfrac{1}{\sqrt{2}} \right)$.

(b)   For $e = 1 - \sqrt{2}$, $x = \left(-\dfrac{1}{2}, \dfrac{1}{\sqrt{2}}, -\dfrac{1}{2}\right)$.

(c)   For $e = 1 + \sqrt{2}$, $x = \left(\dfrac{1}{2}, \dfrac{1}{\sqrt{2}}, \dfrac{1}{2}\right)$.

Note that if we group the eigenvectors as columns of a 3×3 matrix

$$\mathbf{T} = \begin{pmatrix} \dfrac{1}{\sqrt{2}} & -\dfrac{1}{2} & \dfrac{1}{2} \\ 0 & \dfrac{1}{\sqrt{2}} & \dfrac{1}{\sqrt{2}} \\ -\dfrac{1}{\sqrt{2}} & -\dfrac{1}{2} & \dfrac{1}{2} \end{pmatrix}, \tag{A.14}$$

then the matrix is *orthogonal*. This means that the transpose of the matrix equals its inverse. The transpose $\tilde{\mathbf{T}}$ is defined by $\tilde{\mathbf{T}}_{ij} = \mathbf{T}_{ji}$ and is

$$\tilde{\mathbf{T}} = \begin{pmatrix} \dfrac{1}{\sqrt{2}} & 0 & -\dfrac{1}{\sqrt{2}} \\ -\dfrac{1}{2} & \dfrac{1}{\sqrt{2}} & -\dfrac{1}{2} \\ \dfrac{1}{2} & \dfrac{1}{\sqrt{2}} & \dfrac{1}{2} \end{pmatrix}. \tag{A.15}$$

One can easily verify that $\tilde{\mathbf{T}}\mathbf{T} = \mathbf{T}^{-1}\mathbf{T} = \mathbf{I}$, where $\mathbf{I}$ is the identity matrix [Eq. 13.1)]. Orthogonality is a general property of the eigenvectors of a *symmetric* matrix (i.e., $\tilde{\mathbf{A}} = \mathbf{A}$).

B

$$I = \frac{Z^6}{\pi^2} \int d^3\mathbf{r}_1 \int d^3\mathbf{r}_2 e^{-2Zr_1} e^{-2Zr_2} \frac{1}{r_{12}}$$

Two-Electron Repulsion Integral

From Eqs. (7.6) and (7.8), the two-electron repulsion integral is

$$I = \left\langle \phi_{100}(r_1)\phi_{100}(r_2) \left| \frac{1}{r_{12}} \right| \phi_{100}(r_1)\phi_{100}(r_2) \right\rangle. \tag{B.1}$$

Substituting the 1s functions from Table 6.2, we have

$$I = \frac{Z^6}{\pi^2} \int d^3\mathbf{r}_1 \int d^3\mathbf{r}_2 e^{-2Zr_1} e^{-2Zr_2} \frac{1}{r_{12}}. \tag{B.2}$$

To evaluate this integral, we use spherical coordinates to express $d^3\mathbf{r}_1$ and $d^3\mathbf{r}_2$. In addition, we use the expansion

$$\frac{1}{r_{12}} = \sum_{\ell=0}^{\infty} \sum_{m=-\ell}^{\ell} \frac{4\pi}{2\ell+1} \frac{r_<^\ell}{r_>^{\ell+1}} Y_{\ell m}^*(\theta_1, \phi_1) Y_{\ell m}(\theta_2, \phi_2), \tag{B.3}$$

where $r_<$ is the smaller of r_1 and r_2, and $r_>$ is the greater of r_1 and r_2. Substituting Eq. (B.3) into Eq. (B.2) and recalling that $Y_{00} = (4\pi)^{-1/2}$, we get

$$I = 16Z^6 \sum_{\ell=0}^{\infty} \sum_{m=-\ell}^{\ell} \frac{1}{2\ell+1} \int_0^{\infty} \int_0^{\infty} e^{-2Zr_1} e^{-2Zr_2}$$

$$\times \frac{r_<^\ell}{r_>^{\ell+1}} r_1^2 dr_1 r_2^2 dr_2$$

$$\times \int_0^{2\pi} \int_0^{\pi} Y_{\ell m}^*(\theta_1, \phi_1) Y_{00}(\theta_1, \phi_1) \sin\theta_1 d\theta_1 d\phi_1 \qquad \text{(B.4)}$$

$$\times \int_0^{2\pi} \int_0^{\pi} Y_{00}^*(\theta_2, \phi_2) Y_{\ell m}(\theta_2, \phi_2) \sin\theta_2 d\theta_2 d\phi_2.$$

The Y_{00}'s that appear in the last two lines are just constants $\left(1/\sqrt{4\pi}\right)$ that have been inserted (and factored out in front) to give us the usual orthogonality relation for the spherical harmonics. Invoking this orthogonality gives

$$I = 16Z^6 \int_0^{\infty} \int_0^{\infty} e^{-2Zr_1} e^{-2Zr_2} \frac{r_1^2 r_2^2}{r_>} dr_1 dr_2. \qquad \text{(B.5)}$$

This integral may be evaluated by splitting the r_2 integration range into two parts, one from 0 to r_1, and the other from r_1 to ∞. In this case, $r_> = r_1$ in the first part, and $r_> = r_2$ in the second. This gives

$$I = 16Z^6 \left\{ \int_0^{\infty} e^{-2Zr_1} r_1 \left(\int_0^{r_1} e^{-2Zr_2} r_2^2 dr_2 \right) dr_1 \qquad \text{(B.6)} \right.$$

$$\left. + \int_0^{\infty} e^{-2Zr_1} r_1^2 \left(\int_{r_1}^{\infty} e^{-2Zr_2} r_2 dr_2 \right) dr_1 \right\}.$$

Straightforward evaluation of these integrals yields

$$I = \frac{5}{8} Z. \qquad \text{(B.7)}$$

C

$C_{\infty v}$	E	$2C_\infty{}^\Phi$	...	$\infty\sigma_v$
Σ^+	1	1	...	1
Σ^-	1	1	...	−1
Π	2	$2\cos\Phi$	...	0
Δ	2	$2\cos 2\Phi$	...	0
Φ	2	$2\cos 3\Phi$	...	0
...	...	...	...	...

Character Tables

C_1	E
A	1

C_s	E	σ_h		
A'	1	1	x, y, R_z	x^2, y^2, z^2, xy
A''	1	−1	z, R_x, R_y	yz, xz

C_i	E	i		
A_g	1	1	R_x, R_y, R_z	x^2, y^2, z^2
A_u	1	−1	x, y, z	xy, xz, yz

C_2	E	C_2		
A	1	1	z, R_z	x^2, y^2, z^2, xy
B	1	−1	x, y, R_x, R_y	yz, xz

C_3	E	C_3	$C_3{}^2$		$\varepsilon = \exp(2\pi i/3)$
A	1	1	1	z, R_z	$x^2 + y^2, z^2$
B	$\begin{cases}1 \\ 1\end{cases}$	$\begin{matrix}\varepsilon \\ \varepsilon^*\end{matrix}$	$\left.\begin{matrix}\varepsilon^* \\ \varepsilon\end{matrix}\right\}$	$(x, y)(R_x, R_y)$	$(x^2 - y^2, xy)(yz, xz)$

C_4	E	C_4	C_2	C_4^3		
A	1	1	1	1	z, R_z	$x^2 + y^2,\ z^2$
B	1	-1	1	-1		$x^2 - y^2,\ xy$
E	1	-i	-1	-i	$(x, y)(R_x, R_y)$	(yz, xz)
	1	-i	-1	i		

C_5	E	C_5	C_5^2	C_5^3	C_5^4		$\varepsilon = \exp(2\pi i/5)$
A	1	1	1	1	1	z, R_z	$x^2 + y^2,\ z^2$
E_1	1	ε	ε^2	ε^{2*}	ε^*	$(x, y)(R_x, R_y)$	(yz, xz)
	1	ε^*	ε^{2*}	ε^2	ε		
E_2	1	ε^2	ε^*	ε	ε^{2*}		$(x^2 - y^2,\ xy)$
	1	ε^{2*}	ε	ε^*	ε^2		

C_6	E	C_6	C_3	C_2	C_3^2	C_6^5		$\varepsilon = \exp(2\pi i/6)$
A	1	1	1	1	1	1	z, R_z	$x^2 + y^2,\ z^2$
B	1	-1	1	-1	1	-1		
E_1	1	ε	$-\varepsilon^*$	-1	$-\varepsilon$	ε^*	(x, y)	(xz, yz)
	1	ε^*	$-\varepsilon$	-1	$-\varepsilon^*$	ε	(R_x, R_y)	
E_2	1	$-\varepsilon^*$	$-\varepsilon$	1	$-\varepsilon^*$	$-\varepsilon$		$(x^2 - y^2,\ xy)$
	1	$-\varepsilon$	ε^*	1	$-\varepsilon$	$-\varepsilon^*$		

D_2	E	$C_2(z)$	$C_2(y)$	$C_2(x)$		
A	1	1	1	1		$x^2,\ y^2,\ z^2$
B_1	1	1	-1	-1	z, R_z	xy
B_2	1	-1	1	-1	y, R_y	xz
B_3	1	-1	-1	1	x, R_x	yz

D_3	E	$2C_3$	$3C_2$		
A_1	1	1	1		$x^2 + y^2,\ z^2$
B_2	1	1	-1	z, R_z	
E	2	-1	0	$(x, y)(R_x, R_y)$	$(x^2 - y^2,\ xy)(xz, yz)$

D_4	E	$2C_4$	$C_2(=C_4^2)$	$2C_2'$	$2C_2''$		
A_1	1	1	1	1	1		$x^2 + y^2,\ z^2$
A_2	1	1	1	-1	-1	z, R_z	
B_1	1	-1	1	1	-1		$x^2 + y^2$
B_2	1	-1	1	-1	1		xy
E	2	0	-2	0	0	$(x, y)(R_x, R_y)$	(xz, yz)

D_5	E	$2C_5$	$2C_5^2$	$5C_2$		
A_1	1	1	1	1		$x^2 + y^2,\ z^2$
A_2	1	1	1	-1	z, R_z	
E_1	2	$2\cos 72°$	$2\cos 144°$	0	$(x, y)(R_x, R_y)$	(xz, yz)
E_2	2	$2\cos 144°$	$2\cos 72°$	0		$(x^2 + y^2,\ xy)$

D_6	E	$2C_6$	$2C_3$	C_2	$3C_3'$	$3C_3''$		
A_1	1	1	1	1	1	1		$x^2 + y^2, z^2$
A_2	1	1	1	1	-1	-1	z, R_z	
B_1	1	-1	1	-1	1	-1		
B_2	1	-1	1	-1	-1	1		
E_1	2	1	-1	-2	0	0	$(x, y)(R_x, R_y)$	(xz, yz)
E_2	2	-1	-1	2	0	0		$(x^2 - y^2, xy)$

C_{2v}	E	C_2	$\sigma_v(xz)$	$\sigma_v(yz)$		
A_1	1	1	1	1	z	x^2, y^2, z^2
A_2	1	1	-1	-1	R_z	xy
B_1	1	-1	1	-1	x, R_y	xz
B_2	1	-1	-1	1	y, R_x	yz

C_{3v}	E	$2C_3$	$3\sigma_v$		
A_1	1	1	1	z	$x^2 + y^2, z^2$
A_2	1	1	-1	R_z	
E	2	-1	0	$(x, y)(R_x, R_y)$	$(x^2 - y^2, xy)(xz, yz)$

C_{4v}	E	$2C_4$	C_2	$2\sigma_v$	$2\sigma_v$		
A_1	1	1	1	1	1	z	$x^2 + y^2, z^2$
A_2	1	1	1	-1	-1	R_z	
B_1	1	-1	1	1	-1		$x^2 + y^2$
B_2	1	-1	1	-1	1		xy
E	2	0	-2	0	0	$(x, y)(R_x, R_y)$	(xz, yz)

C_{5v}	E	$2C_5$	$2C_5^2$	$5\sigma_v$		
A_1	1	1	1	1	z	$x^2 + y^2, z^2$
A_2	1	1	1	-1	R_z	
E_1	2	$2\cos 72°$	$2\cos 144°$	0	$(x, y)(R_x, R_y)$	(xz, yz)
E_2	2	$2\cos 144°$	$2\cos 72°$	0		$(x^2 - y^2, xy)$

C_{6v}	E	$2C_6$	$2C_3$	C_2	$3\sigma_v$	$3\sigma_d$		
A_1	1	1	1	1	1	1	z	$x^2 + y^2, z^2$
A_2	1	1	1	1	-1	-1	R_z	
B_1	1	-1	1	-1	1	-1		
B_2	1	-1	1	-1	-1	1		
E_1	2	1	-1	-2	0	0	$(x, y)(R_x, R_y)$	(xz, yz)
E_2	2	-1	-1	2	0	0		$(x^2 - y^2, xy)$

C_{2h}	E	C_2	i	σ_h		
A_g	1	1	1	1	R_z	x^2, y^2, z^2, xy
B_g	1	-1	1	-1	R_x, R_y	xz, yz
A_u	1	1	-1	-1	z	
B_u	1	-1	-1	1	x, y	

C_{3h}	E	C_3	C_3^2	σ_h	S_3	S_3^5		$\varepsilon = \exp(2\pi i/3)$
A'	1	1	1	1	1	1	R_z	x^2+y^2, z^2
E'	$\begin{cases}1\\1\end{cases}$	$\begin{matrix}\varepsilon\\\varepsilon^*\end{matrix}$	$\begin{matrix}\varepsilon^*\\\varepsilon\end{matrix}$	$\begin{matrix}1\\1\end{matrix}$	$\begin{matrix}\varepsilon\\\varepsilon^*\end{matrix}$	$\begin{matrix}\varepsilon^*\\\varepsilon\end{matrix}$	(x, y)	(x^2-y^2, xy)
A''	1	1	1	-1	-1	-1	z	
E''	$\begin{cases}1\\1\end{cases}$	$\begin{matrix}\varepsilon\\\varepsilon^*\end{matrix}$	$\begin{matrix}\varepsilon^*\\\varepsilon\end{matrix}$	$\begin{matrix}-1\\-1\end{matrix}$	$\begin{matrix}-\varepsilon\\-\varepsilon^*\end{matrix}$	$\begin{matrix}-\varepsilon^*\\-\varepsilon\end{matrix}$	(R_x, R_y)	(xz, yz)

C_{4h}	E	C_4	C_2	C_4^3	i	S_4^3	σ_h	S_4		
A_g	1	1	1	1	1	1	1	1	R_z	x^2+y^2, z^2
B_g	1	-1	1	-1	1	-1	1	-1		(x^2-y^2, xy)
E_g	$\begin{cases}1\\1\end{cases}$	$\begin{matrix}i\\-i\end{matrix}$	$\begin{matrix}-1\\-1\end{matrix}$	$\begin{matrix}-i\\i\end{matrix}$	$\begin{matrix}1\\1\end{matrix}$	$\begin{matrix}i\\-i\end{matrix}$	$\begin{matrix}-1\\-1\end{matrix}$	$\begin{matrix}-i\\i\end{matrix}$	(R_x, R_y) z	(xz, yz)
A_u	1	1	1	1	-1	-1	-1	-1	z	
B_u	1	-1	1	-1	-1	1	-1	1		
E_u	$\begin{cases}1\\1\end{cases}$	$\begin{matrix}i\\-i\end{matrix}$	$\begin{matrix}-1\\-1\end{matrix}$	$\begin{matrix}-i\\i\end{matrix}$	$\begin{matrix}-1\\-1\end{matrix}$	$\begin{matrix}-i\\i\end{matrix}$	$\begin{matrix}1\\1\end{matrix}$	$\begin{matrix}i\\-i\end{matrix}$	(x, y)	

C_{5h}	E	C_5	C_5^2	C_5^3	C_5^4	σ_h	S_5	S_5^7	S_5^3	S_5^9		$\varepsilon = \exp(2\pi i/5)$
A'	1	1	1	1	1	1	1	1	1	1	R_z	x^2+y^2, z^2
E_1'	$\begin{cases}1\\1\end{cases}$	$\begin{matrix}\varepsilon\\\varepsilon^*\end{matrix}$	$\begin{matrix}\varepsilon^2\\\varepsilon^{2*}\end{matrix}$	$\begin{matrix}\varepsilon^{2*}\\\varepsilon^2\end{matrix}$	$\begin{matrix}\varepsilon^*\\\varepsilon\end{matrix}$	$\begin{matrix}1\\1\end{matrix}$	$\begin{matrix}\varepsilon\\\varepsilon^*\end{matrix}$	$\begin{matrix}\varepsilon^2\\\varepsilon^{2*}\end{matrix}$	$\begin{matrix}\varepsilon^{2*}\\\varepsilon^2\end{matrix}$	$\begin{matrix}\varepsilon^*\\\varepsilon\end{matrix}$	(x, y)	
E_2'	$\begin{cases}1\\1\end{cases}$	$\begin{matrix}\varepsilon^2\\\varepsilon^{2*}\end{matrix}$	$\begin{matrix}\varepsilon^*\\\varepsilon\end{matrix}$	$\begin{matrix}\varepsilon\\\varepsilon^*\end{matrix}$	$\begin{matrix}\varepsilon^{2*}\\\varepsilon^2\end{matrix}$	$\begin{matrix}1\\1\end{matrix}$	$\begin{matrix}\varepsilon^2\\\varepsilon^{2*}\end{matrix}$	$\begin{matrix}\varepsilon^*\\\varepsilon\end{matrix}$	$\begin{matrix}\varepsilon\\\varepsilon^*\end{matrix}$	$\begin{matrix}\varepsilon^{2*}\\\varepsilon^2\end{matrix}$		(x^2-y^2, xy)
A''	1	1	1	1	1	-1	-1	-1	-1	-1	z	
E_1''	$\begin{cases}1\\1\end{cases}$	$\begin{matrix}\varepsilon\\\varepsilon^*\end{matrix}$	$\begin{matrix}\varepsilon^2\\\varepsilon^{2*}\end{matrix}$	$\begin{matrix}\varepsilon^{2*}\\\varepsilon^2\end{matrix}$	$\begin{matrix}\varepsilon^*\\\varepsilon\end{matrix}$	$\begin{matrix}-1\\-1\end{matrix}$	$\begin{matrix}-\varepsilon\\-\varepsilon^*\end{matrix}$	$\begin{matrix}-\varepsilon^2\\-\varepsilon^{2*}\end{matrix}$	$\begin{matrix}-\varepsilon^{2*}\\-\varepsilon^2\end{matrix}$	$\begin{matrix}-\varepsilon^*\\\varepsilon\end{matrix}$	(R_x, R_y)	(xz, yz)
E_2''	$\begin{cases}1\\1\end{cases}$	$\begin{matrix}\varepsilon^2\\\varepsilon^{2*}\end{matrix}$	$\begin{matrix}\varepsilon^*\\\varepsilon\end{matrix}$	$\begin{matrix}\varepsilon\\\varepsilon^*\end{matrix}$	$\begin{matrix}\varepsilon^{2*}\\\varepsilon^2\end{matrix}$	$\begin{matrix}-1\\-1\end{matrix}$	$\begin{matrix}-\varepsilon^2\\-\varepsilon^{2*}\end{matrix}$	$\begin{matrix}-\varepsilon^*\\-\varepsilon\end{matrix}$	$\begin{matrix}-\varepsilon\\-\varepsilon^*\end{matrix}$	$\begin{matrix}-\varepsilon^{2*}\\-\varepsilon^2\end{matrix}$		

C_{6h}	E	C_6	C_3	C_2	C_3^2	C_6^5	i	S_3^5	S_6^5	σ_h	S_6	S_3		$\varepsilon = \exp(2\pi i/6)$
A_g	1	1	1	1	1	1	1	1	1	1	1	1	R_z	x^2+y^2, z^2
B_g	1	-1	1	-1	1	-1	1	-1	1	-1	1	-1		
E_{1g}	$\begin{cases}1\\1\end{cases}$	$\begin{matrix}-\varepsilon\\\varepsilon^*\end{matrix}$	$\begin{matrix}-\varepsilon^*\\-\varepsilon\end{matrix}$	$\begin{matrix}-1\\-1\end{matrix}$	$\begin{matrix}-\varepsilon\\-\varepsilon^*\end{matrix}$	$\begin{matrix}\varepsilon^*\\\varepsilon\end{matrix}$	$\begin{matrix}1\\1\end{matrix}$	$\begin{matrix}\varepsilon\\\varepsilon^*\end{matrix}$	$\begin{matrix}-\varepsilon^*\\-\varepsilon\end{matrix}$	$\begin{matrix}-1\\-1\end{matrix}$	$\begin{matrix}-\varepsilon\\-\varepsilon^*\end{matrix}$	$\begin{matrix}\varepsilon^*\\\varepsilon\end{matrix}$	(R_x, R_y)	(xz, yz)
E_{2g}	$\begin{cases}1\\1\end{cases}$	$\begin{matrix}-\varepsilon^*\\-\varepsilon\end{matrix}$	$\begin{matrix}-\varepsilon\\-\varepsilon^*\end{matrix}$	$\begin{matrix}1\\1\end{matrix}$	$\begin{matrix}-\varepsilon^*\\-\varepsilon\end{matrix}$	$\begin{matrix}-\varepsilon\\-\varepsilon^*\end{matrix}$	$\begin{matrix}1\\1\end{matrix}$	$\begin{matrix}-\varepsilon^*\\-\varepsilon\end{matrix}$	$\begin{matrix}-\varepsilon\\-\varepsilon^*\end{matrix}$	$\begin{matrix}1\\1\end{matrix}$	$\begin{matrix}-\varepsilon^*\\-\varepsilon\end{matrix}$	$\begin{matrix}-\varepsilon\\-\varepsilon^*\end{matrix}$		(x^2-y^2, xy)
A_u	1	1	1	1	1	1	-1	-1	-1	-1	-1	-1	z	
B_u	1	-1	1	-1	1	-1	-1	1	-1	1	-1	1		
E_{1u}	$\begin{cases}1\\1\end{cases}$	$\begin{matrix}\varepsilon\\\varepsilon^*\end{matrix}$	$\begin{matrix}-\varepsilon^*\\-\varepsilon\end{matrix}$	$\begin{matrix}-1\\-1\end{matrix}$	$\begin{matrix}-\varepsilon\\-\varepsilon^*\end{matrix}$	$\begin{matrix}\varepsilon^*\\\varepsilon\end{matrix}$	$\begin{matrix}-1\\-1\end{matrix}$	$\begin{matrix}-\varepsilon\\-\varepsilon^*\end{matrix}$	$\begin{matrix}\varepsilon^*\\\varepsilon\end{matrix}$	$\begin{matrix}1\\1\end{matrix}$	$\begin{matrix}\varepsilon\\\varepsilon^*\end{matrix}$	$\begin{matrix}-\varepsilon^*\\-\varepsilon\end{matrix}$	(x, y)	
E_{2u}	$\begin{cases}1\\1\end{cases}$	$\begin{matrix}-\varepsilon^*\\-\varepsilon\end{matrix}$	$\begin{matrix}-\varepsilon\\-\varepsilon^*\end{matrix}$	$\begin{matrix}1\\1\end{matrix}$	$\begin{matrix}-\varepsilon^*\\-\varepsilon\end{matrix}$	$\begin{matrix}-\varepsilon\\-\varepsilon^*\end{matrix}$	$\begin{matrix}-1\\-1\end{matrix}$	$\begin{matrix}\varepsilon^*\\\varepsilon\end{matrix}$	$\begin{matrix}\varepsilon\\\varepsilon^*\end{matrix}$	$\begin{matrix}-1\\-1\end{matrix}$	$\begin{matrix}\varepsilon^*\\\varepsilon\end{matrix}$	$\begin{matrix}\varepsilon\\\varepsilon^*\end{matrix}$		

D_{2h}	E	$C_2(z)$	$C_2(y)$	$C_2(x)$	i	$\sigma(xy)$	$\sigma(xz)$	$\sigma(yz)$		
A_g	1	1	1	1	1	1	1	1		x^2, y^2, z^2
B_{1g}	1	1	-1	-1	1	1	-1	-1	R_z	xy
B_{2g}	1	-1	1	-1	1	-1	1	-1	R_y	xz
B_{3g}	1	-1	-1	1	1	-1	-1	1	R_x	yz
A_u	1	1	1	1	-1	-1	-1	-1		
B_{1u}	1	1	-1	-1	-1	-1	1	1	z	
B_{2u}	1	-1	1	-1	-1	1	-1	1	y	
B_{3u}	1	-1	-1	1	-1	1	1	-1	x	

D_{3h}	E	$2C_6$	$2C_3$	C_2	$3\sigma_v$	$3\sigma_d$		
A_1'	1	1	1	1	1	1		$x^2 + y^2, z^2$
A_2'	1	1	-1	1	1	-1	R_z	
E'	2	-1	0	2	-1	0	(x, y)	$(x^2 - y^2, xy)$
A_1''	1	1	1	-1	-1	-1		
A_2''	1	1	-1	-1	-1	1	z	
E''	2	-1	0	-2	1	0	(R_x, R_y)	(xz, yz)

D_{4h}	E	$2C_4$	C_2	$2C_2'$	$2C_2''$	i	$2S_4$	σ_h	$2\sigma_v$	$2\sigma_d$		
A_{1g}	1	1	1	1	1	1	1	1	1	1		$x^2 + y^2, z^2$
A_{2g}	1	1	1	-1	-1	1	1	1	-1	-1	R_z	
B_{1g}	1	-1	1	1	-1	1	-1	1	1	-1		$x^2 - y^2$
B_{2g}	1	-1	1	-1	1	1	-1	1	-1	1		xy
E_g	2	0	-2	0	0	2	0	-2	0	0	(R_x, R_y)	(xz, yz)
A_{1u}	1	1	1	1	1	-1	-1	-1	-1	-1		
A_{2u}	1	1	1	-1	-1	-1	-1	-1	1	1	z	
B_{1u}	1	-1	1	1	-1	-1	1	-1	-1	1		
B_{2u}	1	-1	1	-1	1	-1	1	-1	1	-1		
E_u	2	0	-2	0	0	-2	0	2	0	0	(x, y)	

D_{5h}	E	$2C_5$	$2C_5^2$	$5C_2$	σ_h	$2S_5$	$2S_5^3$	$5\sigma_v$		
A_1'	1	1	1	1	1	1	1	1		$x^2 + y^2, z^2$
A_2'	1	1	1	-1	1	1	1	-1	R_z	
E_1'	2	$2\cos 72°$	$2\cos 144°$	0	2	$2\cos 72°$	$2\cos 144°$	0	(x, y)	
E_2'	2	$2\cos 144°$	$2\cos 72°$	0	2	$2\cos 144°$	$2\cos 72°$	0		$(x^2 - y^2, xy)$
A_1''	1	1	1	1	-1	-1	-1	-1		
A_2''	1	1	1	-1	-1	-1	-1	1	z	
E_1''	2	$2\cos 72°$	$2\cos 144°$	0	-2	$-2\cos 72°$	$-2\cos 144°$	0	(R_x, R_y)	(xz, yz)
E_2''	2	$2\cos 144°$	$2\cos 72°$	0	-2	$-2\cos 144°$	$-2\cos 72°$	0		

D_{6h}	E	$2C_6$	$2C_3$	C_2	$3C_2'$	$3C_2''$	i	$2S_3$	$2S_6$	σ_h	$3\sigma_d$	$3\sigma_v$		
A_{1g}	1	1	1	1	1	1	1	1	1	1	1	1		x^2+y^2, z^2
A_{2g}	1	1	1	1	-1	-1	1	1	1	1	-1	-1	R_z	
B_{1g}	1	-1	1	-1	1	-1	1	-1	1	-1	1	-1		
B_{2g}	1	-1	1	-1	-1	1	1	-1	1	-1	-1	1		
E_{1g}	2	1	-1	-2	0	0	2	1	-1	-2	0	0	(R_x,R_y)	(xz,yz)
E_{2g}	2	-1	-1	2	0	0	2	-1	-1	2	0	0		(x^2-y^2,xy)
A_{1u}	1	1	1	1	1	1	-1	-1	-1	-1	-1	-1		
A_{2u}	1	1	1	1	-1	-1	-1	-1	-1	-1	1	1	z	
B_{1u}	1	-1	1	-1	1	-1	-1	1	-1	1	-1	1		
B_{2u}	1	-1	1	-1	-1	1	-1	1	-1	1	1	-1		
E_{1u}	2	1	-1	-2	0	0	-2	-1	1	2	0	0	(x,y)	
E_{2u}	2	-1	-1	2	0	0	-2	1	1	-2	0	0		

D_{2d}	E	$2S_4$	C_2	$2C_2'$	$2\sigma_d$		
A_1	1	1	1	1	1		x^2+y^2, z^2
A_2	1	1	1	-1	-1	R_z	
B_1	1	-1	1	1	-1		x^2-y^2
B_2	1	-1	1	-1	1	z	xy
E	2	0	-2	0	0	$(x,y)(R_x,R_y)$	(xz,yz)

D_{3d}	E	$2C_3$	$3C_2$	i	$2S_6$	$3\sigma_d$		
A_{1g}	1	1	1	1	1	1		x^2+y^2, z^2
A_{2g}	1	1	-1	1	1	-1	R_z	
E_g	2	-1	0	2	-1	0	(R_x,R_y)	$(x^2-y^2,xy)(xz,yz)$
A_{1u}	1	1	1	-1	-1	-1		
A_{2u}	1	1	-1	-1	-1	1	z	
E_u	2	-1	0	-2	1	0	(x,y)	

D_{4d}	E	$2S_8$	$2C_4$	$2S_8^{\ 3}$	C_2	$4C_2'$	$4\sigma_d$		
A_1	1	1	1	1	1	1	1		x^2+y^2, z^2
A_2	1	1	1	1	1	-1	-1	R_z	
B_1	1	-1	1	-1	1	1	-1		
B_2	1	-1	1	-1	1	-1	1	z	
E_1	2	$\sqrt{2}$	0	$-\sqrt{2}$	-2	0	0	(x,y)	
E_2	2	0	-2	0	2	0	0		(x^2-y^2,xy)
E_3	2	$-\sqrt{2}$	0	$\sqrt{2}$	-2	0	0	(R_x,R_y)	(xz,yz)

D_{5d}	E	$2C_5$	$2C_5^{\ 2}$	$5C_2$	i	$2S_{10}^{\ 3}$	$2S_{10}$	$5\sigma_d$		
A_{1g}	1	1	1	1	1	1	1	1		x^2+y^2, z^2
A_{2g}	1	1	1	-1	1	1	1	-1	R_z	
E_{1g}	2	$2\cos 72°$	$2\cos 144°$	0	2	$2\cos 72°$	$2\cos 144°$	0	(R_x,R_y)	(xz,xy)
E_{2g}	2	$2\cos 144°$	$2\cos 72°$	0	2	$2\cos 144°$	$2\cos 72°$	0		(x^2-y^2,xy)
A_{1u}	1	1	1	1	-1	-1	-1	-1		
A_{2u}	1	1	1	-1	-1	-1	-1	1	z	
E_{1u}	2	$2\cos 72°$	$2\cos 144°$	0	-2	$-2\cos 72°$	$-2\cos 144°$	0	(x,y)	
E_{2u}	2	$2\cos 144°$	$2\cos 72°$	0	-2	$-2\cos 144°$	$-2\cos 72°$	0		

D_{6d}	E	$2S_{12}$	$2C_6$	$2S_4$	$2C_3$	$2S_{12}{}^5$	C_2	$6C_2'$	$6\sigma_d$		
A_1	1	1	1	1	1	1	1	1	1		$x^2 + y^2, z^2$
A_2	1	1	1	1	1	1	1	-1	-1	R_z	
B_1	1	-1	1	-1	1	-1	1	1	-1		
B_2	1	-1	1	-1	1	-1	1	-1	1	z	
E_1	2	$\sqrt{3}$	1	0	-1	$-\sqrt{3}$	-2	0	0	(x, y)	
E_2	2	1	-1	-2	-1	1	2	0	0		$(x^2 - y^2, xy)$
E_3	2	0	-2	0	2	0	-2	0	0		
E_4	2	-1	-1	2	-1	-1	2	0	0		
E_5	2	$-\sqrt{3}$	1	0	-1	$\sqrt{3}$	-2	0	0	(R_x, R_y)	(xz, yz)

S_4	E	S_4	C_2	$S_4{}^3$		
A	1	1	1	1	R_z	$x^2 + y^2, z^2$
B	1	-1	1	-1	z	$x^2 - y^2, xy$
E	$\begin{cases} 1 \\ 1 \end{cases}$	$\begin{matrix} i \\ -i \end{matrix}$	$\begin{matrix} -1 \\ -1 \end{matrix}$	$\begin{matrix} -i \\ i \end{matrix}$	$(x, y) (R_x, R_y)$	(xz, yz)

S_6	E	C_3	$C_3{}^2$	i	$S_6{}^5$	S_6		$\varepsilon = \exp(2\pi i/3)$
A_g	1	1	1	1	1	1	R_z	$x^2 + y^2, z^2$
E_g	$\begin{cases} 1 \\ 1 \end{cases}$	$\begin{matrix} \varepsilon \\ \varepsilon^* \end{matrix}$	$\begin{matrix} \varepsilon^* \\ \varepsilon \end{matrix}$	$\begin{matrix} 1 \\ 1 \end{matrix}$	$\begin{matrix} \varepsilon \\ \varepsilon^* \end{matrix}$	$\begin{matrix} \varepsilon^* \\ \varepsilon \end{matrix}$	(R_x, R_y)	$(x^2 - y^2, xy)(xz, yz)$
A_u	1	1	1	-1	-1	-1	z	
E_u	$\begin{cases} 1 \\ 1 \end{cases}$	$\begin{matrix} \varepsilon \\ \varepsilon^* \end{matrix}$	$\begin{matrix} \varepsilon^* \\ \varepsilon \end{matrix}$	$\begin{matrix} -1 \\ -1 \end{matrix}$	$\begin{matrix} -\varepsilon \\ -\varepsilon^* \end{matrix}$	$\begin{matrix} -\varepsilon^* \\ -\varepsilon \end{matrix}$	(x, y)	

T	E	$4C_3$	$4C_3{}^2$	$3C_2$		$\varepsilon = \exp(2\pi i/3)$
A	1	1	1	1		$x^2 + y^2 + z^2$
E	$\begin{cases} 1 \\ 1 \end{cases}$	$\begin{matrix} \varepsilon \\ \varepsilon^* \end{matrix}$	$\begin{matrix} \varepsilon^* \\ \varepsilon \end{matrix}$	$\begin{matrix} 1 \\ 1 \end{matrix}$		$(2z^2 - x^2 - y^2, x^2 - y^2)$
T	3	0	0	-1	$(R_x, R_y, R_z)(x, y, z)$	(xy, xz, yz)

T_h	E	$4C_3$	$4C_3{}^2$	$3C_2$	i	$4S_6$	$4S_6{}^5$	$3\sigma_h$		$\varepsilon = \exp(2\pi i/3)$
A_g	1	1	1	1	1	1	1	1		$x^2 + y^2 + z^2$
A_u	1	1	1	1	-1	-1	-1	-1		
E_g	$\begin{cases} 1 \\ 1 \end{cases}$	$\begin{matrix} \varepsilon \\ \varepsilon^* \end{matrix}$	$\begin{matrix} \varepsilon^* \\ \varepsilon \end{matrix}$	$\begin{matrix} 1 \\ 1 \end{matrix}$	$\begin{matrix} 1 \\ 1 \end{matrix}$	$\begin{matrix} \varepsilon \\ \varepsilon^* \end{matrix}$	$\begin{matrix} \varepsilon^* \\ \varepsilon \end{matrix}$	$\begin{matrix} 1 \\ 1 \end{matrix}$		$(2z^2 - x^2 - y^2, x^2 - y^2)$
E_u	$\begin{cases} 1 \\ 1 \end{cases}$	$\begin{matrix} \varepsilon \\ \varepsilon^* \end{matrix}$	$\begin{matrix} \varepsilon^* \\ \varepsilon \end{matrix}$	$\begin{matrix} 1 \\ 1 \end{matrix}$	$\begin{matrix} -1 \\ -1 \end{matrix}$	$\begin{matrix} -\varepsilon \\ -\varepsilon^* \end{matrix}$	$\begin{matrix} -\varepsilon^* \\ -\varepsilon \end{matrix}$	$\begin{matrix} -1 \\ -1 \end{matrix}$		
T_g	3	0	0	-1	1	0	0	1	(R_x, R_y, R_z)	(xz, yz, xy)
T_u	3	0	0	-1	-1	0	0	1	(x, y, z)	

T_d	E	$8C_3$	$3C_2$	$6S_4$	$6\sigma_d$		
A_1	1	1	1	1	1		$x^2 + y^2 + z^2$
A_2	1	1	1	-1	-1		
E	2	-1	2	0	0		$(2z^2 - x^2 - y^2, x^2 - y^2)$
T_1	3	0	-1	1	-1	(R_x, R_y, R_z)	
T_2	3	0	-1	-1	1	(x, y, z)	(xy, xz, yz)

O	E	$6C_4$	$3C_2(=C_4^2)$	$8C_3$	$6C_2$		
A_1	1	1	1	1	1		$x^2 + y^2 + z^2$
A_2	1	-1	1	1	-1		
E	2	0	2	-1	0		$(2z^2 - x^2 - y^2, x^2 - y^2)$
T_1	3	1	-1	0	-1	$(R_x, R_y, R_z); (x, y, z)$	
T_2	3	-1	-1	0	1		(xy, xz, yz)

O_h	E	$8C_3$	$6C_2$	$6C_4$	$3C_2(=C_4^2)$	i	$6S_4$	$8S_6$	$3\sigma_h$	$6\sigma_d$		
A_{1g}	1	1	1	1	1	1	1	1	1	1		$x^2 + y^2 + z^2$
A_{2g}	1	1	-1	-1	1	1	-1	1	1	-1		
E_g	2	-1	0	0	2	2	0	-1	2	0		$(2z^2 - x^2 - y^2, x^2 - y^2)$
T_{1g}	3	0	-1	1	-1	3	1	0	-1	-1	(R_x, R_y, R_z)	
T_{2g}	3	0	1	-1	-1	3	-1	0	-1	1		(xz, yz, xy)
A_{1u}	1	1	1	1	1	-1	-1	-1	-1	-1		
A_{2u}	1	1	-1	-1	1	-1	1	-1	-1	1		
E_u	2	-1	0	0	2	-2	0	1	-2	0		
T_{1u}	3	0	-1	1	-1	-3	-1	0	1	1	(x, y, z)	
T_{2u}	3	0	1	-1	-1	-3	1	0	1	-1		

$C_{\infty v}$	E	$2C_\infty^\Phi$	$\cdots$	$\infty\sigma_v$		
Σ^+	1	1	$\cdots$	1	z	$x^2 + y^2, z^2$
Σ^-	1	1	$\cdots$	-1	R_z	
Π	2	$2\cos\Phi$	$\cdots$	0	$(x, y); (R_x, R_y)$	(xz, yz)
Δ	2	$2\cos 2\Phi$	$\cdots$	0		$(x^2 - y^2, xy)$
Φ	2	$2\cos 3\Phi$	$\cdots$	0		
$\cdots$	$\cdots$	$\cdots$	$\cdots$	$\cdots$		

$D_{\infty h}$	E	$2C_\infty^\Phi$	$\cdots$	$\infty\sigma_v$	i	$2S_\infty^\Phi$	$\cdots$	∞C_2		
Σ_g^+	1	1	$\cdots$	1	1	1	$\cdots$	1		$x^2 + y^2, z^2$
Σ_g^-	1	1	$\cdots$	-1	1	1	$\cdots$	-1	R_z	
Π_g	2	$2\cos\Phi$	$\cdots$	0	2	$-2\cos\Phi$	$\cdots$	0	(R_x, R_y)	(xz, yz)
Δ_g	2	$2\cos 2\Phi$	$\cdots$	0	2	$2\cos 2\Phi$	$\cdots$	0		$(x^2 - y^2, xy)$
$\cdots$	$\cdots$	$\cdots$	$\cdots$	$\cdots$	$\cdots$	$\cdots$	$\cdots$	$\cdots$		
Σ_u^+	1	1	$\cdots$	1	-1	-1	$\cdots$	-1	z	
Σ_u^-	1	1	$\cdots$	-1	-1	-1	$\cdots$	1		
Π_u	2	$2\cos\Phi$	$\cdots$	0	-2	$2\cos\Phi$	$\cdots$	0	(x, y)	
Δ_u	2	$2\cos 2\Phi$	$\cdots$	0	-2	$-2\cos 2\Phi$	$\cdots$	0		
$\cdots$	$\cdots$	$\cdots$	$\cdots$	$\cdots$	$\cdots$	$\cdots$	$\cdots$	$\cdots$		

D

PHYSICAL CONSTANTS[a]

Constant	Symbol	SI Value
Speed of light	c	2.99792458×10^8 m/s
Charge on proton	e	$1.6021764 \times 10^{-19}$ C
Permittivity of vacuum	ε_0	$8.8541878 \times 10^{-12}$ J^{-1}C^{2}m^{-1}
Avogadro's number	N_A	6.022142×10^{23} mol^{-1}
Rest mass of electron	m_e	9.109382×10^{-31} kg

Atomic Units, Energy Conversion Factors, and Physical Constants

ATOMIC UNITS

Quantity	Symbol or SI Expression	CGS Equivalent or SI Equivalent	Important Related Properties
Mass	m_e	9.10938×10^{-28} g 9.10938×10^{-31} kg	mass of electron
Charge	e	4.803204×10^{-10} stat C 1.602176×10^{-19} C	$-e$ = charge on electron
Angular momentum	$\hbar$	1.05457×10^{-27} erg s 1.05457×10^{-34} Js	
Length (bohr)	$a_0 = \dfrac{4\pi\varepsilon_0\hbar^2}{m_e e^2}$	0.5291772×10^{-8} cm $0.5291772 \times 10^{-10}$ m	radius of lowest energy Bohr orbit
Energy (hartree)	$E_h = \dfrac{e^2}{4\pi\varepsilon_0 a_0}$	4.35974×10^{-11} erg 4.35974×10^{-18} J	$\frac{1}{2}E_h$ = ionization energy of hydrogen
Time	$\tau_0 = \dfrac{(4\pi\varepsilon_0)^2\hbar^3}{m_e e^4}$	2.41888×10^{-17} s	$2\pi\tau_0$ = period of lowest Bohr orbit
Frequency	$\dfrac{m_e e^4}{(4\pi\varepsilon_0)^2\hbar^3}$	4.13414×10^{16} s^{-1}	
Velocity	$\dfrac{e^2}{4\pi\varepsilon_0\hbar}$	2.18770×10^8 cm/s 2.18770×10^6 m/s	velocity of electron in lowest Bohr orbit

(continued)

ATOMIC UNITS

Quantity	Symbol or SI Expression	CGS Equivalent or SI Equivalent	Important Related Properties
Force	$\dfrac{e^2}{4\pi\varepsilon_0 a_0^2}$	8.23872×10^{-3} dynes 8.23872×10^{-8} N	
Electric field	$\dfrac{e}{4\pi\varepsilon_0 a_0^2}$	1.71526×10^7 stat C/cm^2 5.14221×10^{11} V/m	field acting on electron in lowest Bohr orbit
Electric potential	$\dfrac{e}{4\pi\varepsilon_0 a_0}$	9.07675×10^{-2} stat C/cm 27.2114 V	
Fine structure constant	$\alpha = \dfrac{e^2}{4\pi\varepsilon_0 \hbar c}$	1/137.036	
Power	$\dfrac{m_e^2 e^8}{(4\pi\varepsilon_0)^4 \hbar^5}$	1.80239×10^6 erg/s 0.180239 W	
Magnetic moment	$\beta_e = \dfrac{e\hbar}{2m_e}$	9.27399×10^{-21} erg/gauss 9.27399×10^{-24} J/T	magnetic moment of orbital motion of electron in lowest Bohr orbit
Magnetic field	$\dfrac{eh}{m_e a_0^3}$	7.86455×10^5 gauss 7.86455×10^1 T	Magnetic field at nucleus due to orbital motion of electron in first Bohr orbit

ENERGY CONVERSION FACTORS

	K	cm^{-1}	kJ/mol	kcal/mol	eV	hartree
K	1	0.695036	8.31447×10^{-3}	1.98721×10^{-3}	8.61734×10^{-5}	3.16682×10^{-6}
cm^{-1}	1.43878	1	1.19627×10^{-2}	2.85914×10^{-3}	1.23984×10^{-4}	4.55634×10^{-6}
kJ/mol	120.272	83.5935	1	2.39006×10^{-1}	1.03643×10^{-2}	3.80880×10^{-4}
kcal/mol	503.219	349.755	4.184	1	4.33641×10^{-2}	1.59360×10^{-3}
eV	11,604.5	8,065.54	96.4853	23.0606	1	3.67493×10^{-2}
hartree	315,775.	219,475.	2,625.50	627.509	27.2114	1

PHYSICAL CONSTANTS[a]

Constant	Symbol	SI Value	CGS Value
Speed of light	c	2.99792458×10^8 m/s	$2.99792458 \times 10^{10}$ cm/s
Charge on proton	e	$1.6021764 \times 10^{-19}$ C	4.803204×10^{-10} statC
Permittivity of vacuum	ε_0	$8.8541878 \times 10^{-12}$ J^{-1}C^{2}m^{-1}	
Avogadro's number	N_A	6.022142×10^{23} mol^{-1}	6.022142×10^{23} mol^{-1}
Rest mass of electron	m_e	9.109382×10^{-31} kg	9.109382×10^{-28} g
Proton mass	m_p	1.672622×10^{-27} kg	1.672622×10^{-24} g
Planck constant	h	6.62607×10^{-34} Js	6.62607×10^{-27} erg s
Bohr radius	a_0	5.291772×10^{-11} m	0.5291772×10^{-8} cm
Bohr magneton	β_e	9.27401×10^{-24} J/T	
Nuclear magneton	β_N	5.05078×10^{-27} J/T	
Electron g value	g_e	2.0023193044	2.0023193044
Gas constant	R	8.31447 J/mol K	8.31447×10^7 erg/mol K
Boltzmann constant	k	1.380650×10^{-23} J/K	1.380650×10^{-16} erg/K

[a] Adapted from values given on the web page: physics. nist. gov/constants.
Source: P. J. Mohr and B. N. Taylor, *Rev. Mod. Phys.* 72 (2) 2000.

Solutions to Odd-Numbered Problems

Chapter 1

1.1 Here we use SI units, except for the product $\bar{\nu}c = (4395.2 \text{ cm}^{-1})(3 \times 10^{10} \text{ cm/s}) = 1.31 \times 10^{14} \text{ s}^{-1}$. Then the force constant is

$$k = \mu\omega^2 = \mu(2\pi)^2(c\bar{\nu})^2$$

$$= \frac{1}{2}(1.67 \times 10^{-27})(2\pi)^2[(4395.2)(3 \times 10^{10})]^2$$

$$= 5.73 \times 10^2 \text{ kg s}^{-2} = 5.73 \times 10^2 \text{ N m}^{-1}.$$

The force is zero at equilibrium (where $x = 0$). At the turning points, the force is obtained from $F = -kx$, where $E = 1/2\hbar\omega = 1/2kx^2$. This gives $F = -(k\hbar\omega)^{1/2}$, so that

$$F = -\sqrt{(5.73 \times 10^2 \text{ N/m})(6.625 \times 10^{-34} \text{ Js})(4392.3)(3 \times 10^{10})\text{s}^{-1}}$$

$$= -7.07 \times 10^{-9} \text{ N}.$$

The velocity would be obtained from

$$v = \sqrt{\frac{2hc\bar{\nu}}{2\mu}} = \sqrt{\frac{2(6.625 \times 10^{-34})(4395.2)(3 \times 10^{10})}{2 \cdot \frac{1}{2}(1.67 \times 10^{-27})}}$$

$$= 1.02 \times 10^4 \text{ m/s}$$

The corresponding thermal velocity is lower:

$$v = \sqrt{\frac{3kT}{m}} = \sqrt{\frac{3(1.38 \times 10^{-23})(300)}{2(1.67 \times 10^{-27})}} = 1.93 \times 10^3 \text{ m s}^{-1}.$$

1.3 (a) $R = \dfrac{m_1 \mathbf{r}_1 + m_2 \mathbf{r}_2}{m_1 + m_2}$ $\mathbf{r} = \mathbf{r}_1 - \mathbf{r}_2$

$$\mathbf{r}_1 = \mathbf{R} + \frac{m_2 \mathbf{r}}{m_1 + m_2}$$

$$\mathbf{r}_2 = \mathbf{R} - \frac{m_1 \mathbf{r}}{m_1 + m_2}$$

$$\mathbf{r} \equiv (x, y, z); \quad \mathbf{R} \equiv (X, Y, Z)$$

$$H = \frac{1}{2} m_1 (\dot{x}_1^2 + \dot{y}_1^2 + \dot{z}_1^2) + \frac{1}{2} m_2 (\dot{x}_2^2 + \dot{y}_2^2 + \dot{z}_2^2) + V$$

$$= \frac{1}{2} m_1 \left(\left(\dot{R}_x + \frac{m_2}{m_1 + m_2} \dot{r}_x \right)^2 + (\quad)^2 + (\quad)^2 \right)$$

$$+ \frac{1}{2} m_2 \left(\left(\dot{R}_x - \frac{m_1}{m_1 + m_2} \dot{r}_x \right)^2 + (\quad)^2 + (\quad)^2 \right) + V$$

$$= \frac{1}{2} (m_1 + m_2)(\dot{R}_x^2 + \dot{R}_y^2 + \dot{R}_z^2)$$

$$+ \frac{1}{2} \left\{ \frac{m_1 m_2^2}{(m_1 + m_2)^2} + \frac{m_2 m_1^2}{(m_1 + m_2)^2} \right\} \times (\dot{r}_x^2 + \dot{r}_y^2 + \dot{r}_z^2) + V$$

$$= \frac{1}{2} M \dot{\mathbf{R}}^2 + \frac{1}{2} \mu \dot{\mathbf{r}}^2 + V(r)$$

(b) $m_1 \ddot{x}_1 = -\dfrac{\partial V}{\partial x_1}$ $m_2 \ddot{x}_2 = -\dfrac{\partial V}{\partial x_2}$

reduces to

$$M \ddot{R}_x = -\frac{\partial V}{\partial R_x} = 0 \qquad \mu \ddot{r}_x = -\frac{\partial V}{\partial r} \frac{\partial r}{\partial r_x} = -\frac{\partial V}{\partial r} \left(\frac{x}{r} \right)$$

R = linear function of time, uncoupled to r.

1.5 The Bragg equation, $n\lambda = d \sin\theta$ indicates that the larger λ is, the larger θ is (for fixed n). Since electrons have the smaller mass, they will have the larger λ and hence the larger θ.

Chapter 2

2.1 (a) No   **(b)** No   **(c)** No   **(d)** No
(e) No   **(f)** Yes   **(g)** Yes   **(h)** No

2.3 $[p_x, x^2] = x[p_x, x] + [p_x, x]x = -2i\hbar x$, where $[p_x, x] = -i\hbar$

2.5 (a) T   **(b)** F   **(c)** F

  (d) F   **(e)** T   **(f)** T

2.7 (a) $\displaystyle\int_{-1}^{1}\psi_1^*\psi_2\,dx = \int_{-1}^{1} x\,dx = 0$ by symmetry

(b) $\displaystyle\int_{-1}^{1}\psi_1^*\psi_3\,dx = \int (x^2 + bx + c)\,dx = \frac{2}{3} + 2c = 0$

$$c = -\frac{1}{3}$$

$$\int \psi_2^*\psi_3\,dx = \int (x^3 + bx^2 + cx)\,dx = \frac{2}{3}b = 0 \quad b = 0$$

$$\int \psi_1\psi_4\,dx = \int (x^3 + dx^2 + ex + f)\,dx = \frac{2}{3}d + 2f \quad f = -\frac{1}{3}d$$

$$\int \psi_2\psi_4\,dx = \int (x^4 + dx^3 + ex^2 + fx)\,dx = \frac{2}{5} + e\frac{2}{3} \quad e = -\frac{3}{5}$$

$$\int \psi_3\psi_4\,dx = \int \left(x^2 - \frac{1}{3}\right)(x^3 + dx^2 + ex + f)\,dx$$

$$= \int \left(x^5 + dx^4 + ex^3 + fx^2 - \frac{1}{3}x^3 - \frac{1}{3}dx^2 - \frac{1}{3}ex - \frac{1}{3}f\right)dx$$

$$= \frac{2}{5}d + \frac{2}{3}f - \left(\frac{1}{3}\right)\left(\frac{2}{3}\right)d - \frac{2}{3}f = \left(\frac{2}{5} - \frac{2}{9}\right)d = 0$$

These results indicate that $d = 0$ and $f = 0$.

Results: $\psi_1 = 1 \quad \psi_2 = x \quad \psi_3 = x^2 - \frac{1}{3} \quad \psi_4 = x^3 - \frac{3}{5}x$

(c) $\displaystyle N^2\int_{-1}^{1}\psi_1^2\,dx = 2 \qquad \psi_1 = \frac{1}{\sqrt{2}}$

$$N^2\int_{-1}^{1}\psi_2^2\,dx = \int x^2\,dx = \frac{2}{3} \qquad \psi_2 = \sqrt{\frac{3}{2}}\,x$$

$$N^2\int_{-1}^{1}\psi_3^2\,dx = \int \left(x^4 - \frac{2}{3}x^2 + \frac{1}{9}\right)dx$$

$$= \frac{2}{5} - \left(\frac{2}{3}\right)\left(\frac{2}{3}\right) + \frac{2}{9} = \frac{2(9-5)}{45} = \frac{8}{45}$$

$$\psi_3 = \sqrt{\frac{45}{8}}\left(x^2 - \frac{1}{3}\right) = \sqrt{\frac{5}{2}}\left(\frac{3x^2 - 1}{2}\right)$$

$$N^2 \int_{-1}^{1} \psi_4^2 dx = \int \left(x^6 - \frac{6}{5} x^4 + \frac{9}{25} x^2 \right) dx$$

$$= \frac{2}{5} - \left(\frac{6}{5} \right) \left(\frac{2}{5} \right) + \left(\frac{3}{25} \right) \left(\frac{2}{1} \right)$$

$$= \frac{2}{7} - \frac{6}{25} = \frac{2(25-21)}{7(25)} = \frac{8}{7(25)}$$

$$\psi_4 = \sqrt{\frac{7(25)}{8}} \left(x^3 - \frac{3}{5} x \right) = \sqrt{\frac{7}{2}} \frac{5x^3 - 3x}{2}$$

2.9 $\langle p_x \rangle = \dfrac{\int \psi^* \hat{p}_x \psi dx}{\int \psi^* \psi dx} = \dfrac{\int e^{-ipx/\hbar} \dfrac{\hbar}{i} \dfrac{\partial}{\partial x} e^{ipx/\hbar} dx}{\int \psi^* \psi dx}$

$$= \frac{p \int \psi^* \psi dx}{\int \psi^* \psi dx} = p$$

Probability density $= \psi^*(x)\psi(x) = e^{-ipx/\hbar} e^{ipx/\hbar} = 1$

$$\langle p_x \rangle = p$$

$$\langle p_x^2 \rangle = p^2 \quad \text{(easily shown)}$$

Therefore, $\quad \Delta p = (\langle p_x^2 \rangle - \langle p_x \rangle^2)^{1/2} = 0$

What this shows is that $e^{ipx/\hbar}$ is an eigenfunction of $\hat{p}_x$, so there is no uncertainty associated with measuring p_x.

2.11 If the operator associated with H is $i\hbar \partial / \partial t$, then

$$[H, t] = i\hbar \frac{\partial}{\partial t} t - ti\hbar \frac{\partial}{\partial t} = i\hbar + i\hbar t \frac{\partial}{\partial t} - i\hbar t \frac{\partial}{\partial t} = i\hbar$$

$$\Delta t = \hbar / \Delta E = \frac{1.054 \times 10^{-34} \text{ Js}}{(0.2)1.6 \times 10^{-19} \text{ J}} = 3.3 \times 10^{-15} \text{ s}$$

$$= 3.3 \text{ fs where 1 femtosecond} = 10^{-15} \text{ s}$$

Chapter 3

3.1 (a and b)

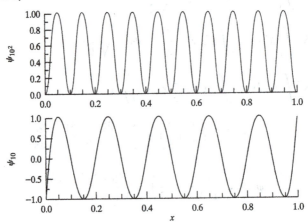

(c) $E_n \propto 1/m$

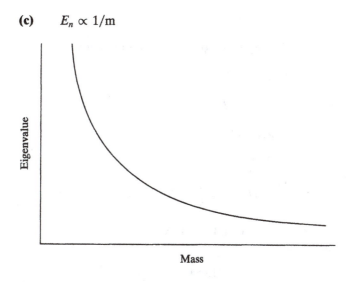

3.3 (a) $\psi(0) = 0 \quad \psi(a) = 0$

$\psi\left(x = \dfrac{a}{2}\right)$ is continuous. $\psi'\left(x = \dfrac{a}{2}\right)$ is continuous.

(b) $\psi(x) = A \sin kx \qquad 0 \le x \le \dfrac{a}{2} \qquad k = \sqrt{\dfrac{2mE}{\hbar^2}}$

(c) $\psi(x) = B \sin k'(x - a) \qquad \dfrac{a}{2} \le x \le a \qquad k' = \sqrt{\dfrac{2m}{\hbar^2}(E - V_s)}$

(d) at $x = \dfrac{a}{2}, \quad A \sin k\dfrac{a}{2} = B \sin k'\left(-\dfrac{a}{2}\right) = -B \sin k'\dfrac{a}{2}$

$Ak \cos k\dfrac{a}{2} = Bk' \cos k'\left(-\dfrac{a}{2}\right) = Bk' \cos k'\dfrac{a}{2}$

Take the ratio to get

$$\frac{1}{k} \tan k\frac{a}{2} = -\frac{1}{k'} \tan k'\frac{a}{2}$$

We need to solve this numerically to get the allowed E's.

3.5 (a) $\qquad \psi(x < 0) = 0$

$\psi(0 \le x \le L) = A \sin kx + B \cos kx \quad k = \sqrt{\dfrac{2m}{\hbar^2}E}$

$\psi(x > L) = Ce^{-Kx} - De^{Kx} \quad K = \sqrt{\dfrac{2m}{\hbar^2}(V_0 - E)}$

(b) $\psi(0) = 0 \rightarrow B = 0$

$\psi(x = L)$ is continuous, $\psi'(x = L)$ is continuous

$\psi(x = \infty) \rightarrow 0 \Rightarrow D = 0$

At $x = L$, $\quad A \sin kL = Ce^{-KL}$

$$kA \cos kL = -KCe^{-KL}$$

We take the ratio to get

$$\frac{1}{k} \tan kL = -\frac{1}{K} \text{ (solve numerically to get } E)$$

(c) $L = 1$ Å, $m = 1$ g/mol, $V_0 = 4$ eV
∞box results:

$$E = \frac{\hbar^2}{2m} \frac{\pi^2}{L^2} n^2 = \frac{(1.054 \times 10^{-34} \text{ Js})^2 \pi^2 n^2}{2(1.67 \times 10^{-27} \text{ kg})(1 \times 10^{-10} \text{ m})^2}$$

$$= 3.3 \times 10^{-21} \text{ J } n^2 = (0.0205 \text{ eV})n^2$$

$E = 0.0205, 0.0821, 0.185, 0.328$ eV

$$\tan \sqrt{\frac{2m}{\hbar^2} EL^2} = -\sqrt{\frac{E}{V_0 - E}}$$

Let $E = (0.0205 \text{ eV})z$. Then the expression becomes $\tan \pi \sqrt{z} = -\sqrt{\dfrac{.0205z}{(4 - .0205z)}}$.
Solutions, obtained graphically, are $E = 0.0196, 0.0784, 0.176$, and 0.313 eV.

3.7 $\psi = \dfrac{1}{\sqrt{2}} (\psi_{1,2,3} - \psi_{3,1,2})$ is an eigenfunction (sum of degenerate solutions)

$$\psi = \psi_{1,1,1} - \psi_{2,2,2} \text{ is not an eigenfunction.}$$

3.9 $k =$ quantum number of HOMO $= 2 \times$ # electrons $=$ # double bonds

$$\frac{hc}{\lambda} = E_{\text{LUMO}} - E_{\text{HOMO}} = \frac{\hbar^2}{2m} \frac{\pi^2}{a^2} [(k + 1)^2 - k^2]$$

$$= \frac{\hbar^2 \pi^2}{2ma^2} [2k + 1]$$

Substitute $a = 1.4 \text{ Å}(k + 2)$ to get

$$\frac{hc}{\lambda} = \frac{\hbar^2 \pi^2}{2m(1.4 \text{ Å})^2} \frac{(2k + 1)}{(k + 2)^2}$$

Need to solve this for k:

$$\lambda = \frac{2m(1.4 \text{ Å})^2}{\hbar^2 \pi^2} (hc) \frac{(k+2)^2}{(2k+1)}$$

$$= \frac{2(9.1 \times 10^{-31} \text{ kg})(1.4 \times 10^{-10} \text{ m})^2}{(1.054 \times 10^{-34} \text{ Js})^2 \pi^2} (6.625 \times 10^{-34} \text{ Js})(3 \times 10^8 \text{ m/s}) \frac{(k+2)^2}{(2k+1)}$$

$$= 6.47 \times 10^{-8} \text{ m} \frac{(k+2)^2}{(2k+1)}$$

$$= 64.7 \text{ nm} \frac{(k+2)^2}{2k+1}$$

λ	k	Need $k = 15$ double bonds
207 nm	2	$CH_2=CH-(CH=CH)_{13}-CH=CH_2$
571	14	
603	15	
635	16	
667	17	
699	18	

Chapter 4

4.1

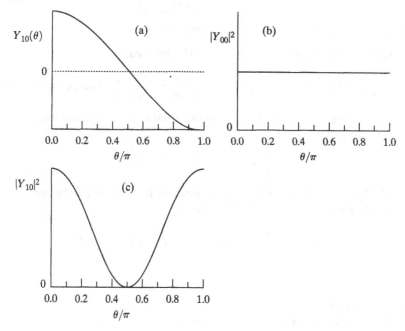

4.3 If the particle is confined to the surface of the cylinder, then the kinetic energy reflects two factors: (a) free-particle motion in the z coordinate, as with a particle in a box, and (b) rotational motion in the ϕ coordinate, as in a rigid rotor. The radius

r would be fixed, so no kinetic energy would be associated with this degree of freedom. Overall, the kinetic-energy expression is $E = E_z + E_\phi$, and the Hamiltonian operator is

$$\hat{H} = \frac{-\hbar^2}{2m} \frac{\partial^2}{\partial z^2} + \frac{-\hbar^2}{2\mu r^2} \frac{\partial^2}{\partial \phi^2}$$

where both the mass m and the mass μ would be approximately equal to the electron mass m_e.

(b) Given the preceding Hamiltonian, the wavefunction is a solution, provided that n and m are both integers, with $n = 1, 2, 3, \ldots$, and $m = 0, \pm 1, \pm 2, \pm 3, \pm 4$.

(c) The energy expression is

$$E_{n,m} = \frac{n^2 \pi^2 \hbar^2}{2m_e \ell^2} + \frac{m^2 \hbar^2}{2m_e a^2}.$$

The energy expression is $E_{n,m} = 0.0026 \text{ eV } n^2 + 0.126 m^2$. The lowest two spacings are 0.0078 and 0.013 eV. 300π electrons with double occupation, and including for the twofold degeneracy of states with m values other than zero, will occupy energies up to $n = 22$ and $m = 2$, and the spacing between that and the $n = 16, m = 3$ level would be 0.035 eV. The energy of the highest occupied state is 1.77 eV.

4.5 (i) $\langle 2p_x | \ell_z | 2p_z \rangle = 0$

(ii) $\langle 2p_x | \ell_x^2 + \ell_y^2 | 2p_z \rangle = \langle 2p_x | \ell^2 - \ell_z^2 | 2p_z \rangle$

$$= (\hbar^2 \ell(\ell + 1)) \langle 2p_x | 2p_z \rangle = 0$$

4.7 $[\ell_+, \ell_x] = [\ell_x + i\ell_y, \ell_x] = i[\ell_y, \ell_x] = i(-i\hbar \ell_z) = \hbar \ell_z$

4.9 (a) From Eq. (4.26), we can write the Hamiltonian:

$$\frac{-\hbar^2}{2m} \left\{ \frac{1}{r} \frac{\partial^2}{\partial r^2} r \right\} \psi(r, \phi, \theta) + \frac{-\hbar^2}{2\mu r^2} \left\{ \frac{1}{\sin\theta} \frac{\partial}{\partial \theta} \sin\theta \frac{\partial}{\partial \theta} + \frac{1}{\sin^2 \theta} \frac{\partial^2}{\partial \phi^2} \right\} \psi(r, \phi, \theta) = E\psi(r, \phi, \theta).$$

(b) The angular eigenfunctions are the usual spherical harmonics.

(c) The radial equation is

$$\frac{1}{r} \frac{\partial^2}{\partial r^2} rR - \frac{\ell(\ell + 1)}{r^2} R = \frac{-2mE}{\hbar^2} R.$$

(d) At $r = a$, we want the wavefunction and its derivative to be zero.

(e) This works provided that $\ell = 0$ and

$$E_k = \frac{\hbar^2 k^2}{2\mu a^2} \quad \text{with } ka = n\pi \text{ and } n = 1, 2, 3, \ldots$$

Chapter 5

5.1 $n = 6$ six nodes

See the accompanying figures for results.

5.3 $\langle 0 | bbb^+ b^+ | 0 \rangle = 2$

$\langle 0 | b^+ bb^+ b | 0 \rangle = 0$

$\langle 0 | b^+ bb^+ | 0 \rangle = 0$

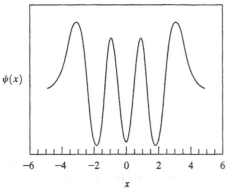

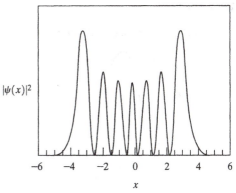

Figure S5.1

5.5 (a) $H = \displaystyle\sum_{k=1}^{3} \hbar\omega_k \left(b_k^+ b_k + \frac{1}{2} \right)$

(b) $E = \displaystyle\sum_{k=1}^{3} \frac{1}{2} \hbar\omega_k$

$$\psi = \prod_{k=1}^{3} \sqrt[4]{\frac{\alpha_k}{\pi}} \, e^{-\alpha_k Q_k^2 / 2} \qquad \alpha_k = \mu_k \omega_k / \hbar$$

$$\mu_k = 1 \text{ in normal coordinates}$$

(c) $V = C_{112} Q_1^2 Q_2$

$$V = C_{112} \left(\sqrt{\frac{\hbar}{2\mu_1 \omega_1}} \right)^2 (b_1 + b_1^+)^2 \sqrt{\frac{\hbar}{2\mu_2 \omega_2}} (b_2 + b_2^+)$$

(d) $E_1 = \langle 0 | V | 0 \rangle = 0.$

(e) $|000\rangle$ mixes with $|210\rangle$.

5.7 $H = \dfrac{J^2}{2I} + \mu E \cos\phi$

(a) $E^{(0)} = \dfrac{\hbar^2}{2I} m^2 \qquad m = 0, \pm 1, \pm 2, \ldots, 2\pi$

(b) $E^{(1)} = \langle J | \mu E \cos\phi | J \rangle = \dfrac{1}{2\pi} \displaystyle\int_0^{2\pi} e^{i(0)} \mu E \cos\phi \, d\phi = 0 \qquad$ No change.

(c) $E^{(2)} = \displaystyle\sum_J \frac{|\langle J | \mu E \cos\phi | 0 \rangle|^2}{E_0 - E_J} = (\mu E)^2 \left\{ \frac{|\langle +1 | \cos\phi | 0 \rangle|^2}{E_0 - E_1} + \frac{|\langle -1 | \cos\phi | 0 \rangle|^2}{E_0 - E_1} \right\}$

$$= (\mu E)^2 \left(\frac{2I}{\hbar^2} \right) \left\{ \frac{1}{4} \frac{1}{(-1)} + \frac{1}{4} \frac{1}{(-1)} \right\} = -(\mu E)^2 \left(\frac{I}{\hbar^2} \right)$$

Chapter 6

6.1 (a) $n = 3, \ell = 2$ $\psi \sim r^2 e^{-Zr/3a_0}$ (Plot follows.)

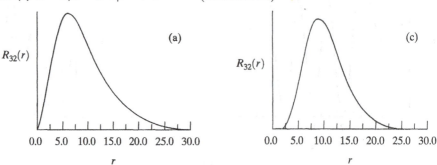

(b) This is the same as the $2p_z$ wavefunction that is plotted in Figure 6.4.

(c) $P_{32} \sim r^2 r^4 e^{-2Zr/3a_0}$.

6.3 Radius $= 0.5 \times 10^{-12}$ cm

$$\text{Prob} = \int_0^{\text{radius}} \frac{1}{\pi} \left(\frac{Z}{a_0} \right)^3 e^{-2Zr/a_0} r^2 dr \int \sin\theta d\theta \int d\phi$$

$$= 4 \left(\frac{Z}{a_0} \right)^3 \int_0^{\text{radius}} e^{-2Zr/a_0} r^2 dr$$

Note: $e^{-2Zr/a_0} \approx 1$.

$$\text{Prob} = \frac{4}{3} (\text{radius})^3 \left(\frac{Z}{a_0} \right)^3$$

$$= \frac{4}{3} \left(\frac{(\text{radius})Z}{a_0} \right)^3 = \frac{4}{3} \left(\frac{0.5 \times 10^{-12}\ \text{cm}}{0.5 \times 10^{-8}\ \text{cm}} \right)^3 = \frac{4}{3} \times 10^{-12}$$

6.5 Maxima are at $Zr = 1a_0$ for $1s$, $Zr = 0.764a_0$ and $5.236a_0$ for $2s$, and $4a_0$ for $2p$. The $2s$ orbital has the outermost maximum and is therefore more available for covalent bond formation in H atoms. The radial probability density is the same for $2p_x, 2p_y, 2p_+, 2p_-,$ and $2p_z$.

6.7 (a) The integral for the case of $(n, l, m) = (2, 1, m)$ is

$$\left\langle \frac{1}{r} \right\rangle = \frac{1}{64\pi} \left(\frac{Z}{a_0} \right)^5 \int_0^{2\pi} d\phi \int_0^\pi \sin^3\theta \, d\theta \int_0^\infty r^3 e^{-Zr/a_0} dr = \frac{Z}{4a_0}.$$

We can get $\langle T \rangle$ from $\langle T \rangle + \langle V \rangle = E = -Z^2/2n^2a_0$. Since $\langle V \rangle = -Z\langle 1/r \rangle = -Z^2/n^2a_0$, we get

$$\langle T \rangle = -\frac{Z^2}{2n^2a_0} + \frac{Z^2}{n^2a_0} = \frac{Z^2}{2n^2a_0} = -\frac{1}{2}\langle V \rangle,$$

which implies that $k = 2$.

(b) For a harmonic oscillator, $\langle T \rangle = \langle V \rangle$, which implies that $k = -1$.

6.9 (a) In first order, want to evaluate $\langle V \rangle = eE\langle z \rangle$. For the orbitals considered, we can show by symmetry that $\langle z \rangle = 0$, so there is no first-order contribution to the energy. In second order, symmetry again dictates that the only nonzero matrix element is $\langle 2s \mid z \mid 2p_z \rangle$, so only the $2s$ and $2p_z$ energies are shifted. This indicates that only $\Delta m = 0$ transitions can be induced by a field along the z-axis.

(b) Since $2p_x$ and $2p_y$ are not perturbed in either the first or the second order, the solutions $2p_x$ and $2p_y$ are still meaningful. If the perturbation had lifted the degeneracy, then it would not be meaningful to take linear combinations of non-degenerate states to define new solutions.

Chapter 7

7.1 $\varepsilon_1 = \langle \phi_1 \mid H \mid \phi_1 \rangle$

$\varepsilon_2 = \langle \phi_2 \mid H \mid \phi_2 \rangle$

$$\varepsilon_{trial} = \frac{\langle \phi_1 + c\phi_2 \mid H \mid \phi_1 + c\phi_2 \rangle}{\langle \phi_1 + c\phi_2 \mid \phi_1 + c\phi_2 \rangle}$$

$$= \frac{\varepsilon_1 + c^2 \varepsilon_2}{1 + c^2}$$

$$\frac{\partial \varepsilon_{trial}}{\partial c} = 0 = \frac{2c}{(1 + c^2)^2} \left[-\varepsilon_1 + \varepsilon_2 \right]$$

Extrema are at $c = 0$ and $c = \infty$. The following are possible results:

1. $\varepsilon_1 > \varepsilon_2$. Then $c = \infty$ gives $\varepsilon_{trial} = \varepsilon_2$, which is lower in energy than ε_1.
2. $\varepsilon_1 < \varepsilon_2$. Then $c = 0$ gives the lowest energy, $\varepsilon_{trial} = \varepsilon_1$.

Either way, $\varepsilon_{trial} \leq \varepsilon_1$.

7.3 (a) $H = -\dfrac{1}{2} \nabla_1^2 - \dfrac{1}{2} \nabla_2^2 - \dfrac{Z}{r_1} - \dfrac{Z}{r_2} + \dfrac{1}{r_{12}}$

$$\varepsilon_{trial} = \frac{1}{N} \int_0^\infty \int_0^\pi \int_0^{2\pi} \int_0^\infty \int_0^\pi \int_0^{2\pi} e^{-\alpha(r_1^2 + r_2^2)} H e^{-\alpha(r_1^2 + r_2^2)}$$

$$r_1^2 dr_1 \sin\theta_1 d\theta_1 d\phi_1 r_2^2 dr_2 \sin\theta_2 d\theta_2 d\phi_2$$

$$N = \int_0^\infty \int_0^\pi \int_0^{2\pi} \int_0^\infty \int_0^\pi \int_0^{2\pi} e^{-2\alpha(r_1^2 + r_2^2)}$$

$$r_1^2 dr_1 \sin\theta_1 d\theta_1 d\phi_1 r_2^2 dr_2 \sin\theta_2 d\theta_2 d\phi_2$$

(b) $E_{trial} = \dfrac{11}{4}\alpha - \dfrac{11\sqrt{\alpha}}{2}$

$$\frac{dE_{trial}}{d\alpha} = \frac{11}{4} - \frac{11}{2}\frac{1}{2\sqrt{\alpha}} = 0$$

$$\alpha = 1$$

$$E_{trial} = \frac{11}{4} - \frac{11}{2} = -\frac{11}{4} = -2.75$$

(c) $\phi = e^{-Z'(r_1 + r_2)}$ gives an optimized energy of -2.838. Since this is lower than E_{trial}, $\phi = e^{-Z'(r_1 + r_2)}$ is the better trial wavefunction.

Chapter 8

8.1 (a) No   **(b)** No   **(c)** No   **(d)** Yes

8.3 $\psi_{2s} = \dfrac{1}{\sqrt{4\pi}} R_{2s}(r)$ $\psi_{3s} = \dfrac{1}{\sqrt{4\pi}} R_{3s}(r)$

$$J_{2s,\,3s} = \int\int \psi_{2s}^2(1)\psi_{3s}^2(2)\,\frac{1}{r_{12}}\,d\tau_1 d\tau_2$$

$$= \int_0^\infty r_1^2 dr_1 \int_0^\infty r_2^2 dr_2 \int_0^\pi \sin\theta_1 d\theta_1 \int_0^{2\pi} d\phi_1 \int_0^\pi \sin\theta_2 d\theta_2 \int_0^{2\pi} d\phi_2$$

$$\frac{1}{(4\pi)^2} R_{2s}(r_1)^2 R_{3s}(r_2)^2 \frac{1}{\sqrt{r_1^2 + r_2^2 - 2r_1 r_2 \cos\chi}}$$

$$\cos\chi = \cos\theta_1 \cos\theta_2 + \sin\theta_1 \sin\theta_2 \cos(\phi_1 - \phi_2)$$

8.5 (a) $\psi = N(\alpha + c\beta)$

$$\langle \psi | \psi \rangle = N^2(1 + c^2) = 1 \qquad N = \frac{1}{\sqrt{1 + c^2}}$$

$$\psi = \frac{1}{\sqrt{1 + c^2}}(\alpha + c\beta)$$

(b) $\langle s_z \rangle = \dfrac{1}{1 + c^2}\langle \alpha + c\beta | s_z | \alpha + c\beta \rangle = \dfrac{1}{2}\hbar\dfrac{(1 - c^2)}{(1 + c^2)}$

(c) $H\psi = (As_z + Bs_x)\dfrac{1}{\sqrt{1 + c^2}}(\alpha + c\beta)$

$$= \frac{1}{\sqrt{1 + c^2}}\left[\frac{1}{2}\hbar A(\alpha - c\beta) + B\frac{\hbar}{2}(\beta + c\alpha)\right]$$

$$H\psi = \frac{1}{\sqrt{1 + c^2}}\frac{1}{2}\hbar\left[(A + cB)\alpha + (-Ac + B)\beta\right] = E\frac{1}{\sqrt{1 + c^2}}(\alpha + c\beta)$$

$$\frac{1}{2}\hbar(A + cB) = E \qquad\qquad \frac{1}{2}\hbar(-Ac + B) = cE$$

$$Bc^2 + 2cA - B = 0 \qquad\qquad c = -\frac{2A \pm \sqrt{4A^2 + 4B^2}}{2B}$$

$$= -\frac{A}{B} \pm \sqrt{\frac{A^2}{B^2} + 1}$$

$$E = \frac{1}{2}\hbar\left(A - A \pm \sqrt{A^2 + B^2}\right) = \pm\frac{1}{2}\hbar\sqrt{A^2 + B^2}$$

8.7 (a) $s_x = \dfrac{1}{2}\hbar\sigma_x$ $s_y = \dfrac{1}{2}\hbar\sigma_y$

$$[s_x, s_y] = \frac{\hbar^2}{4}\left[\begin{pmatrix} 0 & 1 \\ 1 & 0 \end{pmatrix}\begin{pmatrix} 0 & -i \\ i & 0 \end{pmatrix} - \begin{pmatrix} 0 & -i \\ i & 0 \end{pmatrix}\begin{pmatrix} 0 & 1 \\ 1 & 0 \end{pmatrix}\right]$$

$$= \frac{\hbar^2}{4}\left\{\begin{bmatrix} i & 0 \\ 0 & -i \end{bmatrix} - \begin{bmatrix} -i & 0 \\ 0 & i \end{bmatrix}\right\} = \frac{1}{2}\hbar^2\begin{bmatrix} i & 0 \\ 0 & -i \end{bmatrix}$$

$$= i\hbar\begin{bmatrix} \frac{1}{2}\hbar & 0 \\ 0 & -\frac{1}{2}\hbar \end{bmatrix} = i\hbar s_z$$

(b)

$$s^2 = s_x^2 + s_y^2 + s_z^2 = \left(\frac{1}{2}\hbar\right)^2\left\{\begin{pmatrix} 0 & 1 \\ 1 & 0 \end{pmatrix}\begin{pmatrix} 0 & 1 \\ 1 & 0 \end{pmatrix} + \begin{pmatrix} 0 & -i \\ i & 0 \end{pmatrix}\begin{pmatrix} 0 & -i \\ i & 0 \end{pmatrix} + \begin{pmatrix} 1 & 0 \\ 0 & 1 \end{pmatrix}\begin{pmatrix} 1 & 0 \\ 0 & 1 \end{pmatrix}\right\}$$

$$= \frac{3\hbar^2}{4}\left\{\begin{matrix} 1 & 0 \\ 0 & 1 \end{matrix}\right\} = \frac{3}{4}\hbar^2 I$$

$$s_z = \begin{pmatrix} \frac{1}{2}\hbar & 0 \\ 0 & -\frac{1}{2}\hbar \end{pmatrix}$$

(c) $s_x^2 s_y^2 s_z^2 = \left(\dfrac{1}{2}\hbar\right)^6\begin{pmatrix} 1 & 0 \\ 0 & 1 \end{pmatrix} = \dfrac{\hbar^6}{64}\begin{pmatrix} 1 & 0 \\ 0 & 1 \end{pmatrix} = \dfrac{1}{27}s^6$

8.9 From Eq. (8.52), the transition energy is $\Delta E = g_N\beta_N B = h\nu$, so if $\nu = 500$ MHz $= 5 \times 10^8$ s^{-1}, then $B = (6.625 \times 10^{-34}$ Js$)(5 \times 10^8$ s$^{-1})/(5.586)$ $(5.05079 \times 10^{-27}$ J/T$) = 11.7$ T.

8.11 E (fermi) $= Am_s m_N$, so we get the following energy levels:

E (fermi)	m_s	m_N
$\frac{1}{4}A$	1/2	1/2
$-\frac{1}{4}A$	1/2	-1/2
$-\frac{1}{4}A$	-1/2	1/2
$\frac{1}{4}A$	-1/2	-1/2

ΔE (fermi) $= \frac{1}{2}A = 0.02381$ cm^{-1}. The wavelength would be 42.0 cm.

Chapter 9

9.1 $d\tau = r_1^2 dr_1 \sin\theta_1 d\theta_1 d\phi_1 r_2^2 dr_2 \sin\theta_2 d\theta_2 d\phi_2 r_3^2 dr_3 \sin\theta_3 d\theta_3 d\phi_3$.

9.3 $2K = -1.525 + 1.752 = 0.227$ hartree.

9.5 $\langle \alpha_1 \beta_2 | S_{1+} S_{2-} + S_{1-} S_{2+} | \beta_1 \alpha_2 \rangle = \hbar^2$

$\langle \alpha_1 \alpha_2 | S^2 | \alpha_1 \alpha_2 \rangle = \langle \alpha_1 \alpha_2 | S_1^2 + S_2^2 + 2(S_{1z} S_{2z} + S_{1x} S_{2x} + S_{1y} S_{2y}) | \alpha_1 \alpha_2 \rangle$

$= \hbar^2 \left(\dfrac{1}{2} \right) \left(\dfrac{1}{2} + 1 \right) + \hbar^2 \dfrac{1}{2} \left(\dfrac{1}{2} + 1 \right) + 2 \left(\dfrac{1}{2} \hbar \right) \left(\dfrac{1}{2} \hbar \right)$

$= 2\hbar^2$

9.7 **(a)** Nearly good,   **(b)** Nearly good,   **(c)** good,   **(d)** good,   **(e)** neither.

9.9 **(a)** F   **(b)** F

9.11 triplets $^3P_2$ singlets $^1F_3$

9.13 $^3P_2$

Chapter 10

10.1 **(a)** T   **(b)** F

10.3

(a)   (b)

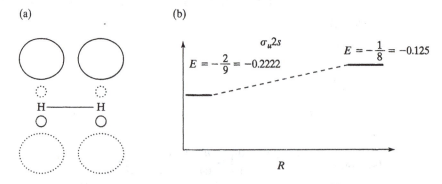

10.5

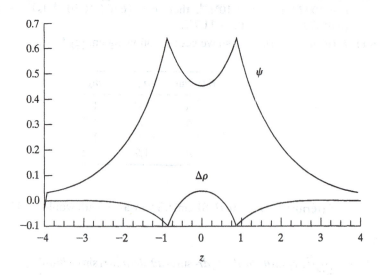

$\Delta\rho$ is positive near the midpoint of the bond, but negative elsewhere.

10.7 $\psi = \dfrac{1}{\sqrt{4!}} \begin{vmatrix} \sigma_g 1s(1)\alpha_1 & \sigma_g 1s(1)\beta_1 & \sigma_u 1s(1)\alpha_1 & \sigma_u 1s(1)\beta_1 \\ \sigma_g 1s(2)\alpha_2 & \sigma_g 1s(2)\beta_2 & \sigma_u 1s(2)\alpha_2 & \sigma_u 1s(2)\beta_2 \\ \sigma_g 1s(3)\alpha_3 & \sigma_g 1s(3)\beta_3 & \sigma_u 1s(3)\alpha_3 & \sigma_u 1s(3)\beta_3 \\ \sigma_g 1s(4)\alpha_4 & \sigma_g 1s(4)\beta_4 & \sigma_u 1s(4)\alpha_4 & \sigma_u 1s(4)\beta_4 \end{vmatrix}$

10.9 VB: more natural description of electron pair bonds (covalent wavefunction)—dissociates correctly. Disadvantage: not a single determinant wavefunction.

10.11 $\psi = \dfrac{1}{2}(\alpha_1\beta_2 + \beta_1\alpha_2)(\sigma_g(1)\sigma_u(2) - \sigma_u(1)\sigma_g(2))$

$= \dfrac{1}{\sqrt{2}}\left\{ \dfrac{1}{\sqrt{2}} \begin{vmatrix} \sigma_g(1)\alpha_1 & \sigma_u(1)\beta_1 \\ \sigma_g(2)\alpha_2 & \sigma_u(2)\beta_2 \end{vmatrix} + \dfrac{1}{\sqrt{2}} \begin{vmatrix} \sigma_g(1)\beta_1 & \sigma_u(1)\alpha_1 \\ \sigma_g(2)\beta_2 & \sigma_u(2)\alpha_2 \end{vmatrix} \right\}$

Chapter 11

11.1 (a) F   **(b)** F   **(c)** F   **(d)** F

11.3 (a) n   **(b)** y   **(c)** y   **(d)** y

11.5 (a) $\dfrac{1}{\sqrt{2}} \begin{vmatrix} 1a(1)\alpha_1 & 1b(1)\beta_1 \\ 1a(2)\alpha_2 & 1b(2)\beta_2 \end{vmatrix} = \dfrac{1}{\sqrt{2}}(1a(1)1b(2)\alpha_1\beta_2 - 1b(1)1a(2)\alpha_2\beta_1)$

(b) Neither. This is a linear combination of $S = 0$ and $S = 1$ with $M_s = 0$ in both cases.

11.7

$$\langle R \rangle = \left\langle \psi \left| \sum_i R_i \right| \psi \right\rangle$$

H_2 ground-state wavefunction:

$$\psi = \sigma_g(1)\sigma_g(2)(\alpha_1\beta_2 - \beta_1\alpha_2)/\sqrt{2}$$

$$\langle R \rangle = \left\langle \sigma_g(1)\sigma_g(2) \left| \sum_i R_i \right| \sigma_g(1)\sigma_g(2) \right\rangle$$

$$= \sum_{i=1}^{2} \langle \sigma_g(i) | R_i | \sigma_g(i) \rangle = 2\langle \sigma_g(a) | R_a | \sigma_g(a) \rangle$$

Now do the same derivation as for the Slater determinant:

$$\left(\dfrac{1}{\sqrt{2}}\right)^2 \left\langle \begin{vmatrix} \sigma_g(1)\alpha_1 & \sigma_g(1)\beta_1 \\ \sigma_g(2)\alpha_2 & \sigma_g(2)\beta_2 \end{vmatrix} \left| \sum_i R_i \right| \begin{vmatrix} \sigma_g(1)\alpha_1 & \sigma_g(1)\beta_1 \\ \sigma_g(2)\alpha_2 & \sigma_g(2)\beta_2 \end{vmatrix} \right\rangle$$

$$= \dfrac{1}{2}\langle \sigma_g(1)\alpha_1 | R_1 | \sigma_g(1)\alpha_1 \rangle \times \text{(integral of cofactor)}$$

$$+ \dfrac{1}{2}\langle \sigma_g(1)\beta_1 | R_1 | \sigma_g(1)\beta_1 \rangle \times \text{(integral of cofactor)}$$

$$+ \text{similar terms involving } R_2$$

$$= 2\langle \sigma_g(a) | R_a | \sigma_g(a) \rangle$$

The general derivation is

$$\langle R \rangle = \frac{1}{N!} \left[\left\langle |\det| \left| \sum_i R_i \right| |\det| \right\rangle \right. = \frac{1}{N!} \langle \phi_1(1)\alpha_1 | R_1 | \phi_1(1)\alpha_1 \rangle \times \text{cofactor}$$

$$+ \langle \phi_1(1)\beta_1 | R_1 | \phi_1(1)\beta_1 \rangle \times \text{cofactor} + \langle \phi_2(1)\alpha_1 | R_1 | \phi_2(1)\alpha_1 \rangle$$

$$\left. \times \text{cofactor} + \cdots + \text{same for } R_2 + \text{same for } R_3 + \cdots \right]$$

Each cofactor integrates to $(N - 1)!$.

$$\langle R \rangle = \frac{1}{N!} (N - 1)! N \big[2\langle \phi_1(1) | R_1 | \phi_1(1) \rangle + 2\langle \phi_2(1) | R_2 | \phi_2(1) \rangle + \cdots \big]$$

$$= 2 \sum_j \langle \phi_j(1) | R_1 | \phi_j(1) \rangle$$

Chapter 12

12.1 (a) y   **(b)** n   **(c)** y   **(d)** n

12.3 (a) y   **(b)** y   **(c)** n   **(d)** n

12.5 (a) n   **(b)** n   **(c)** n   **(d)** y

12.7 (a) y   **(b)** n   **(c)** n   **(d)** n

12.9

$$\begin{vmatrix} \alpha - E & \beta & 0 & \beta \\ \beta & \alpha - E & \beta & 0 \\ 0 & \beta & \alpha - E & \beta \\ \beta & 0 & \beta & \alpha - E \end{vmatrix} = 0$$

$$\begin{vmatrix} x & 1 & 0 & 1 \\ 1 & x & 1 & 0 \\ 0 & 1 & x & 1 \\ 1 & 0 & 1 & x \end{vmatrix} = 0 = x \begin{vmatrix} x & 1 & 0 \\ 1 & x & 1 \\ 0 & 1 & x \end{vmatrix} - \begin{vmatrix} 1 & 1 & 0 \\ 0 & x & 1 \\ 1 & 1 & x \end{vmatrix} - \begin{vmatrix} 1 & x & 1 \\ 0 & 1 & x \\ 1 & 0 & 1 \end{vmatrix}$$

$$= x \big[x^3 - 2x \big] - \big[x^2 + 1 - 1 \big] - \big[1 + x^2 - 1 \big]$$

$$= x^4 - 2x^2 - x^2 - x^2 = 0$$

$$x^4 - 4x^2 = 0$$

$$x^2(x^2 - 4) = 0 \qquad x = 0, 0, \pm 2$$

$$E = \alpha, \alpha, \alpha + 2\beta, \alpha - 2\beta$$

(a) Eigenfunction for $E = \alpha$

$$\begin{pmatrix} \alpha - \alpha & \beta & 0 & \beta \\ \beta & \alpha - \alpha & \beta & 0 \\ 0 & \beta & \alpha - \alpha & \beta \end{pmatrix} \begin{pmatrix} c_1 \\ c_2 \\ c_3 \\ c_4 \end{pmatrix} = 0$$

First Solution: $c_2 + c_4 = 0$

$$c_1 + c_3 = 0$$

Choose $c_1 = 1, c_2 = 0,$ $\psi = \dfrac{1}{\sqrt{2}}(p_1 - p_3)$ $(E = \alpha)$

$$c_3 = -1, c_4 = 0$$

Second Solution: $c_1 = 0, c_2 = 1,$ $\psi = \dfrac{1}{\sqrt{2}}(p_2 - p_4)$ $(E = \alpha)$

$$c_3 = 0, c_4 = -1$$

Third Solution: $E = \alpha + 2\beta$

$$-2c_1 + c_2 + c_4 = 0 \qquad \psi = \frac{1}{2}(p_1 + p_2 + p_3 + p_4) \qquad (E = \alpha + 2\beta)$$
$$c_1 - 2c_2 + c_3 = 0$$
$$c_2 - 2c_3 + c_4 = 0$$

Fourth Solution: $E = \alpha - 2\beta$

$$2c_1 + c_2 + c_4 = 0 \qquad \psi = \frac{1}{2}(p_1 - p_2 + p_3 - p_4) \qquad (E = \alpha - 2\beta)$$

$$c_1 + 2c_2 + c_3 = 0$$
$$c_2 + 2c_3 + c_4 = 0$$

(b) $E_\pi = 2(\alpha + 2\beta) + 2\alpha = 4\alpha + 4\beta.$

(c) π bond order $= 2(1/2)(1/2) = 1/2$ (same for any adjacent pair).
(d) charge $= 2(1/2)(1/2) = 1/2$.
(e) delocalization energy $= 4\alpha + 4\beta - 4(\alpha + \beta) = 0$.
(f) spin multiplicity $= 3$.
12.11 (a) The secular equation is

$$\begin{vmatrix} \alpha - E & \beta & 0 \\ \beta & 0.85\alpha - E & \beta \\ 0 & \beta & \alpha - E \end{vmatrix} = 0$$

The eigenvalues are

$$E_3 = 0.95\alpha - \sqrt{2}\beta\sqrt{1 + 0.0028\frac{\alpha^2}{\beta^2}}$$

$$E_2 = \alpha$$

$$E_1 = 0.95\alpha + \sqrt{2}\beta\sqrt{1 + 0.0028\frac{\alpha^2}{\beta^2}}$$

The total energy is $E_{\text{tot}} = 2[0.95\alpha + \sqrt{2}\beta\sqrt{1 + 0.0028(\alpha^2/\beta^2)}] + \alpha$.

The transition energy is $\Delta E = 0.05\alpha - \sqrt{2}\beta\sqrt{1 + 0.0028(\alpha^2/\beta^2)}$.

(b) Use wavefunctions from Eq. (12.19). The perturbation Hamiltonian is

$$\begin{vmatrix} 0 & 0 & 0 \\ 0 & -0.15\alpha & 0 \\ 0 & 0 & 0 \end{vmatrix} = 0.$$

Plug into first-order perturbation theory to get

$$E_3 = 0.925\alpha - \sqrt{2}\beta$$

$$E_2 = \alpha$$

$$E_1 = 0.925\alpha + \sqrt{2}\beta$$

so that $E_{\text{tot}} = 2(0.925\alpha + \sqrt{2}\beta) + \alpha$ and $\Delta E = 0.075\alpha - \sqrt{2}\beta$.

12.13 (a)

(a)

θ

H_B H_A

$$\begin{array}{c} 1s_A \\ 1s_B \\ 2p_x \\ 2p_y \\ 2p_z \\ 2s \end{array} \begin{vmatrix} \alpha_H - E & 0 & \beta\cos(\theta/2) & \beta\sin(\theta/2) & 0 & \beta' \\ 0 & \alpha_H - E & \beta\cos(\theta/2) & -\beta\sin(\theta/2) & 0 & \beta' \\ \beta\cos(\theta/2) & \beta\cos(\theta/2) & \alpha_0 - E & 0 & 0 & 0 \\ \beta\sin(\theta/2) & -\beta\sin(\theta/2) & 0 & \alpha_0 - E & 0 & 0 \\ 0 & 0 & 0 & 0 & \alpha_0 - E & 0 \\ \beta' & \beta' & 0 & 0 & 0 & \alpha_0' - E \end{vmatrix}$$

(b)

$$\begin{array}{c} (1s_A + 1s_B)/\sqrt{2} \\ 2p_x \\ 2s \\ (1s_A - 1s_B)/\sqrt{2} \\ 2p_y \\ 2p_z \end{array} \begin{vmatrix} \alpha_H - E & \sqrt{2}\beta\cos(\theta/2) & \sqrt{2}\beta' & 0 & 0 & 0 \\ \sqrt{2}\beta\cos(\theta/2) & \alpha_0 - E & 0 & 0 & 0 & 0 \\ \sqrt{2}\beta' & 0 & \alpha_0' - E & 0 & 0 & 0 \\ 0 & 0 & 0 & \alpha_H - E & \sqrt{2}\beta\sin(\theta/2) & 0 \\ 0 & 0 & 0 & \sqrt{2}\beta\sin(\theta/2) & \alpha_0 - E & 0 \\ 0 & 0 & 0 & 0 & 0 & \alpha_0 - E \end{vmatrix}$$

(c) Secular equations

$$\begin{vmatrix} \alpha - E & \sqrt{2}\beta\cos(\theta/2) \\ \sqrt{2}\beta\cos(\theta/2) & \alpha - E \end{vmatrix} = 0$$

$$\begin{vmatrix} \alpha - E & \sqrt{2}\beta\sin(\theta/2) \\ \sqrt{2}\beta\sin(\theta/2) & \alpha - E \end{vmatrix} = 0 \qquad |\alpha - E| = 0$$

Energies: $E = \alpha \pm \sqrt{2}\beta \cos(\theta/2)$ $E = \alpha \pm \sqrt{2}\beta \sin(\theta/2)$ $E = \alpha$

$$\underline{\qquad}\ \alpha - \sqrt{2}\beta\sin(\theta/2)$$

$$\underline{\qquad}\ \alpha - \sqrt{2}\beta\cos(\theta/2)$$

$\underline{\text{#}}\ \alpha$

$\underline{\text{#}}\ \alpha + \sqrt{2}\beta\cos(\theta/2)$

$\underline{\text{#}}\ \alpha + \sqrt{2}\beta\sin(\theta/2)$

$$E_{TOT} = 6\alpha + 2\sqrt{2}\beta\sin(\theta/2) + 2\sqrt{2}\beta\cos(\theta/2)$$

Min is at $\theta = 90°$ (when $\theta > 90°$ is constrained).

(d) Revised secular equation:

$$\begin{vmatrix} \alpha - E & \sqrt{2}\beta\cos(\theta/2) & \sqrt{2}\beta \\ \sqrt{2}\beta\cos(\theta/2) & \alpha - E & 0 \\ \sqrt{2}\beta & 0 & \alpha - E \end{vmatrix} = 0$$

$$(\alpha - E)^3 - (\alpha - E)2\beta^2 - (\alpha - E)2\beta^2\cos^2(\theta/2) = 0$$

$$(\alpha - E)[(\alpha - E)^2 - 2\beta^2(1 + \cos^2(\theta/2))] = 0$$

$$E = \alpha, \alpha \pm \sqrt{2}\beta\sqrt{1 + \cos^2(\theta/2)}$$

$$\underline{\qquad}\ \alpha - \sqrt{2}\beta\sin(\theta/2)$$

$$\underline{\qquad}\ \alpha - \sqrt{2}\beta\sqrt{\cos^2(\theta/2) + 1}$$

$\underline{\text{###}}\ \alpha$

$\underline{\text{#}}\ \alpha + \sqrt{2}\beta\sqrt{\cos^2(\theta/2) + 1}$

$\underline{\text{#}}\ \alpha + \sqrt{2}\beta\sin(\theta/2)$

$$E_{TOT} = 8\alpha + 2\sqrt{2}\beta\sin(\theta/2) + 2\sqrt{2}\beta\sqrt{\cos^2(\theta/2) + 1}$$

Minimum is at $\theta = 180°$.

12.15 Expect that overlap should rise monotonically from $S_{\mu\nu} \to 0$ for $R \to \infty$ to $S_{\mu\nu} \to 1$ for $R \to 0$. Indeed, this is just the behavior of the exact result,

$$S_{ab} = e^{-R}\left\{\frac{R^2}{3} + R + 1\right\}.$$

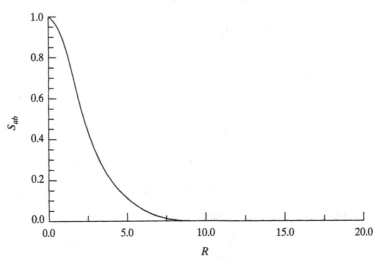

Chapter 13

13.1 The ground term is $^1A'$, while $(e'')^2$ gives rise to $^3A_2'$, $^1E'$, and $^1A_1'$. On the basis of the dipole selection rule, only singlet $\to$ singlet transitions are allowed, and from Table 13.2, only transitions transforming as E' or A_2'' are allowed. This indicates that only the $^1E'$ state is optically connected to the ground state.

13.3

| | I(D_{2h}) | II(D_{4h}) | III(D_{2h}) |

(a) Energy Levels:

	B_{2g}, B_{3u} $\alpha - \beta$	B_{2g} $\alpha - 2\beta$	B_{2g}, B_{1u} $\alpha - \beta$
		E_u α	
	A_g, B_{1u} $\alpha + \beta$	A_{1g} $\alpha + 2\beta$	B_{3u}, A_g $\alpha + \beta$

Many-electron terms: $^1A_{1g}$ $(e_u)^2 = {}^1A_{1g}, {}^1B_{1g}, {}^3A_{2g}, {}^3B_{2g}$ $^1A_{1g}$

Total energy: $4\alpha + 4\beta$ $4\alpha + 4\beta$ 4α

(b) Adiabatic Correlations:

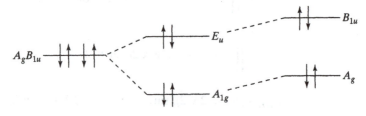

13.5 (a) $\Gamma = 2A_1 + B_2$

$\psi_{a_1^{(1)}} = s_2$

$\psi_{a_1^{(2)}} = 2^{-1/2}(s_1 + s_3)$

$\psi_{b_2} = 2^{-1/2}(s_1 - s_3)$

$E_{a_1} = \alpha + \beta(\gamma \pm (2 + \gamma^2)^{1/2})$

$E_{b_2} = \alpha - 2\beta\gamma$ $\gamma = 1 - \sin(\theta/2)$

Note that this gives the expected results for the limiting cases of 60° and 180°, namely, $\alpha + 2\beta$, $\alpha - \beta$, and $\alpha - \beta$ for 60° and $\alpha - \sqrt{2}\beta$, α, and $\alpha + \sqrt{2}\beta$ for 180°.

(b) $E_{tot} = 2E_{a_1}^{(+)} + E_{b_2}$

$= 3\alpha + 2\beta(2 + \gamma^2)^{1/2}$

Considering $60° \le \theta \le 180°$ only, we see that energy minimizes for $\theta = 60°$. $E_{tot} = 3\alpha + 3\beta$. (However, this is not the physically correct result). For $\theta = 180°$, $E_{tot} = 3\alpha + 2\sqrt{2}\beta$. For H_3^+, a minimum again occurs at 60° (which is physically correct). $E_{tot} = 2\alpha + 4\beta$.

(c) $\psi_{b_2} = 2^{-1/2}(s_1 - s_3)$

$\psi_{a_1} = (2 + \gamma^2)^{-1/4}(s_2/((2 + \gamma^2)^{1/2} \pm \gamma)^{1/2} \pm 1/2(s_1 + s_3)((2 + \gamma^2)^{1/2} \pm \gamma)^{1/2})$

(d)

$D_{\infty h}$	C_{2v}	D_{3h}
σ_g	a_1	a_1'
σ_u	b_2	
σ_g	a_1	e'

(e)

H_3	H_3^+	H_3^-
$(a_1')^2 e'$	$(a_1')^2$	$(a_1')^2(e')^2$
$^2E'$	$^1A_1'$	$^1A_1', {}^3A_2', {}^1E'$

13.7 The representation of the C atom may be obtained by inspection from Table 13.2 (looking at the (x, y) and z entries) and is $E' + A_2''$. For the three H's, we have the following:

D_{3h}	E	$2C_3$	$3C_2$	σ_h	$2S_3$	$3\sigma_v$
Three H's	9	0	−1	3	0	1

This reduces to $A_1' + A_2' + 2E' + A_2'' + E''$, so for all four atoms, we get $A_1' + A_2' + 3E' + 2A_2'' + E''$. Now, subtract the three translations $(E' + A_2'')$ and three rotations $(A_2' + E'')$ to get $A_1' + A_2'' + 2E'$ for the six vibrational modes.

The corresponding treatment for C_{3v} uses the character tables in Appendix C. The C atom reduces to $A_1 + E$ and the three H's reduce to $2A_1 + A_2 + 3E$, so the sum is $3A_1 + A_2 + 4E$. Subtracting the translations $(A_1 + E)$ and the rotations $(A_2 + E)$ gives $2A_1 + 2E$ for the six vibrations. The correlations with D_{3h} are

$$A_1' \rightarrow A_1$$

and

$$A_2'' \rightarrow A_1 \qquad E'' \rightarrow E.$$

Chapter 14

14.1 The results of CIS calculations of the ground and first four excited singlet states, generated with a 6-311G** basis set, are shown in Figure S14.1. Note that polarization functions on hydrogen are necessary if one is to have π orbitals in the basis set, so 6-31G would not be sufficient to give the fourth excited state. The results are in qualitative agreement with expectations (cf. Figure 10.8), with the first excited state corresponding to the $\sigma_u 1s$ or $1\Sigma_u^+$ level, the second to $\sigma_g 2s$ or $2\Sigma_g^+$, the third to $\sigma_u 2s$ or $2\Sigma_u^+$, and the fourth to $\pi_u 2p$ or $1\Pi_u$. Note also that the first and third excited states are repulsive, while the second and fourth have minima. The $\pi_u 2p$ level is below $\sigma_g 2p$, which is typical for a light homonuclear diatomic. Comparison with experiment (Huber and Herzberg) is qualitative. Only the bound excited states are listed in Huber and Herzberg, and of these, the first is the B state (0.42-hartree excitation energy from the ground state), which corresponds to $\sigma_u 2p$. The second bound state is the C state (0.46-hartree excitation energy), correponding to $\pi_u 2p$, and the third is the E state (0.46-hartree excitation energy), corresponding to $\sigma_g 2s$.

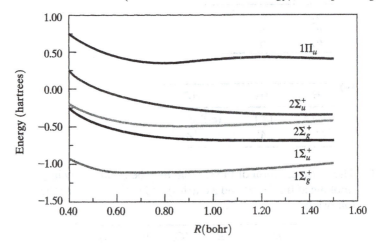

Figure S14.1

14.3 Table S14.1 presents the infrared and Raman spectra of benzene, as obtained from HF, MP2, and B3LYP calculations with a 6-31G basis set. For each method, the frequencies (in wavenumbers) are listed first, then the IR intensities, and then the

Raman intensities. Experimental results are listed at the far right. Note that there are no modes that are both IR and Raman allowed. This reflects the fact that IR spectroscopy is a one-photon process, and the dipole matrix element that determines this intensity requires a change in symmetry of each state in order to be allowed. By contrast, Raman spectroscopy is a two-photon spectroscopy, and as a result, the dipole matrix element is replaced by a polarizability matrix element whose selection rule requires initial and final states that are both even or both odd. In addition, of the $3N - 6 = 30$ possible vibrational normal modes, Table S14.1 shows only four IR lines and seven Raman lines, some of which may be degenerate. (The degeneracy factor is listed in the first column of the table.) The total number of states associated with these 11 levels is 19. The difference between 30 total modes and 19 observed in IR and Raman is accounted for by modes that are both IR and Raman forbidden. The comparison with experiment is very good for the IR intensities, but only fair for the Raman intensities. This demonstrates that a larger basis set is needed to generate accurate Raman intensities than is required to produce IR intensities.

TABLE S14.1 FREQUENCIES (CM^{-1}), INFRARED (IR) AND RAMAN (RAM) INTENSITIES FOR BENZENE

deg	hf freq	hf–ir	hf–ram	mp2 freq	mp2–ir	mp2– ram	b3lyp freq	b3–ir	b3–ram	exp freq	IR (Bertie)	Ram (Koput)
2	700	0.00	2.29	632	0.00	3.59	640	0.00	3.59	608	0	3
1	811	24.59	0.00	672	114.49	0.00	706	102.63	0.00	674	94	0
2	1,038	0.00	6.21	828	0.00	6.26	884	0.00	5.94	847	0	1
1	1,173	0.00	36.93	1,001	0.00	53.53	1,022	0.00	54.94	993	0	36
2	1,226	0.22	0.00	1,068	3.43	0.00	1,078	3.84	0.00	1,038	9	0
2	1,371	0.00	12.66	1,239	0.00	9.07	1,232	0.00	9.65	1,178	0	3
2	1,773	15.17	0.00	1,526	8.75	0.00	1,542	8.71	0.00	1,484	14	0
2	1,933	0.00	21.62	1,626	0.00	11.63	1,654	0.00	12.93	1,610	0	7
2	3,723	0.00	75.88	3,181	0.00	124.58	3,194	0.00	134.24	3,057	0	34
2	3,737	8.81	0.00	3,200	53.23	0.00	3,213	56.45	0.00	3,064	73	0
1	3,748	0.00	191.95	3,214	0.00	345.66	3,227	0.00	369.35	3,074	0	50

14.5 Table S14.2 shows the energy and geometry of ozone, based on 6–31G calculations. For the ground state, HF, MP2, B3LYP, and experimental results are presented. For the singlet excited states, the CIS energies given are the vertical excitation energies using the HF geometry, while the O–O distance and internal angle refer to a CIS calculation in which the minimum-energy geometry is determined. Here we compare the present CIS results with multireference coupled-cluster calculations [N. Vaval and S. Pal, *JCP* 111, 4051 (1999)] and with experiment [J. Steinfeld, S. Golden, and J. Gallagher, *J. Phys. Chem. Ref. Data* 16, 911 (1987)]. The lowest excited state in the CIS calculations has B_1 symmetry and corresponds to excitation of the highest occupied a_1 symmetry orbital to the b_1 (out-of-plane) orbital. The second excited CIS state has A_2 symmetry and corresponds to exciting the highest occupied b_2 orbital to the b_1 unoccupied state. Neither of these states has a D_{3h} minimum, but the second excited state is strongly bent. Note that the order of the excited-state energies is reversed in the higher level calculation and in experiment.

TABLE S14.2   ENERGY AND GEOMETRY OF OZONE

Method	Energy (hartrees)	O–O distance (Å)	Internal angle (degrees)
HF	−224.138626	1.2506	119.6
MP2	−224.618582	1.3567	115.2
B3LYP	−225.323570	1.3215	117.7
Experiment		1.2717	116.7
First excited state:			
CIS	1.80 eV (exc. energy)	1.2904	134.6
Vaval and Pal/Expt	1.98/2.1 eV		
Second excited state:			
CIS	2.85 eV(exc. energy)	1.3173	100.0
Vaval and Pal/Expt	1.87/1.6 eV		

14.7 Table S14.3 compares energies, geometries, and frequencies for the minimum and
for the *cis* and *trans* isomers of HOOH. Each structure was calculated with a

TABLE S14.3   ENERGY AND GEOMETRY OF THE HYDROGEN PEROXIDE MINIMUM AND *CIS* AND
TRANS BARRIERS

Method	Energy (hartrees)	O–O distance (Å)	O–H distance (Å)	HOO angle (degrees)	Dihedral angle (degrees)
HF–min	−150.7647869	1.3964	0.9492	102.07	116.00
HF–*trans*	−150.7632795	1.4058	0.9485	100.56	180.
HF–*cis*	−150.7502011	1.4039	0.9489	106.63	0.
cis–trans	8.2 kcal/mol				
trans–min	0.95 kcal/mol				
MP2–min	−151.1301264	1.4683	0.9756	98.65	121.17
MP2–*trans*	−151.1291702	1.4783	0.9754	97.23	180.
MP2–*cis*	−151.1151022	1.4755	0.9762	104.02	0.
cis–trans	8.8 kcal/mol				
trans–min	0.60 kcal/mol				
MP4–min	−151.1508028	1.4789	0.9768	98.66	121.11
MP4–*trans*	−151.1498713	1.4895	0.9764	97.26	180.
MP4–*cis*	−151.1362919	1.4881	0.9779	103.80	0.
cis–trans	8.5 kcal/mol				
trans–min	0.58 kcal/mol				
B3LYP–min (6–31G)	−151.4910832	1.5313	0.9835	97.92	180.
B3LYP–min (6–311G)	−151.545579	1.5229	0.9779	98.47	174.75
B3LYP–min cc-PVTZ	−151.6116816	1.4520	0.9659	100.39	113.98
B3LYP–min (6–31G*)	−151.533213	1.4558	0.9736	99.66	118.73
B3LYP–*trans*	−151.5320845	1.4665	0.9725	98.10	180.
B3LYP–*cis*	−151.5190728	1.4660	0.9740	104.63	0.
cis–trans	8.2 kcal/mol				
trans–min	0.71kcal/mol				
expt:					
cis–trans	6.0 kcal/mol				
trans–min	1.1 kcal/mol				
Harding–min		1.462	0.964	99.6	113.4
Harding–*trans*		1.472	0.963	98.3	180.
Harding–*cis*		1.471	0.963	104.1	0.

6-31G* basis, and then the B3LYP method was used for several basis sets to demonstrate convergence behavior. Note that 6-31G gives a planar minimum structure (i.e., the *trans* structure is a minimum). Larger basis sets are needed to give a nonplanar minimum. The results show that, for the ground-state properties, the MP2, MP4, and B3LYP structural parameters are all similar to each other. Also, the barriers are all very much the same—even HF. At the bottom of the table are results from experiment and high-level theory (from Harding's paper). The calculated results are in good agreement with experiment and Harding's results.

A CIS calculation at the HF ground-state geometry predicts that the lowest excited state is at 7.81 eV, while the second excited state is at 9.39 eV. The lowest excitation is from the HOMO (an out-of-plane π excited state) to unoccupied in-plane states (nominally, σ^*).

14.9 Table S14.4 compares HF, MP2, MP4, and B3LYP results for acetylene and vinylidene, using a 6-31G* basis set. ΔE is the energy difference between the respective minima of acetylene and vinylidene. At the bottom are results from either experiment (acetylene, and the vinylidene ΔE value from the Lineberger paper) or high-level calculations (the vinylidene structure from the Schaefer paper). The table shows good agreement of the calculated structural properties with each other and with experiment, for both species. The HF ΔE value is off by over 10 kcal/mol, but the other results are closer, and the MP4 and B3LYP results are within the experimental error bars.

TABLE S14.4 ENERGY AND GEOMETRY OF ACETYLENE AND VINYLIDENE

Method	Energy (hartrees)	C–C distance (Å)	C–H distance (Å)	H–C–C angle (degrees)
HF–acetylene	−76.8178265	1.1854	1.0569	180.
HF–vinylidene	−76.7633954	1.2935	1.0779	120.26
ΔE (kcal/mol)	34.2			
MP2–acetylene	−77.0667935	1.2176	1.0661	180.
MP2–vinylidene	−76.9855738	1.3085	1.0879	120.46
ΔE (kcal/mol)	51.0			
MP4–acetylene	−77.0935971	1.2196	1.0695	180.
MP4–vinylidene	−77.0218428	1.3154	1.0919	120.78
ΔE (kcal/mol)	45.0			
B3LYP–acetylene	−77.3256462	1.2050	1.0666	180.
B3LYP–vinylidene	−77.2597943	1.3041	1.0906	120.87
ΔE (kcal/mol)	41.3			
Exp.–acetylene		1.203	1.061	180.
Exp.–vinylidene	$\Delta E = 44 \pm 4$	1.300	1.081	120.0
	(Lineberger)	(Schaefer)	(Schaefer)	(Schaefer)

14.11 Figure S14.2 shows the UHF and RHF energies as a function of R. Note that they are the same for $R = 1.2a_0$ and smaller, but above that they are different, with the UHF result approaching the correct asymptote for large R and the RHF exhibiting unphysical behavior because of the equal mixing of ionic and covalent configurations. The reason for the correct behavior at large R is that UHF allows up and down spins to have different orbitals, and at large R this leads to the localization of one spin around one nucleus and the other spin around the other nucleus.

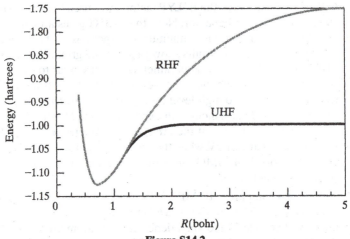

Figure S14.2

Figure S14.3 shows UHF, UMP2, and B3LYP curves for the OH radical. Here, all the curves dissociate to different asymptotes that are determined by the energy associated with O + H for the given level of theory. Table S14.5 lists the comparisons between the different levels of theory for the equilibrium geometry and frequency. Here, the MP2 results are the closest to experiment.

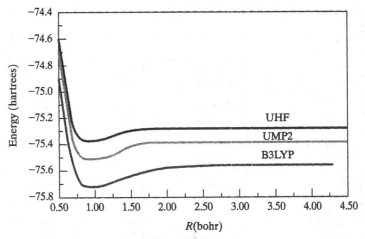

Figure S14.3

TABLE S14.5   ENERGY AND GEOMETRY OF OH

Method	Energy (hartrees)	$r_e(\text{Å})$	$\omega_e(\text{cm}^{-1})$
UHF	−75.3822753	0.9585	3,997
UMP2	−75.5225720	0.9797	3,739
B3LYP	−75.7234548	0.9831	3,643
Exp (Huber&Herzberg)		0.9706	3,735

14.13 The cover of the book shows a plot of energy versus twist angle for the ground state $(^1A_1)$ and the $^1B_1$ excited state based on HF and CIS calculations with a 6-31G basis set. The ground state has a twisted minimum, but the excited state is planar. Thus fluorescence should come from the planar geometry. This result agrees with higher level calculations, and with experiment, although the real picture is much more complex due to the presence of other excited states.

Chapter 15

15.1 From Eq. (15.4), we have

$$\langle \Psi(\mathbf{x}, t) | \Psi(\mathbf{x}, t) \rangle = \left\langle \sum_m C_m(t)\phi_m(\mathbf{x}) \Big| \sum_n C_n(t)\phi_n(\mathbf{x}) \right\rangle$$

$$= \sum_m \sum_n C_m(t)C_n(t)\langle \phi_m(\mathbf{x}) | \phi_n(\mathbf{x}) \rangle$$

$$= \sum_m \sum_n C_m(t)C_n(t)\delta_{m,n}$$

$$= \sum_m |C_m(t)|^2 = 1.$$

This shows that the norm of the wavefunction, which is a constant that we set equal to unity, is equal to the sum of the squares of the coefficients, $|C_m|^2$. Consequently, each of the squared coefficients can be interpreted at the probability of being in state m at time t.

15.3 The ground configuration $(a_2'')^2$ leads to the term $^1A'$, while $(a_2'')^1(e'')^1$ leads to $^{1,3}E'$, and $(e'')^2$ gives rise to $^3A_2'$, $^1E'$, and $^1A_1'$. On the basis of the dipole selection rule, only singlet $\rightarrow$ singlet transitions are allowed, and from Table 13.2, only transitions transforming as E' or A_2'' are allowed. So starting from the ground state, transitions can occur to the two $^1E'$ excited states.

15.5 The second laser would probe at 384.2 cm^{-1}, which corresponds to $\lambda = 26\mu m$. For the $3 \rightarrow 0$ transition, the frequency would be 13,810.6 cm^{-1}, so the wavelength would be 724 nm.

15.7 First calculate $\omega = 2\pi c\bar{\nu} = 4.1 \times 10^{14}$ s^{-1}. From this number, one finds that $(\hbar/2\mu\omega)^{1/2} = 0.034$ Å, so that $g = 3.18$ and $g^2 = 10.1$.

(a) $n = 0$: 4.1×10^{-5}; $n = 1$: 4.1×10^{-4}; $n = 2$: 2.1×10^{-3}.

(b) To maximize the FC overlap, it is easiest to make the replacement $\ln n! \cong n \ln n - n$. If one makes the replacment $h = g^2$, then the maximum can be found by differentiating $\ln h^n/n! = n \ln h - n \ln n + n$. This gives $(n)_{max} = h$, which implies that $(n)_{max} = g^2 = 10.1$. By explicit calculation, one finds that the FC overlap maximizes at $n = 10$. To use the FC principle, one calculates the energy $1/2kx^2$ for $x = 0.107$ Å, using $k = \mu\omega^2$. Since $x = g\{\hbar/2\mu\omega\}^{1/2}$, one finds that

$$\text{energy} = 1/2\mu\omega^2 g^2\{\hbar/2\mu\omega\} = \hbar\omega g^2/4 = 2.5\hbar\omega.$$

This means that the energy associated with a vertical excitation corresponds to 2.5 quanta above the minimum. This is not the same as the value obtained from the explicit overlap calculation, indicating that the classical result is an underestimate. Following is a plot of the FC overlap:

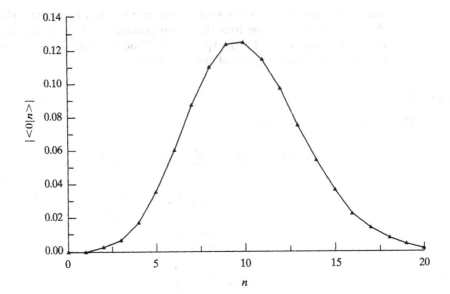

Index